Resource
Center

E. coli Gene Expression Protocols

METHODS IN MOLECULAR BIOLOGY™

John M. Walker, SERIES EDITOR

221. **Generation of cDNA Libraries:** *Methods and Protocols*, edited by *Shao-Yao Ying, 2003*
220. **Cancer Cytogenetics:** *Methods and Protocols,* edited by *John Swansbury, 2003*
219. **Cardiac Cell and Gene Transfer:** *Principles, Protocols, and Applications,* edited by *Joseph M. Metzger, 2003*
218. **Cancer Cell Signaling:** *Methods and Protocols,* edited by *David M. Terrian, 2003*
217. **Neurogenetics:** *Methods and Protocols,* edited by *Nicholas T. Potter, 2003*
216. **PCR Detection of Microbial Pathogens:** *Methods and Protocols,* edited by *Konrad Sachse and Joachim Frey, 2003*
215. **Cytokines and Colony Stimulating Factors:** *Methods and Protocols,* edited by *Dieter Körholz and Wieland Kiess, 2003*
214. **Superantigen Protocols,** edited by *Teresa Krakauer, 2003*
213. **Capillary Electrophoresis of Carbohydrates,** edited by *Pierre Thibault and Susumu Honda, 2003*
212. **Single Nucleotide Polymorphisms:** *Methods and Protocols,* edited by *Piu-Yan Kwok, 2003*
211. **Protein Sequencing Protocols, 2nd ed.,** edited by *Bryan John Smith, 2003*
210. **MHC Protocols,** edited by *Stephen H. Powis and Robert W. Vaughan, 2003*
209. **Transgenic Mouse Methods and Protocols,** edited by *Marten Hofker and Jan van Deursen, 2002*
208. **Peptide Nucleic Acids:** *Methods and Protocols,* edited by *Peter E. Nielsen, 2002*
207. **Recombinant Antibodies for Cancer Therapy:** *Methods and Protocols.* edited by *Martin Welschof and Jürgen Krauss, 2002*
206. **Endothelin Protocols,** edited by *Janet J. Maguire and Anthony P. Davenport, 2002*
205. ***E. coli* Gene Expression Protocols,** edited by *Peter E. Vaillancourt, 2002*
204. **Molecular Cytogenetics:** *Protocols and Applications,* edited by *Yao-Shan Fan, 2002*
203. ***In Situ* Detection of DNA Damage:** *Methods and Protocols,* edited by *Vladimir V. Didenko, 2002*
202. **Thyroid Hormone Receptors:** *Methods and Protocols,* edited by *Aria Baniahmad, 2002*
201. **Combinatorial Library Methods and Protocols,** edited by *Lisa B. English, 2002*
200. **DNA Methylation Protocols,** edited by *Ken I. Mills and Bernie H, Ramsahoye, 2002*
199. **Liposome Methods and Protocols,** edited by *Subhash C. Basu and Manju Basu, 2002*
198. **Neural Stem Cells:** *Methods and Protocols,* edited by *Tanja Zigova, Juan R. Sanchez-Ramos, and Paul R. Sanberg, 2002*
197. **Mitochondrial DNA:** *Methods and Protocols,* edited by *William C. Copeland, 2002*
196. **Oxidants and Antioxidants:** *Ultrastructure and Molecular Biology Protocols,* edited by *Donald Armstrong, 2002*
195. **Quantitative Trait Loci:** *Methods and Protocols*, edited by *Nicola J. Camp and Angela Cox, 2002*
194. **Posttranslational Modifications of Proteins:** *Tools for Functional Proteomics*, edited by *Christoph Kannicht, 2002*
193. **RT-PCR Protocols,** edited by *Joe O'Connell, 2002*
192. **PCR Cloning Protocols, 2nd ed.,** edited by *Bing-Yuan Chen and Harry W. Janes, 2002*
191. **Telomeres and Telomerase**: Methods and Protocols, edited by *John A. Double and Michael J. Thompson, 2002*
190. **High Throughput Screening:** *Methods and Protocols,* edited by *William P. Janzen, 2002*
189. **GTPase Protocols:** *The RAS Superfamily,* edited by *Edward J. Manser and Thomas Leung, 2002*
188. **Epithelial Cell Culture Protocols,** edited by *Clare Wise, 2002*
187. **PCR Mutation Detection Protocols,** edited by *Bimal D. M. Theophilus and Ralph Rapley, 2002*
186. **Oxidative Stress Biomarkers and Antioxidant Protocols,** edited by *Donald Armstrong, 2002*
185. **Embryonic Stem Cells:** *Methods and Protocols,* edited by *Kursad Turksen, 2002*
184. **Biostatistical Methods,** edited by *Stephen W. Looney, 2002*
183. **Green Fluorescent Protein:** *Applications and Protocols,* edited by *Barry W. Hicks, 2002*
182. **In Vitro Mutagenesis Protocols, 2nd ed.**, edited by *Jeff Braman, 2002*
181. **Genomic Imprinting:** *Methods and Protocols,* edited by *Andrew Ward, 2002*
180. **Transgenesis Techniques, 2nd ed.:** *Principles and Protocols,* edited by *Alan R. Clarke, 2002*
179. **Gene Probes:** *Principles and Protocols,* edited by *Marilena Aquino de Muro and Ralph Rapley, 2002*
178. **Antibody Phage Display:** *Methods and Protocols,* edited by *Philippa M. O'Brien and Robert Aitken, 2001*
177. **Two-Hybrid Systems:** *Methods and Protocols,* edited by *Paul N. MacDonald, 2001*
176. **Steroid Receptor Methods:** *Protocols and Assays,* edited by *Benjamin A. Lieberman, 2001*
175. **Genomics Protocols**, edited by *Michael P. Starkey and Ramnath Elaswarapu, 2001*
174. **Epstein-Barr Virus Protocols,** edited by *Joanna B. Wilson and Gerhard H. W. May, 2001*
173. **Calcium-Binding Protein Protocols, Volume 2:** *Methods and Techniques,* edited by *Hans J. Vogel, 2001*
172. **Calcium-Binding Protein Protocols, Volume 1:** *Reviews and Case Histories,* edited by *Hans J. Vogel, 2001*
171. **Proteoglycan Protocols,** edited by *Renato V. Iozzo, 2001*
170. **DNA Arrays:** *Methods and Protocols,* edited by *Jang B. Rampal, 2001*
169. **Neurotrophin Protocols,** edited by *Robert A. Rush, 2001*
168. **Protein Structure, Stability, and Folding,** edited by *Kenneth P. Murphy, 2001*
167. **DNA Sequencing Protocols,** *Second Edition,* edited by *Colin A. Graham and Alison J. M. Hill, 2001*
166. **Immunotoxin Methods and Protocols**, edited by *Walter A. Hall, 2001*
165. **SV40 Protocols**, edited by *Leda Raptis, 2001*
164. **Kinesin Protocols,** edited by *Isabelle Vernos, 2001*
163. **Capillary Electrophoresis of Nucleic Acids, Volume 2:** *Practical Applications of Capillary Electrophoresis,* edited by *Keith R. Mitchelson and Jing Cheng, 2001*
162. **Capillary Electrophoresis of Nucleic Acids, Volume 1:** *Introduction to the Capillary Electrophoresis of Nucleic Acids,*

METHODS IN MOLECULAR BIOLOGY™

E. coli Gene Expression Protocols

Edited by

Peter E. Vaillancourt

Applied Molecular Evolution
San Diego, CA

Humana Press **Totowa, New Jersey**

999 Riverview Drive, Suite 208
Totowa, New Jersey 07512

www.humanapress.com

This publication is printed on acid-free paper. ∞

ANSI Z39.48-1984 (American National Standards Institute) Permanence of Paper for Printed Library Materials.

Cover design by Patricia F. Cleary.

Cover illustration:Principle of a bacterial two-hybrid system (Fig. 1, Chapter 17; *see* full caption and discussion on pp. 251–252).

For additional copies, pricing for bulk purchases, and/or information about other Humana titles, contact Humana at the above address or at any of the following numbers: Tel: 973-256-1699; Fax: 973-256-8341; E-mail: humana@humanapr.com or visit our website at http://humanapress.com

Printed in the United States of America. 10 9 8 7 6 5 4 3 2 1

Library of Congress Cataloging-in-Publication Data

E. coli gene expression protocols / edited by Peter E. Vaillancourt.
p. cm. -- (Methods in molecular biology ; v. 205)
Includes bibliographical references and index.
ISBN 1-58829-008-5 (alk. paper)
1. Gene expression--Laboratory manuals. 2. Escherichia coli--Genetics--Laboratory manuals. I. Vaillancourt, Peter E./ II. Series.

QH 450.E15 2003
572.8'65--dc21

2002020572

Preface

The aim of *E. coli Gene Expression Protocols* is to familiarize and instruct the reader with currently popular and newly emerging methodologies that exploit the advantages of using *E. coli* as a host organism for expressing recombinant proteins. The chapters generally fall within two categories: (1) the use of *E. coli* vectors and strains for production of pure, functional protein, and (2) the use of *E. coli* as host for the functional screening of large collections of proteins or peptides. These methods and protocols should be of use to researchers over a wide range of disciplines. Chapters that fall within the latter category describe protocols that will be particularly relevant for functional genomics studies.

The chapters of *E. coli Gene Expression Protocols* are written by experts who have hands-on experience with the particular method. Each article is written in sufficient detail so that researchers familiar with basic molecular techniques and experienced with handling *E. coli* and its bacteriophages should be able to carry out the procedures successfully. As in all volumes of the Methods in Molecular Biology series, each chapter includes an extensive Notes section, in which practical details peculiar to the particular method are described.

E. coli Gene Expression Protocols is not intended to be all inclusive, but is focused on new tools and techniques—or new twists on old techniques—that will likely be widely used in the coming decade. There are several well-established *E. coli* expression systems (e.g., the original T7 RNA polymerase expression strains and vectors developed by William F. Studier and colleagues; the use of GST and polyhistidine fusion tags for protein purification) that have been extensively described in other methods volumes and peer-reviewed journal articles and are thus not included in this volume, with the exception of a few contributions in which certain of these systems have been adapted for novel applications or otherwise improved upon.

It is my sincerest hope that both novice and seasoned molecular biologists will find *E. coli Gene Expression Protocols* a useful lab companion for years to come. I wish to thank all the authors for their excellent contributions and Prof. John M. Walker for sound advice and assistance throughout the editorial process.

Peter E. Vaillancourt

Contents

Contributors

Ernesto Abel-Santos • *Department of Biochemistry, Albert Einstein College of Medicine, Bronx, NY*
François Baneyx • *Department of Chemical Engineering, University of Washington, Seattle, WA*
Stephen J. Benkovic • *Department of Chemistry, Pennsylvania State University, University Park, PA*
Frank Breitling • *Institut für Molekulare Genetik, Universität Heidelberg, Heidelberg, Germany*
Paul J. Brett • *Quorex Pharmaceuticals, Carlsbad, CA*
Olaf Broders • *Institut für Molekulare Genetik, Universität Heidelberg, Heidelberg, Germany*
Mary N. Burtnick • *Genomics Institute of the Novartis Research Foundation, San Diego, CA*
Dolores J. Cahill • *Max-Planck-Institute for Molecular Genetics, Berlin, Germany; PROT@GEN, Bochum, Germany*
Lisa Campbell • *Department of Biochemistry and Biophysics; Center for Macromolecular Design, Texas A&M University, College Station, TX*
Carsten-Peter Carstens • *Stratagene, La Jolla, CA*
Lisa A. Collins-Racie • *Genetics Institute/Wyeth Research, Cambridge, MA*
Gregory D. Davis • *Clontech Laboratories, Palo Alto, CA*
Heidi A. Davis • *The Center for Reproduction of Endangered Species, San Diego, CA*
Elizabeth A. DiBlasio-Smith • *Genetics Institute/Wyeth Research, Cambridge, MA*
Simon L. Dove • *Division of Infectious Diseases, Children's Hospital, Harvard Medical School, Boston, MA*
Stefan Dübel • *Institut für Molekulare Genetik, Universität Heidelberg, Heidelberg, Germany*
Thomas C. Evans, Jr. • *New England Biolabs, Inc., Beverly, MA*

STEVEN M. FIRESTINE • *Department of Medicinal Chemistry, Mylan School of Pharmacy, Duquesne University, Pittsburgh, PA*
JEFFREY D. FOX • *Macromolecular Crystallography Laboratory, Center for Cancer Research, National Cancer Institute, Frederick, MD*
ROGER G. HARRISON • *School of Chemical Engineering and Materials Science, University of Oklahoma, Norman, OK*
SABINE HORN • *Max-Planck-Institute for Molecular Genetics, Berlin, Germany; PROT@GEN, Bochum, Germany*
JAMES C. HU • *Center for Macromolecular Design, Department of Biochemistry and Biophysics, Texas A&M University, College Station, TX*
KARIN A. HUGHES • *Prolinx Inc., Bothell, WA*
JOHN F. HUNT • *Department of Biological Sciences, Columbia University, New York, NY*
ALBERT JELTSCH • *Institut für Biochemie, Justus-Liebig-Universität, Giessen, Germany*
KAREN JOHNSTON • *Department of Biochemistry and Biophysics, The Wistar Institute, Philadelphia, PA*
WOLFGANG KLEIN • *Institute for Pharmaceutical Biology, Rheinische Friedrich-Wilhelm University Bonn, Bonn, Germany*
THOMAS LANIO • *Justus-Liebig-Universität, Institut für Biochemie, Giessen, Germany*
EDWARD R. LAVALLIE • *Genetics Institute/Wyeth Research, Cambridge, MA*
HANS LEHRACH • *Max-Planck-Institute for Molecular Genetics, Berlin, Germany; PROT@GEN, Bochum, Germany*
ZHIJIAN LU • *Genetics Institute/Wyeth Research, Cambridge, MA*
ANGELIKA LUEKING • *Max-Planck-Institute for Molecular Genetics, Berlin, Germany; PROT@GEN, Bochum, Germany*
LEONARDO MARIÑO-RAMÍREZ • *Center for Macromolecular Design, Department of Biochemistry and Biophysics, Texas A&M University, College Station, TX*
RONEN MARMORSTEIN • *Department of Biochemistry and Biophysics, The Wistar Institute, Philadelphia, PA*
JOHN M. MCCOY • *Biogen, Inc., Cambridge, MA*
SAMU MELKKO • *Institute of Pharmaceutical Sciences, Zurich, Switzerland*
MIRNA MUJACIC • *Department of Chemical Engineering, University of Washington, Seattle, WA*
REBECCA L. MULLINAX • *Stratagene, La Jolla, CA*
DARIO NERI • *Institute of Pharmaceutical Sciences, Zurich, Switzerland*

Kerstein A. Padgett • *Division of Insect Biology, University of California, Berkeley, CA*
Joanne L. Palumbo • *Department of Chemical Engineering, University of Washington, Seattle, WA*
Alfred Pingoud • *Institut für Biochemie, Justus-Liebig-Universität, Giessen, Germany*
Frank Salinas • *Department of Chemistry, Pennsylvania State University, University Park, PA*
Gerhard Sandmann • *Botanisches Institut, Goethe Universität, Frankfurt, Germany*
Charles P. Scott • *Department of Microbiology and Immunology, Thomas Jefferson University, Philadelphia, PA*
Joseph A. Sorge • *Stratagene, La Jolla, CA*
Rhesa D. Stidham • *Department of Physiology and Graduate Program in Molecular Biophysics, The University of Texas Southwestern Medical Center, Dallas, TX*
Philip J. Thomas • *Department of Physiology, The University of Texas Southwestern Medical Center, Dallas, TX*
David S. Waugh • *Macromolecular Crystallography Laboratory, Center for Cancer Research, National Cancer Institute, Frederick, MD*
W. Christian Wigley • *Department of Physiology, The University of Texas Southwestern Medical Center, Dallas, TX*
Jean P. Wiley • *Prolinx Inc., Bothell, WA*
David T. Wong • *GenVault, Carlsbad, CA*
Donald E. Woods • *Department of Microbiology and Infectious Diseases, Faculty of Medicine, University of Calgary Health Sciences Centre; Canadian Bacterial Diseases Network, Calgary, Alberta, Canada*
Ming-Qun Xu • *New England Biolabs, Inc., Beverly, MA*

1

Cold-Inducible Promoters for Heterologous Protein Expression

François Baneyx and Mirna Mujacic

1. Introduction

1.1. Cold Shock Response and Cold Shock Proteins of Escherichia coli

Rapid transfer of exponentially growing *E. coli* cultures from physiological to low temperatures (10–15°C) has profound consequences on cell physiology: membrane fluidity decreases, which interferes with transport and secretion, the secondary structures of nucleic acids are stabilized, which affect the efficiencies of mRNA transcription/translation and DNA replication, and free ribosomal subunits and 70S particles accumulate at the expense of polysomes, negatively impacting translation of most cellular mRNAs ***(1–3)***. It is therefore not surprising that cell growth and the synthesis of the vast majority of cellular proteins abruptly stop upon sudden temperature downshift ***(4)***. However, this lag phase is only transient, and growth resumes with reduced rates after 2–4 h incubation at low temperatures, depending on the genetic background ***(4,5)***. Such remarkable ability to survive drastic changes in environmental conditions is not atypical for *E. coli,* which has evolved multiple, often synergistic, adaptive strategies to handle stress. In the case of cold shock, the need for restoring transcription and translation is handled by an immediate increase in the synthesis of about 16 cold shock proteins (Csps) ***(4)***, while the cell solves the problem of membrane fluidity by raising the concentration of unsaturated fatty acids that are incorporated into membrane phospholipids ***(6)***. Interestingly, translation of the alternative sigma factor σ^S, a global regulator of gene expression in *E. coli*, has been reported to increase at 20°C ***(7)*** suggesting that RpoS-dependent gene products may also play a role in cellular adaptation to mild—but probably not severe—cold shock.

From: *Methods in Molecular Biology, vol. 205, E. coli Gene Expression Protocols*
Edited by: P. E. Vaillancourt © Humana Press Inc., Totowa, NJ

E. coli Csps have been divided into two classes depending on their degree of induction by low temperatures *(8)*. Class II Csps are easily detectable at 37°C but undergo a 2–10 fold increase in synthesis following cold shock. These include the recombination factor RecA, the GyrA subunit of the topoisomerase DNA gyrase, initiation factor IF-2, and HN-S, a nucleoid-associated DNA-binding protein that modulates the expression of many genes at the transcriptional level. By contrast, Class I Hsps are synthesized at low levels at physiological temperatures but experience a more than 10-fold induction following temperature downshift. Two of these, CsdA and RbfA, are associated with the ribosome. CsdA binds to 70S particles and exhibits RNA-unwinding activity *(9)*. RbfA, which only interacts with 30S subunits, has been proposed to function as a late maturation or initiation factor *(2)*, and is required for the efficient translation of most cellular mRNAs at low temperatures *(3)*. Additional Class I Csps include NusA, a transcription termination-antitermination factor, and PNPase, an exonuclease involved in mRNA turnover. The most highly cold-inducible protein, CspA, belongs to a family of nine low molecular mass (≈ 7 kDa) paralogs, four of which—CspA, CspB, CspG and CspI—are upregulated upon temperature downshift with different optimal temperature ranges *(10,11)*. CspA, the best characterized member of the set, has been ascribed an RNA chaperone function based on the observations that it binds single-stranded nucleic acids with low specificity, destabilizes RNA secondary structures *(12)*, and acts as a transcription antiterminator in vivo *(13)*. At present, the function of CspB, CspG and CspI remains unclear, although their high degree of homology to CspA and genetic studies suggests that these proteins may perform similar, albeit complementary roles in cold adaptation *(11,14)*.

1.2. CspA Regulation

CspA, the major *E. coli* cold shock protein, is virtually undetectable at 37°C but more than 10% of the cellular synthetic capacity is devoted to its production during the first hour that follows transfer to 15°C *(15)*. Unlike heat shock genes which rely on specific promoter sequences and alternative sigma factors for transcription, the *cpsA* core promoter is not strikingly different from vegetative promoters (**Fig. 1A**) and is believed to be recognized by the $E\sigma^{70}$ holoenzyme at all temperatures *(16,17)*. An AT-rich UP element, located immediately upstream of the –35 hexamer (**Fig. 1A**) increases the strength of the *cspA* promoter by facilitating transcription initiation *(16,17)*. As a result, large amounts of *cspA* transcripts are synthesized at physiological temperatures. The seemingly inconsistent observation that little CspA is present at 37°C is explained by the presence of a highly structured 159-nt long untranslated region (UTR) at the 5' end of the *cspA* mRNA (**Fig. 1A**; *see* **Note 1**). At 37°C, this extension makes the *cspA* transcript very short-lived ($t_{1/2} \approx 10$ s), thereby preventing its

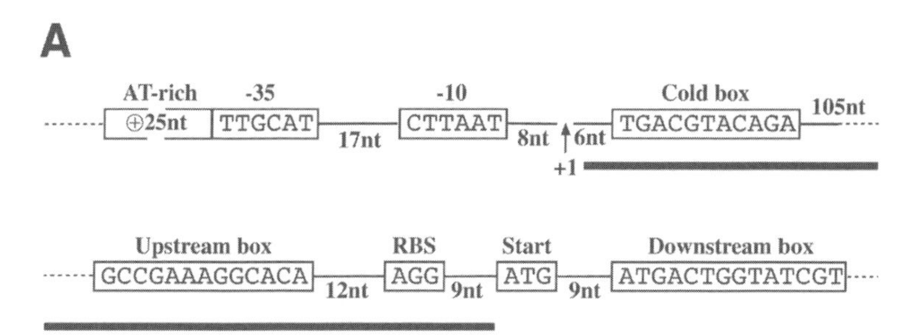

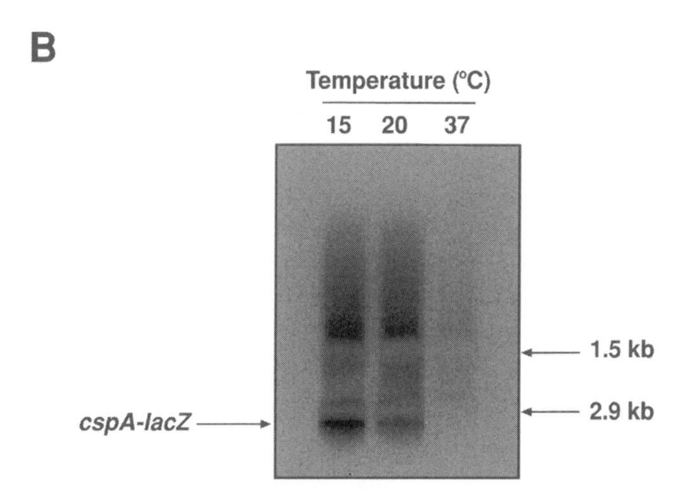

Fig. 1. *cspA* regulatory regions and influence of the downshift temperature on *cspA*-driven transcription. **(A)** Regulatory elements involved in the transcriptional (AT-rich element, -35 and -10 hexamers), posttranscriptional (cold box), and translational (upstream and downstream boxes) control of CspA synthesis are boxed and consensus sequences are given (*see* **Subheading 1.2.** for details). RBS represents the ribosome binding site. The black line spans the length of the 5' UTR. **(B)** JM109 cells harboring pCSBG, a plasmid encoding a *cspA::lacZ* translational fusion ***(21)***, were grown to midexponential phase in LB medium at 37°C and incubated for 45 min at 15, 20, or 37°C. Total cellular RNA was extracted and the *cspA::lacZ* transcript was detected following Northern blotting using a *lacZ*-derived probe. The migration position of the *cspA::lacZ* mRNA and those of the 23S and 16S rRNAs are indicated by arrows (adapted from **ref.** ***(22)***).

efficient translation ***(18–20)***. Of importance for practical applications, the *cspA* UTR is fully portable and fusing it to the 5' terminus of other genes destabilizes the resulting hybrid transcripts at physiological temperatures (*see* **Fig. 1B** and **Note 2** ***[21,22]***).

Following temperature downshift, the *cspA* core promoter is slightly stimulated ***(16)*** but the main contributor to the rapid induction of CspA synthesis is an almost two order of magnitude increase in transcript stability that appears to be related to a conformational change in the 5' UTR (**Fig. 1B**; *see* ***ref. [18–20,23]***). Translational effects also play a role in the induction process. Deletion analysis indicates that a conserved region near the 3' end of the UTR (the so called upstream box; **Fig. 1A**) makes the *cspA* transcript more accessible to the cold-modified translation machinery ***(24)***. In addition, a region complementary to a portion of the 16S rRNA and located 12 bp after the *cspA* start codon (the downstream box; **Fig. 1A**) has been reported to enhance *cspA* translation initiation following cold shock ***(17)***. It should however be noted that the latter feature is not essential to achieve efficient low temperature expression of a variety of heterologous genes fused to the *cspA* promoter-UTR region (unpublished data; *see* ***ref. [21,25,26]***).

After 1–2 h incubation at low temperatures, synthesis of native CspA as well as that of recombinant proteins placed under *cspA* transcriptional control stops. An 11 bp-long element located at the 5' end of the UTR and conserved among cold shock genes (the cold box; **Fig. 1A**), as well as CspA itself, appear to be implicated in this process ***(27–29)***. It has been hypothesized that the cold box is either a binding site for a repressor molecule or a transcriptional pausing site. In the first scenario, binding of the putative repressor (possibly CspA ***[27]***) to the cold box interferes with transcription or destabilizes the mRNA, leading to a shutdown in CspA synthesis. The second model envisions that the putative cold box pausing site is somehow bypassed by RNA polymerase immediately after temperature downshift. However, once CspA reaches a threshold concentration, it binds to its own mRNA, thereby destabilizing the RNA polymerase elongation complex and attenuating transcription ***(30)***.

Repression of CspA synthesis coincides with resumption of cell growth. This phenomenon has been explained by the ribosome adaptation model ***(3,31)*** which states that cold shock proteins RbfA, CsdA, and IF-2 associate with the free ribosomal subunits and 70S particles that accumulate immediately after cold shock to progressively convert them into functional, cold-adapted ribosomes and polysomes capable of translating non cold shock mRNAs. It is possible that these changes in the translational machinery also contribute to the repression of CspA synthesis as suggested by the fact that *rbfA* mutants produce cold shock proteins constitutively following temperature downshift ***(3)***. The fact that *rbfA* cells do not repress the synthesis of Csps at the end of the lag phase is of great practical value and has been exploited to significantly increase the intracellular accumulation of gene products placed under *cspA* transcriptional control in both shake flasks and fermentors ***(5)***.

1.3. Advantages and Drawbacks of Low Temperature Expression

A number of studies have demonstrated that expression in the 15–23°C range often—but not always—improves the folding of recombinant proteins that form inclusion bodies at 37°C (reviewed in *ref. [32]*). Although the mechanistic basis for this observation remains unclear, several non-exclusive possibilities can account for improved folding at low temperatures. First, in contrast to other forces (e.g., H-bonding), hydrophobic interactions weaken with decreasing temperatures. Since hydrophobic effects contribute to the formation and stabilization of protein aggregates, newly synthesized proteins may have a greater chance to escape off-pathway aggregation reactions. Second, because peptide elongation rates decrease with the temperature *(33)*, nascent polypeptides may have a higher probability of forming local elements of secondary structure, thus avoiding unproductive interactions with neighboring partially folded chains. Finally, a decrease in translation rates should increase the likelihood that a protein requiring the assistance of folding helpers to reach a proper conformation is captured and processed by molecular chaperones and foldases based on mass effect considerations.

In addition to improving folding, expression at low temperatures can prove helpful in reducing the degradation of proteolytically sensitive polypeptides (reviewed in *ref. [32]*). Here again, the fundamental reasons underpinning this phenomenon remain obscure. However, it has been reported that cold shock is accompanied by a transient decrease in the synthesis of heat shock proteins (Hsps; *[34]*). Since a number of Hsps are ATP-dependent proteases and at least two of these (Lon and ClpYQ) participate in non-specific protein catabolism *(35)*, a polypeptide synthesized early-on after temperature downshift may have a better chance to bypass the cellular degradation machinery (*see* **Note 3**).

Because aggregation and degradation are two major drawbacks associated with the production of heterologous proteins in *E. coli*, expression at low temperatures is of obvious practical interest. Unfortunately, the vast majority of routinely used promoter systems (e.g., *tac* and T7) experience a decrease in efficiency upon temperature downshift *(26,32)*. Furthermore, following transfer of cultures to 15°C, the absence of a cold shock UTR precludes translation of typical transcripts by the cold-modified translational machinery until the end of the transient lag phase *(25)*. Because of its strength and mechanism of induction, the *cspA* promoter-UTR region is particularly well suited for the production of aggregation-prone and proteolytically sensitive polypeptides at low temperatures *(25,26)*. In addition, by destabilizing elements of secondary structures interfering with ribosome binding, the *cspA* UTR can greatly facilitate the translation of otherwise poorly translated mRNAs. The remainder of this chapter highlights procedures and precautions for *cspA*-driven recombinant protein expression.

2. Materials

2.1. Growth and Maintenance of E. Coli Strains

2.1.1. Strains

1. Routine cloning and plasmid maintenance is in Top10 (F– *mcrA* Δ*(mrr-hsdRMS-mcrBC)* Φ*80* Δ*lacZ*ΔM15 Δ*lacX74 deoR recA1 araD139* Δ*(ara-leu)7697 galU galK* λ– *rpsL endA1 nupG*) (Invitrogen) or any other *endA recA* strain.
2. A good wild type host for low temperature expression is CSH142 ***(5)***.
3. The source of the *rbfA* deletion is CD28 (F^- *ara* Δ*(gpt-lac)5 rbfA::kan*) ***(2)***.

2.1.2. Growth Media

1. Luria-Bertani (LB) broth: Mix 10 g of Difco tryptone peptone, 5 g Difco yeast extract, and 10 g of NaCl in 950 mL of ddH_2O. Shake to dissolve all solids, adjust the pH to 7.4 with 5 *N* NaOH, and the volume to 1 L with ddH_2O; autoclave. If desired, add 5 mL of 20% (wt/vol) glucose from a filter sterilized stock.
2. LB plates: add 15 g of agar (Sigma) per liter of LB before autoclaving.
3. Add antibiotics at a final concentration of 50 μg/mL after the solution has cooled to 40–50°C (*see* **Note 4**).

2.1.3. Antibiotic Stock Solution

1. Prepare stock solutions of carbenicillin (or ampicillin), and neomycin (or kanamycin) at 50 mg/mL by dissolving 0.5 g of powder into 10 mL of ddH_2O and filter-sterilizing the solutions through a 0.2 μm filter. Antibiotics are stored in 1 mL aliquots at –20°C until needed (*see* **Note 5**).

2.1.4. Glycerol Stock

1. Weigh 80 g of glycerol in a graduated cylinder. Fill to 100 mL with ddH_2O. Sterilize through a 0.2 μm filter into a sterile bottle and store at room temperature.

2.2. Cloning Vectors for cspA-Driven Expression and rbfA Strains

2.2.1. Plasmids pCS22 and pCS24

1. These plasmids are available upon request from François Baneyx, Ph.D., Department of Chemical Engineering, The University of Washington, Box 351750, Seattle, WA 98195.

2.2.2. Construction and Phenotypic Verification of rbfA::kan Mutants

1. CD28 donor strain (*see* **Subheading 2.1.1.**), desired recipient strain and P1*vir* lysate.
2. $CaCl_2$ solution in ddH_2O (1 *M*), filter sterilized and stored at room temperature.
3. Soft agar: Add 0.75 g of agar to 100 mL of LB (*see* **Subheading 2.1.2.**) and autoclave. Dispense 3 mL aliquots in sterile 18 mm culture tubes before solidification. Store tubes at room temperature.

4. LB medium: LB and LB-neomycin plates (*see* **Subheading 2.1.2.**).
5. Chloroform.
6. MC buffer: 100 m*M* $MgSO_4$, 5 m*M* $CaCl_2$ and 1 *M* sodium citrate in ddH_2O. Filter-sterilize both solutions and store at room temperature.

2.2.3. Transformation and Storage of rbfA *Mutants*

1. $CaCl_2$ in ddH_2O, 100 m*M*, filter sterilized and stored at room temperature.
2. Glycerol stock solution (*see* **Subheading 2.1.4.**).

2.3. Cold Induction in Shake Flasks and Fermentors

2.3.1. Shake Flasks Cultures

1. Temperature controlled water bath with orbital shaking.
2. Cooling coil accessory.
3. VWR model 1172 refrigeration unit or equivalent.

2.3.2. Fermentations

1. New Brunswick BioFloIII fermentor or equivalent equipped with temperature, agitation, pH, dissolved oxygen and foam control.
2. Antifoam (Sigma 289).
3. Glucose stock solution, 20% w/v, filter sterilized.
4. 1 *M* HCl and 5% NH_4OH (v/v) for pH control.
5. Neslab Coolflow HX-200 cooling unit or equivalent.

3. Methods

3.1. Placing PCR Products under cspA *Transcriptional Control*

The cloning vectors pCS22 and pCS24 (**Fig. 2**; ***ref. [25]***) are pET22b(+) and pET24a(+) (Novagen) derivatives that have been engineered to facilitate the positioning of structural genes downstream of the *cspA* promoter-UTR region. Plamid pCS22 is an ampicillin-resistant construct encoding the ColE1(pMB1) origin of replication, a *pelB* signal sequence, a multiple cloning site (MCS) derived from pET22b(+), a 3' hexahistidine tail, and the phage T7 transcription termination sequence (**Fig. 2A**). Plasmid pCS24 is a kanamycin-resistant ColE1 derivative encoding a MCS derived from pET24a(+), a 3' hexahistidine tail and the phage T7 terminator region (**Fig. 2B**). For cytoplasmic expression, cloning should be carried out as follows:

1. Amplify the desired gene using a forward primer designed to create a *Nde*I site overlaping the ATG initiation codon and a reverse primer selected to introduce one of the unique restriction sites available in the MCS of pCS22 or pCS24 (we typically make use of *Xho*I).
2. Purify the amplified fragment following low melting point (LMP) agarose electrophoresis using the QIAGEN QIAquick gel extraction kit or equivalent. If *Taq*

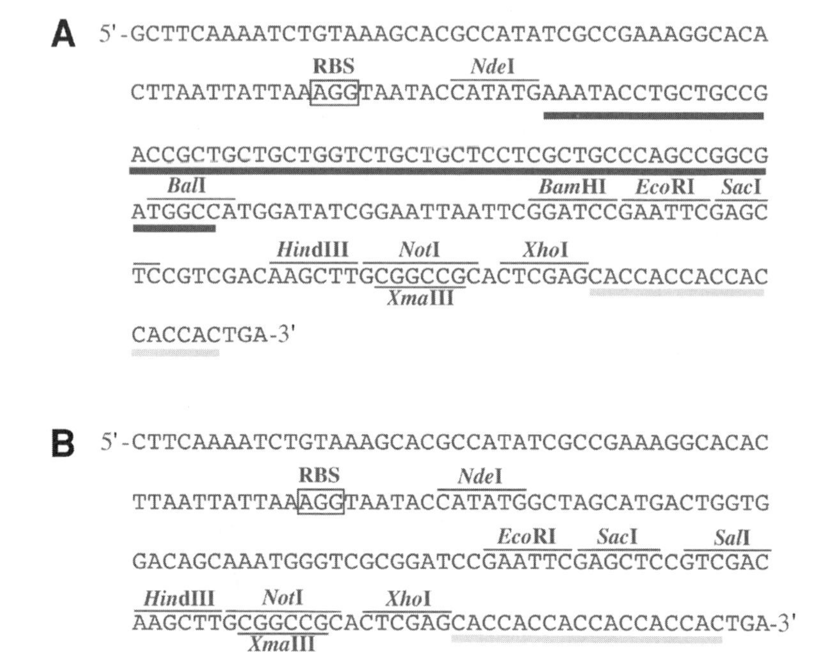

Fig. 2. Cloning regions of pCS22 and pCS24. **(A)** Unique restriction sites in the polylinker of the ampicillin-resistant ColE1 derivative pCS22 are shown. The black line spans the length of the *pelB* signal sequence. The gray line shows the location of the hexahistidine tail. **(B)** Unique restriction sites in the polylinker of the kanamycin-resistant ColE1 derivative pCS24 are shown. The gray line shows the location of the hexahistidine tail. RBS represents the ribosome binding site.

polymerase has been used for amplification, subclone the purified DNA fragment into the Invitrogen TOPO TA cloning vector or equivalent according to the manufacturer's instructions. If the polymerase yields a blunt fragment, subclone into Invitrogen Zero Blunt TOPO cloning vector or equivalent.

3. Digest pCS22 or pCS24 DNA and the plasmid encoding the desired gene with *Nde*I and the appropriate 3' enzyme (*see* **Note 6**). Isolate backbone and insert DNA following LMP agarose electrophoresis as in **step 2**.
4. Ligate at a 3:1 insert to backbone ratio, transform electrocompetent Top10 cells and plate on LB agar supplemented with 50 µg/mL of carbenicillin (for pCS22 derivatives) or 50 µg/mL neomycin (for pCS24 derivatives). Screen the colonies for the presence of the insert.

It is possible to target gene products to the *E. coli* periplasm by taking advantage of the presence of the *pelB* signal sequence in pCS22 (**Fig. 2A**; *see* **Note 7**). However, the *Nco*I site which is typically used to fuse gene products to the *pelB* signal peptide in pET22b(+) is no longer unique in pCS22. Downstream sites (e.g., *Bam*HI, *Eco*RI and *Sac*I) may be used but will lead to an non-native N-terminus following processing of the signal sequence by the leader peptidase. If an intact N-terminus is required, cloning must be accomplished by using partial *Nco*I digestion.

3.2. Increasing the Length of the Production Phase by Using rbfA Mutants

E. coli strains bearing a null mutation in the ribosomal factor RbfA remain able to repress the synthesis of CspA at or above 37°C, but constitutively produce Csps following temperature downshift *(3)*. This phenotype can be exploited to increase the length of time over which recombinant proteins transcribed from the *cspA* promoter-UTR region are synthesized, allowing the accumulation of the target polypeptide to about 15–20% of the total cellular protein compared to approximately 5–10% in $rbfA^+$ genetic backgrounds *(5)*. However, *rbfA* mutants exhibit a cold-sensitive phenotype *(2)* which limits the choice of the downshift temperature to the 23–30°C range and requires that the strains be maintained at or above 37°C.

3.2.1. Construction of rbfA::kan Mutants

The *rbfA::kan* mutation can be easily moved from CD28 to any genetic background by P1 transduction provided that the recipient strain does not already contain a kanamycin or neomycin marker. A P1*vir* lysate is first raised on CD28 as follows (*see* **Note 8**):

1. Use a toothpick or a sterile loop to scrape a few cells from a frozen glycerol stock of CD28 and inoculate 5 mL of LB medium supplemented with 5 µL of 50 mg/mL neomycin and 25 µL of 1 *M* $CaCl_2$ in in an 18-mm culture tube. Grow overnight at 42°C.
2. Combine 0.5 mL of the overnight culture with 100 µL of P1*vir* lysate raised on wild type *E. coli* in a fresh sterile 18 mm tube; incubate at 42°C for 20 min.
3. Melt a 3 mL aliquot of LB soft agar (*see* **Subheading 2.2.2.**) at 50°C, combine it with the cells and P1 lysate at the end of the 42°C incubation period and pour the mixture onto an LB-agar plate preheated at 42°C. Gently swirl the plate to evenly cover the bottom agar and allow to solidify for 10 min at room temperature.
4. Incubate the plate upright (not inverted) at 42°C for 8–12 h. Do not exceed 12 h incubation.
5. At the end of the incubation period (*see* **Note 9**), dip a spatula in ethanol, flame sterilize, and scrape the soft agar into a sterile 30 mL PA tube.
6. Add 200 µL of chloroform and vortex at high speed for 1 min.
7. Centrifuge at 2000*g* for 5 min, recover the supernatant with a sterile pipet (*see* **Note 10**) and use immediately, or store in a sterile Eppendorf tube in the dark at 4°C (*see* **Note 11**).

Transfer of the *rbfA::kan* allele to the desired genetic background is accomplished as follows:

1. Grow an overnight inoculum of the recipient strain in 5 mL of LB medium at 37°C.
2. Transfer 1 mL of culture in a sterile Eppendorf tube, spin at 2000*g* for 5 min in a microfuge, discard the supernatant and resuspend the cell pellet in 1 mL of MC buffer.
3. Place the Eppendorf tube at 37°C and aerate the culture by shaking for at least 20 min.
4. Mix 100 µL of cells from **step 3** with 100 µL of P1 lysate raised on CD28 into a fresh sterile Eppendorf tube. Incubate at 42°C for 20 min or less.
5. Add 100 µL of 1 *M* sodium citrate to chelate calcium ions and stop phage infection.
6. Transfer the mixture to a sterile 18-mm culture tube, add 1 mL of LB, and incubate at 42°C for 1–2 h to allow accumulation of kanamycin phosphotransferase.
7. At the end of the incubation period, centrifuge the culture at 2000*g* for 5 min in a microfuge, resuspend the cells in 100 µL of LB, spread onto a LB-neomycin plate prewarmed at 42°C (*see* **Note 12**) and incubate at 42°C overnight.

3.2.2. *Verification of the* rbfA *Phenotype*

The protocol of **Subheading 3.2.1.** typically yields 3–20 neomycin-resistant colonies following overnight incubation. If this is not the case, the titer of the P1 lysate may be too low (*see* **Note 8**). Transductants should be checked for their ability to grow in liquid medium and for the presence of the *rbfA::kan* allele as follows:

1. Use a sterile loop or toothpick to transfer cells from 5 individual colonies into 18 mm culture tubes containing 5 mL of LB medium supplemented with 5 µL of 50 mg/mL neomycin.
2. Incubate for 14–17 h at 42°C and check for healthy growth.
3. Restreak cells from **step 2** onto four LB-neomycin plates sectored into 6 areas. Use the vacant area to streak *rbfA*$^+$ recipient cells as a positive control.
4. Incubate overnight at room temperature, or in incubators held at either 30, 37, or 42°C. Strains containing the *rbfA::kan* mutation should grow comparably to the wild type at 37 and 42°C, while exhibiting very poor growth (if any) at 30°C and no growth at room temperature.
5. Pick individual colonies from positive transductants using the 42°C plate and prepare overnight cultures (**steps 1–2**).
6. On the next day, mix 800 µL of cells with 200 µL of glycerol stock solution into a cryogenic tube and store at –80°C.

3.2.3. *Transformation of* rbfA *Cells*

rbfA mutants can be readily transformed with pCS22 derivatives or home-made *cspA* cloning vectors that do not contain a kanamycin resistance cartridge. However, plasmid pCS24 or constructs encoding a kanamycin/neomycin resistance gene cannot be stably maintained in *rbfA::kan* cells. Competent cells are prepared by modification of the classic $CaCl_2$ method:

1. Grow an overnight inoculum of *rbfA::kan* cells at 37 or 42°C in 5 mL of LB medium supplemented with 5 µL of a 50 mg/mL neomycin stock solution.
2. Dispense 25 mL of LB and 25 µL of a 50 mg/mL neomycin stock into a 125-mL sterile shake flask, inoculate with 500 µL of seed culture and incubate with shaking at 37 or 42°C to $A_{600} \approx 0.4$.
3. Transfer the culture to a pre-chilled, sterile 30 mL PA tube and centrifuge at 8000*g* for 8 min.
4. Discard the supernatant and gently resuspend the cell pellet in 12.5 mL of 100 m*M* $CaCl_2$.
5. Incubate on ice for 20 min (*see* **Note 13**).
6. Centrifuge at 8000*g* for 8 min, discard the supernatant and resuspend the pellet in 625 µL of 100 m*M* $CaCl_2$.
7. Immediately add 78.1 µL of 80% glycerol stock, dispense into 200-µL aliquots in sterile Eppendorf tubes and store at –80°C until needed.

Transformation with the desired plasmid is carried out as follows:

1. Thaw a 200-µL aliquot of competent cells on ice; add 5 µL of plasmid DNA purified using the QIAGEN QIAprep Spin miniprep kit or equivalent and mix by tapping. Incubate the cells on ice for 30 min.
2. Transfer the cells to a 42°C water bath for a 45- to 60-s heat shock and hold on ice for 2 min.
3. Add 800 µL of LB and incubate at 37 or 42°C for 1.5–2 h.
4. Centrifuge at 8000*g* for 5 min in a microfuge, discard the supernatant, resuspend the pellet in 140 µL of LB, and plate onto a LB plate containing 50 µg/mL neomycin as well as the appropriate selective pressure to maintain the plasmids.
5. Incubate overnight at 42°C (or 37°C) and check transformants for cold sensitivity as in **Subheading 3.2.2.**

3.2.4. Precautions to be Taken with rbfA *Mutants*

The following guidelines should be adhered to when working with *rbfA::kan* cells:

1. Always grow *rbfA* strains in medium containing 50 µg/mL neomycin at 37 or 42°C.
2. Do not store plated or streaked *rbfA* cells at 4°C for future use.
3. Do not use *rbfA* cultures that have been subjected to temperature downshift for inoculum preparation.
4. Do not rapidly cool or warm *rbfA* cells.
5. Periodically check the neomycin-resistant and cold-sensitive phenotypes of glycerol stocks and make new stocks every few months.

3.3. Induction in Shake Flask Cultures

3.3.1. Host Strains

Any *E. coli* strain that does not exhibit a cold sensitive phenotype (the specific case of *rbfA* mutants is discussed in **Subheading 3.3.3.**) may be used as a host for the production of proteins whose genes are under transcriptional control of the *cspA* promoter-UTR region. We have successfully achieved cold-induction in JM109 *(5,21,26)*, CSH142 *(5)*, BL21(DE3) *(25)*, as well as in MC4100 and W3110 derivatives. However, as is the case with other promoter systems-gene combinations, the host genetic background can exert a profound influence on production levels and case-by-case optimization may be necessary. For instance, in the case of *cspA*-driven production of β-galactosidase, almost 10-times more active enzyme was present in CSH142 cells 1 h after transfer from 42 to 23°C compared to JM109 *(5)*.

3.3.2. Leaky Expression

Although the *cspA* UTR efficiently destabilizes transcripts to which it is fused (**Fig. 1B**), repression is by no means complete at physiological temperatures *(5,21,25,26)*. Growth of seed cultures and biomass accumulation at 42°C prior to cold shock help reduce—but do not completely abolish—leaky expression (*5,25*; *see* **Note 14**). It is thus important to bear in mind that the *cspA* system may be unsuitable for the production of proteins that are highly toxic to *E. coli* (*see* **Note 15**)

3.3.3. Choice of the Downshift Temperature

Induction of *cspA*-driven expression can be achieved by temperature downshifts as small as 7°C *(21)*, and recombinant protein production remains possible at temperatures as low as 10°C *(26)*. Thus, a wide range of induction conditions is available. The following issues should be carefully considered when selecting the downshift temperature. (1) Transferring cultures to the 10–15°C temperature range yields the highest levels of target transcript (**Fig. 2**), but causes a reduction in translational efficiency compared to 20–25°C *(21)*. As a result, overall recovery yields are typically comparable at 15 and 23°C *(21,25)*. (2) In wild type cells, the length of the lag phase over which *cspA*-driven transcription takes place increases as the downshift temperature decreases *(5)*. This means that the same amounts of target protein will accumulate faster at 20–25°C relative to 15°C, thereby enhancing productivity (*see* **Note 16**). (3) On the other hand, proper folding of aggregation-prone proteins greatly improves when cultures are transferred to 10°C, but little material accumulates at this temperature *(26)*. We therefore recommend carrying out preliminary studies at both 15 and 23°C as follows.

1. Start an inoculum of the desired culture in 5 mL of LB supplemented with the appropriate antibiotics and grow overnight at 37 or 42°C.
2. Adjust the temperature of the cooling system to 5–7°C and the set point of the water bath to the selected downshift temperature (15 or 23°C). Allow bath temperature to equilibrate overnight.

3. On the next day, inoculate a 125 mL shake flask containing 25 mL of LB supplemented with the appropriate antibiotics using 500 µL of seed culture; grow at 37 or 42°C to $A_{600} \approx 0.5$.
4. Take a 1 mL sample for subsequent analysis and transfer the flask to the chilled water bath (*see* **Note 17**).
5. Collect 1 mL samples 1, 2, 3, and 24 h after temperature downshift. Process and analyze by sodium dodecyl sulfate polyacrylamide gel electrophoresis (SDS-PAGE), immunoblotting or adequate activity assay.

As noted in **Subheading 3.2.**, *rbfA* cells allow continuous expression of recombinant proteins placed under *cspA* transcriptional control. However, they exhibit a cold-sensitive phenotype and die when shifted to temperatures lower than 20°C. Thus, the range of induction conditions is more limited with *rbfA* mutants, and these strains may not be suitable for the production of highly aggregation-prone or proteolytically sensitive proteins. For *rbfA* cells, induction in shake-flask cultures is performed as described for wild type strains except that seed cultures and growth to $A_{600} \approx 0.5$ should be carried out at 42°C, while the downshift temperature should be 23°C or higher.

3.4. Cold-Shock Induction in Fermentors

The *cpsA* system has been shown to be suitable for the production of recombinant proteins in both batch and fed-batch fermentation setups *(5)*. An important consideration in these experiments is the choice of the cooling rate. Optimal production of β-galactosidase in a 2.5-L working volume batch fermentor was observed when the medium was chilled from 37 to 15°C using a cooling rate of 0.5°C/min. Cooling under heat transfer-limiting conditions or the use of a 0.3°C/min cooling profile reduced the accumulation levels of active enzyme by about 30% *(5)*. The higher product yield in fermentors cooled at intermediate rates likely reflects an optimal situation in which more efficient translation compensates for lower levels of transcript synthesis.

Multiple induction of the *cspA* promoter can be achieved by temperature cycling between 15 and 25°C, or by using stepwise temperature downshifts between 37, 29, 21, and 13°C. However, re-induction is inefficient in temperature cycling experiments and requires that the cells be held at intermediate temperatures for at least 60 min in stepwise downshift experiments. This is probably owing to a need to dilute out the repressor via biomass increase before high efficiency re-induction can take place *(5)*. Overall, the increase in productivity conferred by fermentation engineering techniques is small, and a single temperature downshift step is probably suitable for the vast majority of applications. A typical batch fermentation is performed as follows (*see* **Note 18**).

1. Dispense 50 mL of LB in a 250-mL shake flask; supplement with the appropriate antibiotics and inoculate with a few cells scraped from a glycerol stock; incubate overnight at 37 or 42°C.

2. Fill the fermentor tank with 2.5 L of LB medium and autoclave with probes in place.
3. After cooling, supplement the medium with glucose (0.2% v/v final concentration), the appropriate antibiotics, and 100 µL of antifoam.
4. Hook up all probes, acid (1 *M* HCl), base (5% NH_4OH), antifoam, and air feed lines.
5. Adjust the temperature of the refrigeration unit to 7°C and slave it to the fermentor.
6. Program the following set points in the control unit: pH = 7.0, impeller speed = 500 rpm, aeration rate = 1 L/min, temperature = 37 or 42°C.
7. Grow the cells to A_{600} = 1.0 (*see* **Note 19**) and initiate cooling to 15°C by programming a cooling rate of 0.5°C per min in the control unit (*see* **Note 20**).
8. Harvest the cells 3 h after temperature downshift.

4. Notes

1. All four of the cold-inducible *csp* genes are transcribed with a 5' UTR, reinforcing the idea that this region plays an important role in regulation. These UTRs are of comparable length (159 nt for *cspA*, 161 nt for *cspB*, 156 nt for *cspG,* and 145 nt for *cspI*) and share a fair degree of homology. However, they appear to confer different cold-inducibility ranges ***(10)***. CspI induction takes place over the lowest and narrower span of temperatures (10–15°C), while CspB and CspG are maximally induced in the range 10–20°C. CspA induction occurs at the highest levels over the broadest and most practically useful temperature range (10–30°C).
2. We have observed that about 350 nt of upstream sequence is necessary for efficient *cspA*-driven protein expression.
3. Cold shock also leads to a decrease in the synthesis of heat-inducible molecular chaperones (e.g., DnaK-DnaJ-GrpE and GroEL-GroES) that may be required for the folding of certain recombinant proteins. However, since the production of most host proteins stops immediately after transfer of exponentially growing cells to 10–15°C, a larger supply of uncomplexed chaperones should be available to provide folding assistance to the few newly translated polypeptides that are synthesized following temperature downshift.
4. Most antibiotics are heat-labile and will lose potency when added to the medium immediately after sterilization, leading to partial or complete loss of selective pressure.
5. Carbenicillin and neomycin are more stable than ampicillin and kanamycin, respectively, and should be used in place of the latter antibiotics to maintain plasmids. Antibiotic stocks should be discarded after 5–10 cycles of thawing/freezing.
6. *Nde*I is inhibited by impurities present in certain DNA preparations and has a short half-life at 37°C ($t_{1/2} \approx$ 15 min). If digestion is inefficient, repurify the DNA and add 5 additional units of enzyme to the digestion mixture after 20 min incubation at 37°C. Allow the digestion to proceed for a total time of at least 1 h.
7. Keep in mind that cold shock affects the efficiency of secretion owing to a decrease in membrane fluidity. Thus, precursor proteins may accumulate in the cytoplasm when cultures are cold shocked at 10–15°C.

8. If the P1 lysate is old or has a low titer, generate a fresh lysate on a wild type strain (e.g., MC4100) by following **steps 1–7**.
9. The soft agar layer should be clear after 8–12 h incubation. A hazy appearance is indicative of a low-titer P1 lysate. If this is the case, raise a fresh P1 stock on wild type cells (*see* **Note 8**).
10. The volume of lysate obtained depends on the moisture level of the soft agar layer. This procedure typically yields 100–500 µL of lysate.
11. Addition of one to two drops of chloroform to the lysate will prevent bacterial growth. If chloroform is added, centrifuge the lysate before use to avoid carrying over any of the solvent.
12. As a recommendation: spread 100 µL of 1 *M* sodium citrate on the agar 1 h before plating the cells to inhibit residual phage growth.
13. It is important not to exceed 20 min incubation on ice since *rbfA* cells are cold-sensitive.
14. In addition to low temperatures, nutritional upshift transiently induces the *cspA* promoter ***(36)***: Thus, inoculation of fresh medium with stationary phase seed cultures may lead to low level accumulation of the target protein at 37 or 42°C. This problem can be partially addressed by using actively growing cells for inoculation.
15. In the case of certain inner membrane proteins, toxicity effects become less pronounced at lower temperatures, allowing the use of *cspA*-driven expression (*see* ***ref. [25]***).
16. Since resumption of cell growth correlates with repression of *cspA*-driven transcription in wild type cells, the target protein concentration will decrease upon prolonged incubation at low temperatures. Although dilution effects are relatively small at 15°C, they become significant at downshift temperatures of 20–30°C. If the latter conditions are used, cells should be harvested at the end of the lag phase, which can be ascertained from growth curves. In JM109 transformants grown at 37°C, the lag phase lasts for more than 3 h following transfer to 15°C, 2 h following transfer to 20°C, and 30 min following transfer to 29°C ***(5)***. Keep in mind that these values depend on the identity of the host.
17. This volume of medium will cool to the temperature of the surroundings within 5 min.
18. This protocol is designed for a 2.5-L–working-volume reactor. Nevertheless, we have shown that typical heat transfer limiting cooling profiles encountered in 60-L vessels are adequate for induction ***(5)***. Although we anticipate that the *cspA* system should perform adequately up to 100 L, heat transfer limitations in larger reactors will likely interfere with efficient induction.
19. Richer media (e.g., Superbroth or Terrific broth) can be used to grow the biomass to higher density (A_{600} = 5–10) before temperature downshift. Alternatively, fed-batch fermentations can be carried out as described ***(5)***.
20. In the case of *rbfA* host cells, accumulate the biomass at 42°C and use a final downshift temperature of 23°C with a cooling rate of 0.5°C/min.

References

1. Broeze, R. J., Solomon, C. J., and Pope, D. H. (1978) Effect of low temperature on in vivo and in vitro protein synthesis in *Escherichia coli* and *Pseudomonas fluorescens*. *J. Bacteriol.* **134,** 861–874.
2. Dammel, C. S. and Noller, H. F. (1995) Suppression of a cold-sensitive mutation in 16S rRNA by overproduction of a novel ribosome-binding factor, RbfA. *Genes Dev.* **9,** 626–637.
3. Jones, P. G. and Inouye, M. (1996) RbfA, a 30S ribosomal binding factor, is a cold-shock protein whose absence triggers the cold-shock response. *Mol. Microbiol.* **21,** 1207–1218.
4. Jones, P. G., VanBogelen, R. A., and Neidhardt, F. C. (1987) Induction of proteins in response to low temperatures in *Escherichia coli*. *J. Bacteriol.* **169,** 2092–2095.
5. Vasina, J. A., Peterson, M. S., and Baneyx, F. (1998) Scale-up and optimization of the low-temperature inducible *cspA* promoter system. *Biotechnol. Prog.* **14,** 714–721.
6. Shaw, M. K. and Ingraham, J. L. (1967) Synthesis of macromolecules by *Escherichia coli* near the minimal temperature for growth. *J. Bacteriol.* **94,** 157–164.
7. Sledjeski, D. D., Gupta, A., and Gottesman, S. (1996) The small RNA, DsrA, is essential for the low temperature expression of RpoS during exponential phase growth in *Escherichia coli*. *EMBO J.* **15,** 3993–4000.
8. Thieringer, H. A., Jones, P. G., and Inouye, M. (1998) Cold shock and adaptation. *Bioessays* **20,** 49–57.
9. Jones, P. G., Mitta, M., Kim, Y., Jiang, W., and Inouye, M. (1996) Cold-shock induces a major ribosomal-associated protein that unwinds double-stranded RNA in *Escherichia coli*. *Proc. Natl. Acad. Sci. USA* **93,** 76–80.
10. Wang, N., Yamanaka, K., and Inouye, M. (1999) CspI, the ninth member of the CspA family of *Escherichia coli*, is induced upon cold shock. *J. Bacteriol.* **181,** 1603–1609.
11. Yamanaka, K., Fang, L., and Inouye, M. (1998) The CspA family in *Escherichia coli*: multiple gene duplication for stress adaptation. *Mol. Microbiol.* **27,** 247–255.
12. Jiang, W., Hou, Y., and Inouye, M. (1997) CspA, the major cold-shock protein of *Escherichia coli*, is an RNA chaperone. *J. Biol. Chem.* **272,** 196–202.
13. Bae, W., Xia, B., Inouye, M., and Severinov, K. (2000) *Escherichia coli* CspA-family RNA chaperones are transcription antiterminators. *Proc. Natl. Acad. Sci. USA* **97,** 7784–7789.
14. Xia, B., Ke, H., and Inouye, M. (2001) Acquirement of cold sensitivity by quadruple deletion of the *cspA* family and its suppression by PNPase S1 domain in *Escherichia coli*. *Mol. Microbiol.* **40,** 179–188.
15. Goldstein, J., Pollitt, N. S., and Inouye, M. (1990) Major cold shock protein of *Escherichia coli*. *Proc. Natl. Acad. Sci. USA* **87,** 283–287.
16. Goldenberg, D., Azar, I., Oppenheim, A. B., Brandi, A., Pon, C. L., and Gualerzi, C. O. (1997) Role of *Escherichia coli cspA* promoter sequences and adaptation of translational apparatus in the cold shock response. *Mol. Gen. Genet.* **256,** 282–290.

17. Mitta, M., Fang, L., and Inouye, M. (1997) Deletion analysis of *cspA* of *Escherichia coli:* requirement of the AT-rich UP element for *cspA* transcription and the downstream box in the coding region for its induction. *Mol. Microbiol.* **26,** 321–335.
18. Brandi, A., Pietroni, P., Gualerzi, C. O., and Pon, C. L. (1996) Post-transcriptional regulation of CspA expression in *Escherichia coli. Mol. Microbiol.* **19,** 231–240.
19. Fang, L., Jiang, W., Bae, W., and Inouye, M. (1997) Promoter-independent cold-shock induction of CspA and its derepression at 37°C by mRNA stabilization. *Mol. Microbiol.* **23,** 355–364.
20. Goldenberg, D., Azar, I., and Oppenheim, A. B. (1996) Differential mRNA stability of the *cspA* gene in the cold-shock response of *Escherichia coli. Mol. Microbiol.* **19,** 241–248.
21. Vasina, J. A. and Baneyx, F. (1996) Recombinant protein expression at low temperatures under the transcriptional control of the major *Escherichia coli* cold shock promoter *cspA. Appl. Environ. Microbiol.* **62,** 1444–1447.
22. Vasina, J. A. (1997), Ph.D. Thesis, University of Washington.
23. Jiang, W., Jones, P., and Inouye, M. (1993) Chloramphenicol induces the transcription of the major cold shock gene of *Escherichia coli, cspA. J. Bacteriol.* **175,** 5824–5828.
24. Yamanaka, K., Mitta, M., and Inouye, M. (1999) Mutation analysis of the 5' untranslated region of the cold shock *cspA* mRNA of *Escherichia coli. J. Bacteriol.* **181,** 6284–91.
25. Mujacic, M., Cooper, K. W., and Baneyx, F. (1999) Cold-inducible cloning vectors for low-temperature protein expression in *Escherichia coli*: application to the production of a toxic and proteolytically sensitive fusion protein. *Gene* **238,** 325–332.
26. Vasina, J. A. and Baneyx, F. (1997) Expression of aggregation-prone recombinant proteins at low temperatures: a comparative study of the *Escherichia coli cspA* and *tac* promoter systems. *Protein Express. Purif.* **9,** 211–218.
27. Bae, W., Jones, P. G., and Inouye, M. (1997) CspA, the major cold shock protein of *Escherichia coli*, negatively regulates its own expression. *J. Bacteriol.* **179,** 7081–7088.
28. Fang, L., Hou, Y., and Inouye, M. (1998) Role of the cold-box region in the 5' untranslated region of the *cspA* mRNA in its transient expression at low temperature in *Escherichia coli. J. Bacteriol.* **180,** 90–95.
29. Jiang, W., Fang, L., and Inouye, M. (1996) The role of the 5'-end untranslated region of the mRNA for CspA, the major cold-shock protein of *Escherichia coli*, in cold-shock adaptation. *J. Bacteriol.* **178,** 4919–4925.
30. Phadtare, S., Alsina, J., and Inouye, M. (1999) Cold-shock response and cold-shock proteins. *Curr. Opin. Microbiol.* **2,** 175–180.
31. VanBogelen, R. A. and Neidhardt, F. C. (1990) Ribosomes as sensors of heat and cold shock in *Escherichia coli. Proc. Natl. Acad. Sci. USA* **87,** 5589–5593.
32. Baneyx, F. (1999) In vivo folding of recombinant proteins in *Escherichia coli in Manual of industrial microbiology and biotechnology,* 2nd edn. (Demain, A. L., Davies, J. E., Altas, R. M., et al., eds.), ASM Press, Washington, D. C., pp. 551–565.
33. Farewell, A. and Neidhardt, F. C. (1998) Effect of temperature on in vivo protein synthetic capacity in *Escherichia coli. J. Bacteriol.* **180,** 4704–4710.

34. Taura, T., Kusukawa, N., Yura, T., and Ito, K. (1989) Transient shut off of *Escherichia coli* heat shock protein synthesis upon temperature shift down. *Biochem. Biophys. Res. Commun.* **163,** 438–443.
35. Gottesman, S. (1996) Proteases and their targets in *Escherichia coli. Annu. Rev. Genet.* **30,** 465–506.
36. Brandi, A., Spurio, R., Gualerzi, C. O., and Pon, C. L. (1999) Massive presence of the *Escherichia coli* "major cold shock protein" CspA under non-stress conditions. *EMBO J.* **18,** 1653–1659.

2

Dual-Expression Vectors for Efficient Protein Expression in Both *E. coli* and Mammalian Cells

Rebecca L. Mullinax, David T. Wong, Heidi A. Davis, Kerstein A. Padgett, and Joseph A. Sorge

1. Introduction

In the near future, the nucleotide sequence of the genomes from many different organisms will be available. The next and more challenging step will be to characterize the biological role of each gene and the way in which the encoded protein functions in the cell. Dual-expression vectors for expression of proteins encoded by these genes in mammalian and bacterial cells can be used for this characterization. Typically, eukaryotic genes are expressed in mammalian cells to characterize biological functions and in bacterial cells to facilitate isolation of the protein. This generally requires the use of more than one vector. In contrast, use of a dual-expression vector eliminates the need to subclone from one vector system to another by combining the essential features of both eukaryotic and prokaryotic vectors in a single vector.

The pDual® GC expression system was designed for high-level protein expression in mammalian and bacterial cells (*see* **Fig. 1A**; *[1,2]*). cDNA inserts encoding proteins are inserted into the vector using the unique seamless cloning method (*see* **Fig. 1B**; *[5]*). This method is advantageous because it can result in the expression of the protein without extraneous amino acids encoded by restriction sites at the termini. As an alternative, the method allows for the optional expression of vector-encoded protein sequences that can be used to detect and purify the protein.

All pDual GC clones can express a fusion protein consisting of the cDNA, a thrombin cleavage site, three copies of the c-myc epitope tag, and a single copy of the 6xHis epitope and purification tag. The c-myc epitope is derived from the human c-*myc* gene and contains 10 amino acid residues (EQKLISEEDL;

From: *Methods in Molecular Biology, vol. 205, E. coli Gene Expression Protocols*
Edited by: P. E. Vaillancourt © Humana Press Inc., Totowa, NJ

A

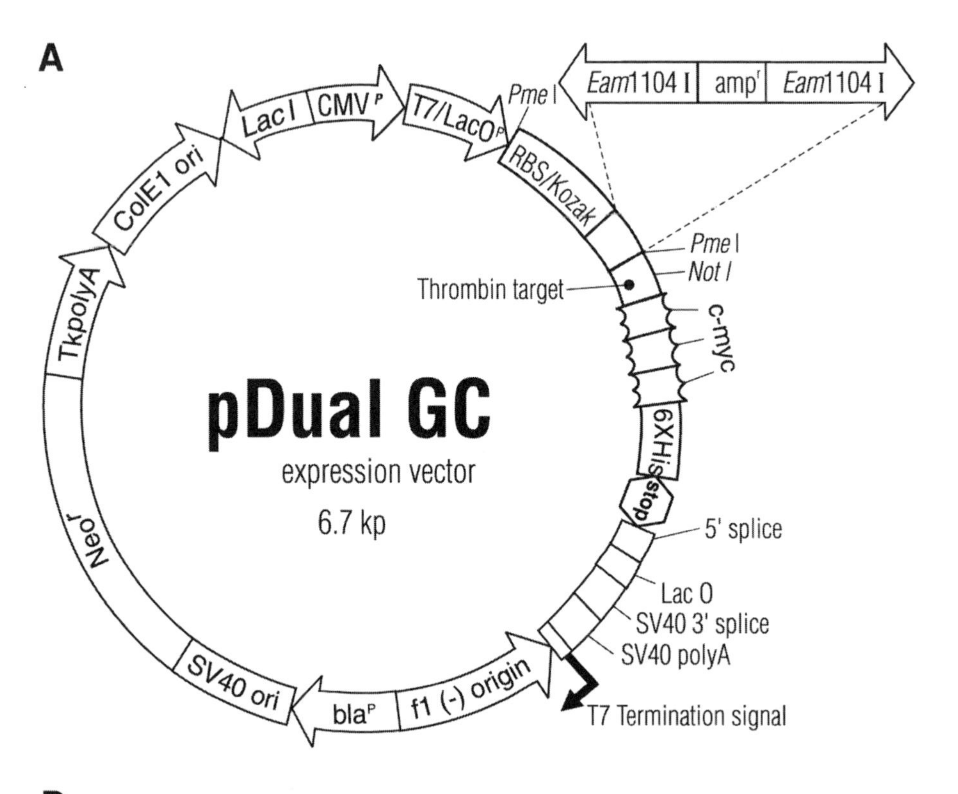

B

*Eam*1104 I — NN**CTCTTC**NATG — cDNA insert — CTTN**GAAGAG**NN — *Eam*1104 I — PCR product
NN**GAGAAG**NTAC — cDNA insert — GAAN**CTTCTC**NN

pDUAL GC --- ATGN**GAAGAG** (*Eam*1104 I) ----------- **CTTCTC**NCTT (*Eam*1104 I) --- pDUAL GC — pDUAL GC
pDUAL GC --- TACN**CTTCTC** ----------- **GAAGAG**NGAA --- pDUAL GC

digest with *Eam*1104 I

----- cDNA insert ------ CTT — PCR product
TAC cDNA insert

pDUAL GC --- ATG --- pDUAL GC — pDUAL GC
pDUAL GC --- GAA --- pDUAL GC

anneal and ligate

pDUAL GC --- ATG ----- cDNA insert -------- CTT --- pDUAL GC — expression clone
pDUAL GC --- TAC ----- cDNA insert -------- GAA --- pDUAL GC

[6]). This allows for sensitive detection and immunoprecipitation of expressed proteins with anti–c-myc antibody. The 6xHis epitope and purification tag consists of six histidine residues and allows for quick and easy detection of expressed proteins with anti-6xHis antibody and purification of the fusion protein from bacterial cells using a nickel-chelating resin *(7)*. A thrombin cleavage site between the protein encoded by the cDNA and the c-myc and 6xHis tags allows for the removal of both tags when desired, for example, following protein purification.

A *Not*I recognition site is located between the cDNA insertion site and sequences encoding the thrombin cleavage site. This site allows for the insertion of nucleotides encoding protein domains that would be expressed as a C-terminal fusion to the expressed protein. An example would be to insert nucleotides encoding hrGFP *(8)* followed by a translational stop codon. The clone would then express

---

Fig. 1. **(A)** The vector contains a mutagenized version of the promoter and enhancer region of the human cytomegalovirus (CMV) immediate early gene for constitutive expression of the clones in either transiently or stably transfected mammalian cells. Inducible gene expression in prokaryotes is directed from the hybrid T7/*lacO* promoter. The vector carries a copy of the *lac* repressor gene (*laqI*^{q}), which mediates tight repression of protein expression in the absence of the inducer, isopropyl-β-D-thiogalactopyranoside (IPTG). Expression is therefore regulated using IPTG in bacteria that express T7 polymerase under the regulation of the *lac* promoter. A tandem arrangement of the bacterial Shine-Dalgarno *(3)* and mammalian Kozak *(4)* ribosomal binding sites (RBS) allows for efficient expression of the open reading frame (ORF) in both bacterial and mammalian systems. In both bacterial and mammalian cells, the dominant selectable marker is the neomycin phosphotransferase gene, which is under the control of the β-lactamase promoter in bacterial cells and the SV40 promoter in mammalian cells. Expression of the neomycin phosphotransferase gene in mammalian cells allows stable clone selection with G418, whereas in bacteria the gene confers resistance to kanamycin. The *beta-lactamase* gene (amp^{r}), which confers resistance to ampicillin in bacteria, is removed during preparation of the expression clone. **(B)** The PCR product and pDUAL GC vector contain Eam*1104* I restriction sites (bold). Digestion of the PCR product and pDUAL GC vector with Eam*1104* I create complementary 3-base overhanging ends (underlined). Directional annealing of the complementary bases followed by ligation results in an expression clone capable of expressing the encoded cDNA in bacterial and mammalian cells. ATG, encoding methionine, is the first codon of the cDNA protein. CTT, encoding leucine, follows the last codon of the protein encoded by the cDNA insert and allows for expression of the downstream thrombin cleavage site, three copies of the c-myc epitope tag, and a single copy of the 6xHis epitope and purification tag. Alternatively, nucleotides encoding a stop codon follow the last codon of the protein encoded by the cDNA insert thereby terminating protein expression.

a fusion protein consisting of the cDNA and hrGFP. Expression of this fusion protein in mammalian cells allows for subcellular detection of the fusion protein.

A wide variety of proteins have been expressed using the pDUAL GC expression vector. To date, over 500 different eukaryotic proteins have been expressed in mammalian cells and detected using the c-myc epitope tag ***(9)***. In addition, over 100 different eukaryotic proteins have been expressed in bacterial cells and detected using the 6xHIS epitope tag (Ed Marsh, personal communication). These proteins are members of many different classes of proteins including kinases, DNA-binding proteins, transferases, transporters, oncogenes, cytochromes, proteases, inflammatory response proteins, cellular matrix proteins, metabolic proteins, synthases, esterases, zinc-finger proteins, and ribosomal proteins. Potential uses for these expressed proteins include analyzing protein function, defining both protein-protein and protein-DNA interactions, elucidating pathways, studying protein degradation, studying catalytic activity, determining the effects of over-expression, and preparing antigen.

2. Materials

2.1. Preparation of Plasmid Expressing Protein of Interest

2.1.1. Preparation of cDNA Insert

1. PCR primers containing *Eam*1104 I recognition sites.
2. DNA template encoding gene of interest.
3. Pfu DNA polymerase.
4. 10X Cloned Pfu polymerase buffer: 100 m*M* KCl, 100 m*M* $(NH_4)_2SO_4$, 200 m*M* Tris-HCl, pH 8.8, 20 m*M* $MgSO_4$, 1% Triton X-100, and 1 mg/mL of nuclease-free bovine serum albumin (BSA).
5. 5-Methyldeoxycytosine (^{m5}dCTP), optional.
6. *Eam*1104 I restriction enzyme.
7. 10X Universal buffer: 1 *M* potassium acetate (KOAc), 250 m*M* Tris-acetate, pH 7.6, 100 m*M* magnesium acetate ($Mg(OAc)_2$), 5 m*M* β-mercaptoethanol, and 100 μg/mL BSA. Autoclave.

2.1.2. Preparation of pDUAL GC Expression Vector

1. pDual® GC expression vector.
2. *Eam*1104 I restriction enzyme.
3. 10X Universal buffer: 1 *M* KOAc, 250 m*M* Tris-acetate, pH 7.6, 100 m*M* $Mg(OAc)_2$, 5 m*M* β-mercaptoethanol, and 100 μg/mL BSA. Autoclave.

2.1.3. Ligation of cDNA insert and pDUAL GC Expression Vector

1. T4 DNA ligase.
2. 10X Ligase Buffer: 500 m*M* Tris-HCl, pH 7.5, 70 m*M* $MgCl_2$, and 10 m*M* dithiothreitol (DTT).

3. T4 DNA ligase dilution buffer (1X ligase buffer).
4. 10 m*M* rATP.
5. Epicurian Coli® XL1-Blue supercompetent cells MRF' (Stratagene).
6. β-Mercaptoethanol.
7. SOB medium per liter: 20.0 g of tryptone, 5.0 g of yeast extract, and 0.5 g of NaCl. Autoclave. Add 10 mL of 1 *M* $MgCl_2$ and 10 mL of 1 *M* $MgSO_4$.
8. SOC medium per 100 mL: 1 mL of a 2 *M* filter-sterilized glucose solution or 2 mL of 20% (w/v) glucose. Adjust to a final volume of 100 mL with SOB medium. Filter sterilize.
9. Luria-Bertani (LB) agar per liter: 10 g of NaCl, 10 g of tryptone, 5 g of yeast extract, and 20 g of agar. Add dH_2O to a final volume of 1 L. Adjust pH to 7.0 with 5 N NaOH. Autoclave. Pour into Petri dishes (~25 mL/100-mm Petri dish).
10. LB-kanamycin agar per liter: Prepare 1 L LB agar. Autoclave. Cool to 55°C and add 5 mL of 10 mg/mL-filter-sterilized kanamycin. Pour into Petri dishes (~25 mL/100-mm plate).
11. LB broth per liter: 10 g of NaCl, 10 g of tryptone, and 5 g of yeast extract. Add deionized H_2O to a final volume of 1 L. Adjust pH to 7.0 with 5 *N* NaOH. Autoclave.
12. LB-kanamycin broth per liter: Prepare 1 L of LB broth. Autoclave. Cool to 55°C. Add 5 mL of 10 mg/mL-filter-sterilized kanamycin.
13. Tris-EDTA (TE) buffer: 10 m*M* Tris-HCl, pH 7.5, and 1 m*M* ethylenediaminetetraacetic acid (EDTA). Autoclave.

2.2. Protein Expression in Bacterial Cells

1. *Escherichia coli* competent cells that express T7 polymerase in the presence of IPTG.
2. IPTG (1 *M*): 238.3 mg/mL in distilled water.
3. 2X Sodium dodecyl sulfate (SDS) gel sample buffer: 100 m*M* tris-HCl, pH 6.5, 4% (w/v) SDS (electrophoresis grade), 0.2% (w/v) bromophenol blue, and 20% (v/v) glycerol. Add dithiothreitol (DTT) to a final concentration of 200 m*M* before use.

3. Methods

Additional information regarding these techniques that is beyond the scope of this chapter can be found in **ref. *10***.

3.1. Design of Primers Used to Amplify cDNA Insert

The cDNA inserts are generated by PCR amplification with primers that contain *Eam*1104 I recognition sites and a minimal flanking sequence at their 5' termini. The ability of *Eam*1104 I to cleave several bases downstream of its recognition site allows the removal of superfluous, terminal sequences from the amplified DNA insert. The elimination of extraneous nucleotides and the generation of unique, nonpalindromic sticky ends permit the formation of directional seamless junctions during the subsequent ligation to the pDual GC expression vector.

The cDNA insert is amplified using PCR primers to introduce *Eam*1104 I recognition sites in each end of the cDNA insert to position the cDNA in the pDUAL GC expression vector for optimal protein expression. *Eam*1104 I is a type IIS restriction enzyme that has the capacity to cut outside its recognition sequence (5'-CTCTTC-3'). The cleavage site extends one nucleotide on the upper strand in the 3' direction and four nucleotides on the lower strand in the 5' direction. Digestion with *Eam*1104 I generates termini that feature three nucleotides in their 5' overhangs. A minimum of two extra nucleotides must precede the 5'-CTCTTC-3' recognition sequence in order to ensure efficient cleavage of the termini. The bases preceding the recognition site can be any of the four nucleotides.

The forward primer must be designed with one extra nucleotide (N) located between the *Eam*1104 I recognition sequence and the gene's translation initiation codon, in order to generate the necessary 5'-ATG overhang that is complementary to the pDUAL GC expression vector sequence. The forward primer should be designed to look as follows: 5'-NNCTCTTCNATG(X)$_{15}$-3'; where N denotes any of the four nucleotides, X represents gene-specific nucleotides, and the underlined nucleotides represent the *Eam*1104 I recognition site.

The reverse primer must be designed with one nucleotide (N) located between the *Eam*1104 I recognition sequence and the AAG triplet that comprises the 5' overhang complementary to the vector sequence. Depending on whether or not the c-myc and 6xHIS tags are desired as fusion partners, the reverse primer should be designed to look as follows: (1) Reverse primer design to allow the expression of the c-myc and 6xHIS fusion tags: 5'-NNCTCTTCNAAG(X)$_{15}$-3'; where N denotes any of the four nucleotides and X represents the gene-specific nucleotides. (2) The reverse primer design that does not allow expression of the c-myc and 6xHIS fusion tags: 5'-NNCTCTTCNAAG*TTA*(X)$_{15}$-3'; where N denotes any of the four nucleotides and X represents the gene-specific nucleotides. The necessary stop codon is shown in italics.

The primer should be complementary to a minimum of 15 nucleotides of the template on the 3' end of the PCR primer in addition to the *Eam*1104 I recognition sequence. The estimated T_m [$T_m \approx 2°C (A + T) + 4°C (G + C)$] of the homologous portion of the primer should be 55°C or higher, with a G-C ratio of 60% or more.

3.2. PCR Amplification of cDNA Insert0

If the insert contains an internal *Eam*1104 I recognition site, the amplification reaction should be performed in the presence of 5-methyldeoxycytosine triphosphate (^{m5}dCTP) for the last five cycles of the PCR (*see* **Note 1**). Incorporation of ^{m5}dCTP during the PCR amplification protects already-existing internal *Eam*1104 I sites from subsequent cleavage by the endonuclease ***(1,2,5)***.

The primer-encoded *Eam*1104 I sites are not affected by the modified nucleotide because the newly synthesized strand does not contain cytosine residues in the recognition sequence.

1. Combine the following components in a 500–µL thin-walled tube (*see* **Note 2**). Add the components in the order given. Mix the components well before adding the Pfu DNA polymerase (*see* **Note 3**): 81.2 µL distilled water, 10.0 µL 10X Pfu DNA polymerase buffer, 0.8 µL 25 m*M* each dNTP, 1.0 µL 1–100 ng/µL plasmid DNA template, 2.5 µL 10 µM primer #1, 2.5 µL 10 µ*M* primer #2, and 2.0 µL 2.5 U/µL cloned Pfu DNA polymerase.
2. Recommended cycling parameters
   a. For inserts that do not contain internal *Eam*1104 I restriction sites (*see* **Note 4**): 1 cycle at 94–98°C, 45 s; 25–30 cycles at 94–98°C, 45 s; primer T_m –5°C, 45 s; and 72°C for 1–2 min/kb of PCR target; and 1 cycle at 72°C, 10 min.
   b. For inserts that contain internal *Eam*1104 I restriction sites: (*see* **Notes 4** and **5**) 1 cycle at 94–98°C, 45 s; 20–25 cycles at 94–98°C, 45 s; primer T_m –5°C, 45 s; and 72°C for 1–2 min/kb of PCR target and 1 cycle at 72°C, 10 min. After the first PCR, add 1 µL 25 m*M* ^{m5}dCTP. Perform a second PCR of 5 cycles at 98°C, 45 s; primer T_m –5°C, 45 s; and 72°C for 1–2 min/kb of PCR target and 1 cycle at 72°C, 10 min.
3. Analyze the PCR amplification products on a 0.7–1.0% (w/v) agarose gel.

3.3. PCR Product Purification

Before digestion, the PCR product must be removed from unincorporated PCR primers, unincorporated nucleotides, and the thermostable polymerase. Suitable purification methods include phenol:chloroform extraction, selective precipitation gel purification, or spin-cup purification. To prepare the insert for ligation, treat the PCR product with *Eam*1104 I (≥24 units/µg PCR product).

1. Mix the following components in a 1.5-mL microcentrifuge tube: dH_2O for a final volume of 30 µL, 1–5 µL of PCR product, 3 µL of 10X universal buffer, and 3 µL of 8 U/µL *Eam*1104 I restriction enzyme.
2. Mix the digestion reaction gently and incubate at 37°C for 1 h.
3. Purify the digested PCR product by gel purification (*see* **Note 6**) and resuspend in TE buffer.

3.4. Eam*1104 I Digestion of pDual GC Expression Vector*

The cloning region of the pDual GC expression vector is characterized by the presence of two *Eam*1104 I recognition sequences (5'-CTCTTC-3') directed in opposite orientations and separated by a spacer region. The sites are positioned for maximal protein expression and optional expression of the downstream epitope and purification tags. Digestion with *Eam*1104 I restriction enzyme creates 3-nucleotide 5' overhangs that are directionally ligated to the 5' overhangs of the cDNA insert. Because one of the sticky ends in the pDUAL

GC expression vector is complementary to the ATG of the cDNA insert, protein expression begins with the gene's own translation initiation codon. Digestion of the pDUAL GC expression vector creates two nonpalindromic, nonidentical overhanging ends and results in directional ligation of the cDNA insert.

To generate a ligation-ready vector for PCR cloning, the pDual GC expression vector is digested with *Eam*1104 I.

1. Mix the following components in a 1.5-mL microcentrifuge tube: dH_2O for a final volume of 30 µL, ≤1 µg pDUAL GC expression vector (*see* **Note 7**), 3 µL of 10X universal buffer and 3 µL of 8 U/µL *Eam*1104 I restriction enzyme.
2. Mix the digestion reaction gently and incubate at 37°C for 2 h.
3. Purify the digested vector by gel purification and resuspend in TE buffer to a final concentration of 100 ng/µL.

3.5. Ligation of Digested Vector and Insert

The vector and insert are directionally ligated at the compatible overhanging ends.

1. Combine the following in a 1.5-mL microcentrifuge tube: 1 µL 100 ng/µL digested pDUAL GC expression vector, x µL digested insert (3:1 molar ratio of insert to vector, *see* **Note 8**), 2 µL 10X ligase buffer, 2 µL of 10 m*M* rATP, 1 µL of (4 U/µL) T4 DNA ligase, and dH_2O to a final volume of 20 µL.
2. Mix the ligation reactions gently and then incubate for 1 h at room temperature or overnight at 16°C.
3. Store the ligation reactions on ice until ready to use for transformation into *E. coli* competent cells.

3.6. Transformation of Ligated DNA

Methylation of nucleic acids has been found to affect transformation efficiency. If the cDNA insert was amplified in the presence of methylated dCTP (^{m5}dCTP), use an *E. coli* strain that does not have an active restriction system that restricts methylated cytosine sequences, such as Epicurian Coli® XL1-Blue MRF' supercompetent cells (Stratagene).

1. Prepare competent cells and keep on ice.
2. Gently mix the cells by hand. Aliquot 100 µL of the cells into a prechilled 15-mL Falcon 2059 polypropylene tube.
3. Add 1.7 µL of the 14.2 *M* β-mercaptoethanol to 100 µL of bacteria.
4. Swirl the contents of the tube gently. Incubate the cells on ice for 10 min, swirling gently every 2 min.
5. Add 5 µL of the ligation reaction to the cells and swirl gently.
6. Incubate the tubes on ice for 30 min.
7. Prepare and equilibrate SOC medium to 42°C.

8. Heat-pulse the tubes in a 42°C water bath for 45 s. The length of time of the heat pulse is critical for obtaining the highest efficiencies.
9. Incubate the tubes on ice for 2 min.
10. Add 0.9 mL of equilibrated SOC medium and incubate the tubes at 37°C for 1 h with shaking at 225–250 rpm (*see* **Note 9**).
11. Using a sterile spreader, plate 5–10% of the transformation reactions onto separate LB-kanamycin agar plates.
12. Incubate the plates overnight at 37°C.
13. Identify colonies containing the desired clone by isolation of miniprep DNA from individual colonies followed by restriction enzyme analysis. Determining the nucleotide sequence of the cDNA insert is highly recommended.

3.7. Protein Expression and Detection in Bacterial Cells

For expression of the fusion protein in bacteria, transform mini-prep DNA into *E. coli* cells which express T7 polymerase when induced with IPTG (*see* **Note 10**). The following is a small-scale protocol intended for the analysis of individual transformants.

1. Prepare competent cells.
2. Transform competent cells with pDUAL GC clone.
3. Identify colonies containing pDUAL GC clone.
4. Inoculate 1-mL aliquots of LB broth (containing 100 μg/mL ampicillin) with single colonies (*see* **Note 11**). Incubate at 37°C overnight with shaking at 220–250 rpm.
5. Transfer 100 μL of each overnight culture into fresh 1-mL aliquots of LB broth without antibiotics. Incubate at 37°C for 2 h with shaking at 220–250 rpm.
6. Transfer 100 μL of each 2-h culture into a clean microfuge tube and place the tube on ice until needed for gel analysis. These samples will be the noninduced control samples.
7. Add IPTG to a final concentration of 1 m*M* to the remaining 2-h cultures. Incubate at 37°C for 4 h with shaking at 220–250 rpm (*see* **Notes 12–13**).
8. At the end of the incubation period, place the induced cultures on ice.
9. Pipet 20 μL of each induced culture into a clean microcentrifuge tube. Add 20 μL of 2X SDS gel sample buffer to each tube.
10. Harvest the cells by centrifugation at 4000*g* for 15 min.
11. Decant the supernatant and store the cell pellet at –70°C if desired or process immediately to purify the induced protein.
12. Mix the tubes containing the non-induced cultures to resuspend the cells. Pipet 20 μL from each tube into a fresh microcentrifuge tube. Add 20 μL of 2X SDS gel sample buffer to each tube.
13. Heat all tubes to 95°C for 5 min and place on ice. Load samples on 6% SDS-PAGE gel with the noninduced samples and induced samples in adjacent lanes. Separate the proteins by electrophoresis at 125 V until the bromophenol blue reaches the bottom of the gel.

14. Stain the separated proteins in the gel using Coomassie® Brilliant Blue.
15. The amount of induced protein should be greater in the induced cultures than in the noninduced cultures.

3.8. Detection and Isolation of Protein from Bacterial Cells

Expression of the fusion protein containing the 6xHIS tag can be detected by Western blot analysis *(10)* using an anti-6xHIS antibody and isolated from the induced cultures by nickel metal affinity chromatography *(7)*. The most commonly used reagent for isolating 6xHIS-tagged proteins is Ni-NTA (nickel nitrilotriacetic acid, QIAGEN).

3.9. Protein Expression and Detection in Mammalian Cells

Transfection of genes into mammalian cells for protein expression is a fundamental tool for the analysis of gene function. There are many well-established protocols that result in a high number of viable cells expressing the protein. These protocols include diethyl amino ethyl (DEAE)-dextran- phosphate *(11)* and calcium-mediated transfection *(12)*. Many factors contribute to transient and stable transfection efficiency; however, the primary factor is the cell type. Different cell lines vary by several orders of magnitude in their ability to take up and express protein from plasmids. Other factors that effect efficiency include the use of highly purified plasmid DNA, optimal cell density, optimal transfection reagent to DNA ratio, and the optimal time the transfection reagent is in contact with the cells prior to dilution with growth medium. These conditions vary with each cell type.

Expression of the fusion protein containing the c-myc epitope can be detected in mammalian cell lysates by Western blot analysis *(10)* using an anti-c-myc antibody.

4. Notes

1. The addition of the ^{m5}dCTP is delayed until the final five cycles of amplification to avoid the possible deamination of the ^{m5}dCTP by extended exposure to cycles of heating and cooling.
2. The use of thin-wall tubes is highly recommended for optimal thermal transfer during PCR.
3. The use of a high fidelity polymerase, such as Pfu DNA polymerase, in the amplification reaction is highly recommended to eliminate mutations that could be introduced during the PCR. In addition, Pfu DNA polymerase is very thermostable and is not inactivated by the high temperatures used in this protocol.
4. Critical optimization parameters for successful amplification of the template DNA include the use of an extension time that is adequate for full-length DNA synthesis, sufficient enzyme concentration, optimization of the reaction buffer, adequate primer-template purity, and concentration and optimal primer design. Extension time is the most critical parameter affecting the yield of PCR product

obtained using Pfu DNA polymerase. The minimum extension time should be 1–2 min/kilobase pair of amplified template.

5. Thermal cycling parameters should be chosen carefully to ensure the shortest denaturation times to avoid enzyme inactivation, template damage and deamination of the ^{m5}dCTP, adequate extension times to achieve full-length target synthesis, and the use of annealing temperatures near the primer melting temperature to improve specificity of the PCR product.
6. Gel purification of the digested insert is optional but will reduce the number of colonies containing vector without insert following transformation.
7. Dephosphorylation of the vector is not required because nonidentical, nonpalindromic sticky ends are generated by the type IIS *Eam*1104 I restriction endonuclease.
8. The ideal insert-to-vector DNA ratio is variable; however, a reasonable starting point is 3:1 (insert-to-vector molar ratio), measured in available picomole ends. This is calculated as follows:

$$\text{picomole ends/microgram of DNA} = \frac{2 \times 106}{\text{number of base pair} \times 660}$$

9. Expression of the kanamycin gene by incubation of transformed cells in LB broth for at least 1 h prior to selection on LB-kanamycin agar plates is essential for efficient transformation.
10. The use of bacterial cells that express tRNA that are rare in *E. coli* but frequent in mammalian proteins is also highly recommended. Use of bacterial cells that are deficient in proteases, such as Lon and OmpT proteases, is highly recommended. These proteases can cause degradation of the over-expressed protein. The BL21-CodonPlus® competent cells contain extra copies of the argU, ileY, leuW, and/or proL tRNA, and are Lon and OmpT protease deficient. These tRNA are frequently of low abundance in *E. coli* cells but may be required for efficient translation of mammalian proteins. BL21-CodonPlus cells express T7 polymerase whose expression is induced in the presence of IPTG.
11. If BL21-CodonPlus cells are used, add 50 µg/mL chloramphenicol to the LB with ampicillin. Chloramphenicol is required to maintain the pACYC plasmid, which expresses the tRNA, in the BL21-Codon Plus strain.
12. The IPTG concentration and the induction time are starting values and may require optimization for each gene expressed.
13. The volume of induced culture required is determined by the protein expression level, protein solubility, and purification conditions. For proteins that are expressed at low levels, the minimum cell culture volume should be 50 mL.

References

1. Mullinax, R. L., Davis, H. A., Wong, D. T., et al. (2000) Sequence-validated and expression-tested human cDNA in a dual expression vector. *Strategies* **13,** 41–43.
2. Davis, H. A., Wong, D. T., Padgett, K. A., Sorge, J. A., and Mullinax, R. L. (2000) High-level dual mammalian and bacterial protein expression vector. *Strategies* **13,** 136–137.

3. Shine, J. and Dalgarno, L. (1974) The 3'-terminal sequence of *Escherichia coli* 16S ribosomal RNA: complementarity to nonsense triplets and ribosome binding sites. *Proc. Natl. Acad. Sci. USA* **71,** 1342–1346.
4. Kozak, M. (1986) Point mutations define a sequence flanking the AUG initiator codon that modulates translation by eukaryotic ribosomes. *Cell* **44,** 283–292.
5. Padgett, K. A. and Sorge, J. A. (1996) Creating seamless junctions independent of restriction sites in PCR cloning. *Gene* **168,** 31–35.
6. Evan, G. I., Lewis, G. K., Ramsay, G., and Bishop, J. M. (1985) Isolation of monoclonal antibodies specific for human c-myc proto-oncogene product. *Mol. Cell Biol.* **5,** 3610–3616.
7. Hochuli, E., Dobeli, H., and Schacher, A. (1987) New metal chelate adsorbent selective for proteins and peptides containing neighbouring histidine residues. *J. Chromatogr.* **411,** 177–184.
8. Felts, F., Rogers, B., Chen, K., Ji, H., Sorge, J., and Vaillancourt, P. (2000) Recombinant *Renilla reniformis* GFP displays low toxicity. *Strategies* **13,** 85–87.
9. Wynne, K., Wong, D. T., Nioko, V., et al. (2000) Sequence-validated and protein expression-tested human cDNA clones now available. *Strategies* **13,** 133–134.
10. Sambrook, J., Fritsch, E. F., and Maniatis, T., eds. (1989), *Molecular Cloning a Laboratory Manual,* Cold Spring Harbor Laboratory Press, Cold Spring Harbor, New York.
11. Wigler, M., Silverstein, S., Lee, L. S., Pellicer, A., Cheng, Y. and Axel, R. (1977) Transfer of purified herpes virus thymidine kinase gene to cultured mouse cells. *Cell* **11,** 223–232.
12. McCutchan, J. H. and Pagano, J. S. (1968) Enchancement of the infectivity of simian virus 40 deoxyribonucleic acid with diethylaminoethyl-dextran. *J. Natl. Cancer Inst.* **41,** 351–357.

3

A Dual-Expression Vector Allowing Expression in *E. coli* and *P. pastoris*, Including New Modifications

Angelika Lueking, Sabine Horn, Hans Lehrach, and Dolores J. Cahill

1. Introduction

Heterologous gene expression is often treated empirically and a number of host organisms are systematically tested. Early successes in the expression of recombinant proteins were achieved using the well-studied bacterium *Escherichia coli* ***(1)***. This prokaryotic expression system is simple to handle, cost-effective, and produces large amounts of heterologous proteins ***(2)***. However, when expressing many different genes, especially eukaryotic genes, this often leads to the production of aggregated and denatured proteins, localized in inclusion bodies, and only a small fraction matures into the desired native form ***(3–5)***. Alternatively, eukaryotic expression systems have been developed to obtain more soluble protein, which in addition, may undergo some eukaryotic post-translational modifications. Yeast expression systems, including the methylotrophic yeast *Pichia pastoris*, have been used over the last few years as powerful expression systems for a number of heterologous genes ***(6–10)***. However, both eukaryotic and prokaryotic systems have their advantages and disadvantages. Therefore, choosing a suitable expression system for a particular protein is a compromise, depending primarily on the properties of the protein, the amounts required, and its intended purpose.

To avoid labor-intensive and costly sub-cloning procedures, we have chosen two commonly used hosts, namely *E. coli* and *P. pastoris*, and generated one vector for inducible protein expression in both systems ***(11)***. Similar to *E. coli*, *P. pastoris* is known for its ability for rapid growth at high cell density and when combined with a strong promotor, has, in a number of cases, yielded up to several grams of the heterologous protein per liter of culture ***(6,12)***. The

From: *Methods in Molecular Biology, vol. 205, E. coli Gene Expression Protocols*
Edited by: P. E. Vaillancourt © Humana Press Inc., Totowa, NJ

dual expression vector combines eukaryotic and prokaryotic promotor elements. Phage T7 promoter, including the ribosomal binding site of the major capsid protein, promotes the efficient bacterial expression and is placed downstream from the *P. pastoris* promoter. The previously described dual shuttle vector consists of the strong alcohol oxidase promoter (*AOX*) that is tightly regulated, since protein expression is completely repressed when grown on glucose and maximally induced when grown on methanol ***(13)***. Our recently developed modification (pZPARS-T7Cup32NST-BT) carries the *CUPI* promotor of *Saccharomyces cerevisiae* ***(14)***, which has been shown to reduce the induction time greatly ***(15)***. Owing to the use of a common selection marker zeocin, the size of the shuttle vector remains small (3.1 and 3.5 kb, respectively), hence it remains convenient for handling, cloning and transformation. By integration of a *Pichia* specific autonomous replicating sequence (PARS1) into this vector, linearization is no longer required and the transformation efficiency is increased up to 10^5 transformants/μg DNA ***(6)***. Additionally, plasmids can be easily recovered from *P. pastoris.* All modifications of the dual expression vector include a double tag consisting of an $RGS(H)_6$ epitope and an in vivo biotinylation sequence ***(16)*** for sensitive detection and rapid purification, respectively. Due to the strong affinity of biotin to avidin, capture and screening assays are enabled.

We will first describe protocols for use of the original dual expression vector in *E. coli* and *P. pastoris*, then we will describe recent modifications.

2. Materials

2.1. Strains and Plasmids

1. Strains: For sub-cloning strategies common *E. coli* strains, such as XL1Blue, DH5α or SCS1 (Invitrogen; Gibco-BRL; Stratagene) are used. The expression in *E. coli* when using the dual expression vector, requires the *E. coli* strain BL21(D3) pLysS (Novagen; Invitrogen) that carries the gene coding for phage T7 polymerase enabling T7 promoter induced transcription of the following cDNA.
   Commonly used *P. pastoris* host strains are GS115 and KM71. The more recently available protease-deficient strain SMD1168 results in a marginal decrease in transformation efficiency and protein expression levels when compared to GS115.
2. Plasmids: pZPARS-T7RGSHis32NST-BT and pZPARS-T7Cup32NST-BT (**Fig. 1**).

2.2. Transformation

2.2.1. E. coli

1. Electro-competent cells, for example XL1Blue (Stratagene), are prepared or obtained from the supplier, with a transformation efficiency of at least 10^9 transformants/μg DNA.

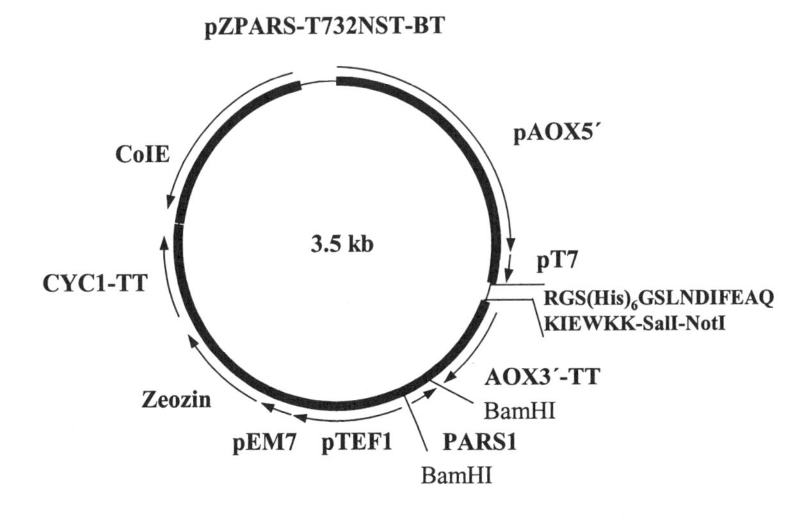

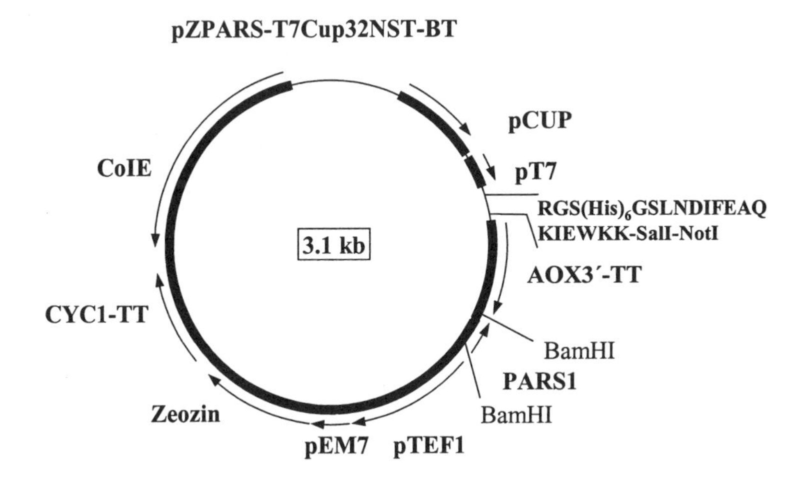

Fig. 1. Schematic map of the dual expression vectors.

2. Strain BL21(D3)pLysS (Novagen; Invitrogen).
3. 40% Glucose in water; autoclaved or sterile-filtrated to sterilize.
4. Antibiotic stock solution: 100 mg/mL zeocin; 34 mg/ mL chloramphenicol.
5. Luria-Bertani (LB) medium per liter: 5 g yeast extract, 10 g NaCl, and 10 g bacto-tryptone. Adjust to pH 7.0 and autoclave. Add 15 g agar for plates. For selection and growth of transformands, add 250 µL of zeocin stock solution (25 µg/mL final concentration), and in the case of BL21(D3)pLysS, add additional 1 mL chloramphenicol (34 µg/mL final concentration).

6. 100% Glycerol, autoclaved.
7. TFBI: 30 m*M* KOAc, 50 m*M* $MnCl_2$, 100 m*M* RbCl, 10 m*M* $CaCl_2$, 15% glycerol, pH 5.8, sterile filtrated.
8. TFBII: 10 m*M* Na-MOPS, 75 m*M* $CaCl_2$, 10 m*M* RbCl, 15% glycerol, pH 7.0, sterile filtered.

2.2.2. P. pastoris

1. Zeocin stock solution: 100 mg/mL.
2. YPD medium: 10 g yeast extract and 10 g bacto-tryptone per liter, autoclave, and add 50 mL of sterile 40% glucose stock solution. For YPD plates: add 15 g agar per liter.
3. 5 *M* Betaine in water, sterile-filtrated.
4. 100% Glycerol, sterile-filtrated.
5. Sterile, distilled water, cooled to 4°C.
6. 1 *M* HEPES, pH 8.0, sterile-filtrated.
7. 1 *M* DTT, sterile-filtrated.
8. 1 *M* Sorbitol, autoclaved and cooled to 4°C.
9. YPD agar plates supplemented with 100 µg/mL zeocin.

2.3. Analysis of Transformants

1. LB agar plate supplemented with 25 µg/mL zeocin.
2. YPD agar plate supplemented with 100 µg/mL zeocin.
3. PCR mix: 50 m*M* KCl, 0.1% Tween-20, 1.5 m*M* $MgCl_2$, 35 m*M* Tris-Base, 15 m*M* Tris-HCl, pH 8.8, 0.2 m*M* dNTPs, 3 units Taq.
4. Primer: AOX5': TTGCGACTGG TTCCAATTGA CAAG; 10 pmol/µL.
   AOX3': CATCTCTCAG GCAAATGGCA TTCTG; 10 pmol/µL.
   CUP5': TGTACAATCA ATCAATCAAT CA; 10 pmol/µL.
5. Lyticase (Sigma L2524) 6 mg/mL in water.

2.4. Protein Expression and Purification

2.4.1. E. coli Protein Expression and Lysis

1. LB medium, supplemented with 25 µg/mL zeocin and 34 µg/mL chloramphenicol.
2. 1 *M* IPTG.
3. Phosphate solution: 50 m*M* NaH_2PO_4, 300 m*M* NaCl, pH 8.0.
4. Lysis buffer: 50 m*M* Tris-HCl, 300 m*M* NaCl, pH 8.0, supplemented with 10 m*M* imidazole, 1 m*M* PMSF, 0.25 mg/mL lysozyme, 1 mg/mL RNAse, and 1 mg/mL DNAse.
5. QIAGEN buffer A: 6 *M* guanidine hydrochloride (Gn-HCl), 0.1 *M* NaH_2PO_4, 10 m*M* Tris-HCl, pH 8.0.

2.4.2. P. pastoris Protein Expression and Lysis

1. YPD medium supplemented with 100 µg/mL zeocin.
2. YNB stock solution: dissolve 134 g yeast nitrogen base with ammonium sulfate and without amino acids (Difco) in 1 L water and autoclave.

3. 100% Methanol (when using pZPARS-T732NST-BT).
4. 1 *M* $CuSO_4$, autoclaved (when using pZPARS-T7Cup-32NST-BT).
5. Biotin stock solution: dissolve 20 mg biotin (Sigma B-4639) in 100 mL water and filter sterilize.
6. 100 m*M* Potassium phosphate buffer, pH 6.0: combine 132 mL 1 *M* $KHPO_4$ and 868 mL KH_2PO_4/L, and filter sterilize.
7. BMMY medium: 10 g yeast extract, 10 g bacto-trypton in 700 mL water, autoclave, add 100 mL YNB stock solution, 5 mL 100% methanol, 2 mL biotin stock solution, 100 mL 100 m*M* calcium phosphate buffer, and 100 µg/mL zeocin.
8. Yeast nitrogen base with dextrose (YNBD) medium: Yeast Nitrogen Base (Difco: 0919–07–03) 6.7 g/L water, autoclave, and add 50 mL/L of filter sterilized, or autoclaved, 40% (w/v) glucose.
9. Glass beads (size 0.5 mm; Sigma G-8772).
10. Lysis buffer: 50 m*M* Tris-HCl, 300 m*M* NaCl, pH 8.0, supplemented with 10 m*M* imidazole, 1 m*M* PMSF.

2.4.3. Native Purification

1. Ni-NTA agarose (QIAGEN).
2. 1 *M* Imidazole.
3. Wash buffer: 50 m*M* Tris-HCl, 300 m*M* NaCl, pH 8.0 supplemented with 20 m*M* imidazole.
4. Elution buffer: 50 m*M* Tris-HCl, 300 m*M* NaCl, pH 8.0, supplemented with 250 m*M* imidazole.

2.4.4. Denatured Purification

1. Ni-NTA agarose (QIAGEN)
2. Buffer C: 8 *M* urea, 100 m*M* NaH_2PO_4, 10 m*M* Tris, pH 6.3.
3. Buffer E: 8 *M* urea, 100 m*M* NaH_2PO_4, 10 m*M* Tris, pH 4.5.

3. Methods

3.1. Cloning of Genes into the Dual Shuttle Vector

The standard cloning steps are not considered here in detail. All methods required are described in Sambrook et al. 1989 ***(17)***. Cloning into the dual expression vector will be described. In general, the dual expression vector offers two restriction sites for cloning: *Sal*I and *Not*I respectively. The cloning procedure/strategy requires an upstream primer containing a *Sal*I site and a downstream primer containing a *Not*I site. Specifically, these primers are as follows: the 5' primer (*Sal*I) is 5'-AAAAG TCG ACC- first triplet behind the ATG/translation initiation codon-$(N)_{15–18}$-3' and the 3' primer (*Not*I) is; 5'AAAA GCG GCC GC-TAA-$(N)_{15–18}$-3'. As previously mentioned, the *Sal*I and the *Not*I sites can be exchanged with *Xho*I, *Ava*I, and *Eag*I sites. If the gene of interest contains one of these restriction sites, alternatively compatible cohesive ends can be generated using the enzymes *Xho*I or *Ava*I at the 5' end and *Eag*I at the 3' end. The gene of interest is amplified using specific primers containing the required restriction sites, following restriction of the purified amplicon and ligation to the *Sal*I/*Not*I or appropriately restricted vector. The transformation step requires an *E. coli* strain with high transformation efficiency, such as electro-competent XL1Blue, DH5α, or SCS1. The designed primer must coincide with the open reading frame of the dual expression vector.

*5' primer-schema (*SalI*)*

5'-AAAAG TCG ACC- first triplet behind the ATG/translation initiation codon-$(N)_{15-18}$-3'

*3' primer-schema (*NotI*)*

5'AAAA GCG GCC GC-TAA-$(N)_{15-18}$-3'

As previously mentioned, the *Sal*I and the *Not*I sites can be exchanged with *Xho*I, *Ava*I and *Eag*I sites.

3.2. Transformation (see Note 1)

Due to the reduced transformation efficiency (10^7–10^8) of the rubidium-competent BL21(D3)pLysS, sub-cloning of the vector in an electro-competent *E. coli* strain is recommended, following the manufacture's instructions. When the transformants are confirmed (*see* **Subheading 3.3.**), the corresponding plasmid is isolated and transformed into the *E. coli* expression strain BL21(D3)pLysS and the *P. pastoris* expression strain, GS115 for example.

3.2.1. Preparation of E. coli BL21(D3)pLysS Competent Cells and Transformation

For transformation in the *E. coli* expression strain using a heat-shock method, competent cells of BL21(D3)pLysS are prepared or obtained from the supplier (Novagen; Invitrogen), according to the following protocol:

1. Inoculate 50 mL 2YT medium (supplemented with 34 µg/mL chloramphenicol) with a fresh colony of BL21/(D3)pLysS from an agar plate, and grow overnight at 37°C with shaking (250 rpm).
2. Inoculate 500 mL 2YT medium without antibiotics with the 5 mL overnight culture and grow it to an OD_{600} = 0.4–0.5 at 37°C with shaking (250 rpm).
3. Cool the culture on ice for 20 min.
4. Harvest the culture by centrifugation at 1300*g* at 4°C for 10 min, and resuspend the cells in 15 mL TFBI on ice.
5. Harvest the culture by centrifugation at 1300*g* at 4°C for 10 min, and resuspend the cells in 4 mL TFBII on ice.
6. The cells can be used directly for transformation or stored in 100 µL aliquots at –70°C until use.

7. Dilute 100–500 ng DNA sample in 5–10 μL total volume of sterile distilled water, add 100 μL pre-cooled competent cells, and incubate on ice for 20 min.
8. Incubate cells in a 42°C water-bath for 1.3 min and cool the cells immediately.
9. Add 1 mL fresh pre-warmed 2YT medium and regenerate the cells for 1 h at 37°C with shaking (250 rpm).
10. Spread aliquots onto agar plates, containing 2YT medium supplemented with 25 μg/mL zeocin and 34 μg/mL chloramphenicol, and incubate overnight at 37°C (*see* **Note 2**).

3.2.2. Preparation of P. pastoris *Electro-competent Cells and Transformation*

For transformation in *P. pastoris* using electroporation, electro-competent cells are prepared, as described below, which can be used directly, or stored at –70°C.

1. Inoculate 10 mL YPD medium with a single fresh colony of *P. pastoris* from an agar plate, and grow overnight at 30°C with shaking (250 rpm).
2. Inoculate 500 mL YPD medium with the 10 mL overnight culture (OD_{600} = 0.1) and grow it to an OD_{600} = 1.3–1.5 at 30°C with shaking (250 rpm).
3. Harvest the culture by centrifugation at 2000*g* at 4°C for 10 min, and suspend the cells in 100 mL YPD supplemented with 20 mL HEPES and 2.5 mL 1 *M* DTT. Incubate the cells for 15 min at 30°C without shaking.
4. Add cold water to 500 mL and harvest the cells by centrifugation at 2000*g* at 4°C for 10 min.
5. Wash the cells with 250 mL cold water and collect the cells by centrifugation at 2000*g* at 4°C for 10 min.
6. Wash the cells with 20 mL cold 1 *M* sorbitol and centrifuge at 2000*g* at 4°C.
7. Resuspend the cells in 500 μL cold 1 *M* sorbitol. The cells can be used directly for transformation, or can be stored in aliquots at –70°C until use.
8. Dilute 100 ng DNA sample in 5 μL total volume of sterile distilled water, add 40 μL competent cells and transfer into a 2-mm gap electroporation cuvet, pre-cooled on ice.
9. Pulse cells according to the following parameters, when a Gene-Pulser (Bio-Rad) is used: 1500 V, 200 Ω, 25 μF. For other electroporation instruments, follow the manufacturer's recommendations with respect to yeast transformation.
10. Immediately add 1 mL cold 1 *M* sorbitol, transfer into a sterile 1.5 mL Eppendorf tube and regenerate cells for at least 30 min at 30°C with shaking.
11. Spread aliquots onto agar plates containing YPD supplemented with 100 μg/mL zeocin, and incubate for two days at 30°C (*see* **Note 2**). When using plasmids containing the PARS replicating sequence, a transformation efficiency of 10^5 transformants/μg DNA is expected.

3.3. Analysis of Transformants

In general, transformants growing on selection medium of both *E. coli* and *P. pastoris* were analyzed by PCR amplification of the specific gene insert using the same primer pair combination: AOX5' and AOX3' (pZPARS-

T7RGSHis32NST-BT or Cup5' and AOX3' (pZPARST7-CupRGSHisNST-BT) respectively. Due to the different stability and composition of the cell wall of *E. coli* and yeast, PCR amplification requires different conditions for cell disruption. *E. coli* cells are disrupted by heating (94°C for 4 min) where the DNA is exposed for amplification, whereas the *P. pastoris* cell wall is enzymatically digested (zymolyase or lyticase at 37°C for 30 min) leading to protoplasts that are more susceptible to heat or detergents. Then, following a heating step (94°C for 4 min), DNA is exposed for amplification.

3.3.1. E. coli

1. Prepare a PCR mix (50 m*M* KCl, 0.1% Tween-20, 1.5 m*M* $MgCl_2$, 35 m*M* Tris-Base, 15 m*M* Tris-HCl, 0.2 m*M* dNTPs, 3 units Taq) sufficient for an appropriate number of transformants. To analyze, add 1 µL of each primer AOX5' or CUP5'/AOX3'; 10 pmol/µL for each transformant.
2. Distribute 30–50 µL per sample in PCR tubes.
3. With a toothpick, pick into a single colony and transfer the cells first onto a fresh LB agar plate supplemented with zeocin, and then into the corresponding PCR tube.
4. The agar plate is incubated at 37°C overnight
5. The PCR is performed under the following conditions: 4 min at 94°C (1 cycle), 45 s at 94°C, 20 s at 55°C, and 1 min 20 s at 72°C (24 cycles).
6. The PCR products are electrophoretically separated and analyzed.

3.3.2. P. pastoris

1. Prepare a PCR mix (50 m*M* KCl, 0.1% Tween-20, 1.5 m*M* $MgCl_2$, 35 m*M* Tris-Base, 15 m*M* Tris-HCl, 0.2 m*M* dNTPs, 3 units Taq) sufficient for an appropriate number of transformants to analyze an add 1 µL of each Primer AOX5' or CUP5'/AOX3'; 10 pmol/µL) for each transformant.
2. Add 0.1 µg/µL lyticase per sample.
3. Distribute 30–50 µL/sample in PCR tubes.
4. With a toothpick, pick into a single colony and transfer the cells first onto a fresh YPD agar plate supplemented with zeocin, and then into the corresponding PCR tube.
5. The agar plate is incubated at 30°C overnight.
6. The PCR is performed under the following conditions: 30 min at 37°C, 4 min at 94°C (1 cycle), 45 s at 94°C, 20 s at 55°C, and 2 min 30 s at 72°C (30 cycles), 10 min at 72°C (1 cycle).
7. The PCR products are electrophoretically separated and analyzed.

3.4. Protein Expression and Purification (see Note 3)

It is recommended that small-scale expression and purification be used to determine if the protein is expressed, and from which host the protein can be solubily purified, in *E. coli* or *P. pastoris*, respectively. When the host and condi-

Table 1
Quantities of Solutions and Materials for Small- and Large-Scale Expression and Purification

	E. coli			*P. pastoris*		
Step	Small scale		Large scale	Small scale		Large scale
Inoculation	200 µL		20 mL	0.5–1 mL		50–200 mL
Induction	+ 1800 µL		+ 200 mL	+ 4 mL		+ 200–800 mL
	denat.	native		denat.	native	
Split	1 mL	1 mL		2.5 mL	2.5 mL	
Lysis Buffer	200 µL	200 µL	0.5–1 mL	200 µL	200 µL	1– 5 mL
Ni-NTA	50 µL	50 µL	200 µL	20 µL	20 µL	50–100 µL
Wash Buffer	200 µL	200 µL	2 mL	200 µL	200 µL	2 mL
Elution Buffer	35 µL	35 µL	100 µL	35 µL	35 µL	100 µL

tions are determined, large-scale expression and purification can be performed, in order to produce sufficient amounts of proteins for following applications.

The quantities of solutions and material of the different scales are listed in **Table 1**.

3.4.1. E. coli *Expression and Lysis*

1. Inoculate LB medium, supplemented with 2% glucose and 25 µg/mL zeocin, with a fresh colony of the transformant, and grown overnight at 37°C with shaking (200 rpm).
2. Inoculate fresh LB medium, supplemented with 25 µg/mL zeocin, with the overnight culture (10% final concentration of cell suspension), and grow at 37°C with shaking to an OD_{600} = 0.6–1.0.
3. Add IPTG to a final concentration of 1 m*M* to induce protein expression, and grow at 37°C with shaking for further 3–5 h
4. For evaluation of small-scale cultures, cultures are divided into two. Cultures were harvested by centrifugation at 4000*g* at 4°C, and frozen for at least 20 min at –70°C.
5. Thaw cell pellets. For evaluation, the two cell pellets from the small-scale culture are re-suspended in either lysis buffer (native lysis) or QIAGEN buffer A (denatured lysis). Cell pellets of the large cultures are resuspended in the appropriate buffer, either lysis buffer or QIAGEN buffer A.
6. Cells re-suspended in lysis buffer are lysed either at 4°C overnight, or 30 min on ice, followed by sonication. Cells resuspended in QIAGEN buffer A are incubated at room temperature for at least 1 h, with shaking (*see* **Note 3**).
7. Lysates were cleared by centrifugation at 10,000*g* for 10 min at 4°C (native lysis) or at room temperature (denatured lysis).

3.4.2. *P. pastoris* Expression and Lysis (see **Note 3**)

1. Inoculate YPD medium, supplemented 100 μg/mL zeocin, with a fresh colony of the transformants, and grow overnight at 30°C with shaking (250 rpm).
2. Inoculate fresh BMMY medium supplemented with 100 μg/mL zeocin (pZPARS-T732NST-BT) or YNBD (pZPARS-T7Cup-32NST-BT) supplemented with 100 μg/mL zeocin and 40 mg/mL histidine, with the overnight culture (10% final concentration of cell suspension), and grow at 30°C with shaking to an OD_{600} = 1.0.
3. Add methanol to final concentration of 0.5% (v/v) (pZPARS-T732NST-BT) or 0.1 m*M* $CuSO_4$ (pZPARS-T7Cup-32NST-BT) to induce protein expression, and grow at 30°C with shaking (250 rpm) for 2–3 d (pZPARS-T732NST-BT) or 1–2 h (pZPARS-T7Cup-32NST-BT).
4. For evaluation of small-scale cultures, cultures are divided into two parts. Cultures were harvest by centrifugation at 2000*g* at 4°C, then frozen for at least 20 min at –70°C.
5. Thaw cell pellets. For evaluation, the two cell pellets of the small-scale culture are resuspended in either lysis buffer (native lysis) or QIAGEN buffer A (denatured lysis). Cell pellets of the large cultures are resuspended in the appropriate buffer, either lysis buffer or QIAGEN buffer A.
6. Add 0.5–1 vol of glass beads and perform 5–7 cycles of 1 min vortex, 1 min incubation on ice.
7. The lysates are cleared by centrifugation at 10,000*g* for 10 min at 4°C (native lysis) or at room temperature (denatured lysis).

3.4.3. Native Purification

1. Add Ni-NTA agarose to the lysate, mix gently and incubate on a rotary shaker for 1 h at 4°C. The appropriate volume of Ni-NTA depends partly of the expression level of the protein. High expressed proteins require more purification matrix, and for less expressed proteins, the volume of Ni-NTA has to be reduced to ensure a good quality of purification.
2. Load the suspension of lysate and Ni-NTA slurry onto a column and collect flow-through.
3. Wash 3× with wash buffer.
4. Elute the protein 4× with buffer E and collect through-flow fractions. Fractions can be analysed by SDS-PAGE and western blot analysis. To increase protein concentration, as well as decrease the elution volume, only one volume can be applied, and incubated for 10 min onto the column without flow through.

3.4.4. Denatured Purification

1. Add Ni-NTA agarose to the lysate, mix gently and incubate on a rotary shaker for 1 h at room temperature. The appropriate volume of Ni-NTA depends partly of the expression level of the protein. High expressed proteins require more purification matrix and for less expressed proteins the volume of Ni-NTA has to be reduced to ensure a good quality of purification.

2. Load the suspension of lysate and Ni-NTA slurry onto a column and collect flow-through.
3. Wash 3× with buffer C.
4. Elute the protein 4× with buffer E and collect through-flow fractions. Fractions can be analysed by SDS-PAGE and western blot analysis. To increase protein concentration, as well as decrease the elution volume, only one volume can be applied, and incubated for 10 min onto the column without flow through.

4. Notes

1. When expressing proteins in *E. coli* using this system, it is important to use an *E. coli* strain that contains a T7 promoter, such as BL21(D3)pLysS or SCS-1.
2. It is also important to note the difference in the concentration of Zeocin antibiotic used in the different expression systems. In *E. coli*, less is used (25 μg/mL) as in Pichia (100 μg/mL Zeocin). In addition, when the *E. coli* strain BL21 is used for expression, it is necessary to add 34 μg/mL Chloramphenicol for selection from pLys.
3. For protein purification in *E. coli*, the lysate should be ultrasonicated longer. Otherwise, the lysate is mucilaginous and may clog the purification column.
4. For planning yeast protein expression experiments, it is important to note that in yeast, the expression takes 2–3 d longer than with *E. coli* when the AOX promoter is used, but not the Cup-promoter. This is because the transformants require 2 d to grow (1 d in *E. coli*) and the induction takes at least 2 d (again with *E. coli*, it is only 1 d).

References

1. Itakura, K., Hirose, T., Crea, R., et al. (1977) Expression in *Escherichia coli* of a chemically synthesized gene for the hormone somatostatin. *Science* **198,** 1056–1063.
2. Baneyx, F. (1999) Recombinant protein expression in Escherichia coli. *Curr. Opin. Biotechnol.* **10,** 411–421.
3. Hannig, G. and Makrides, S. C. (1998) Strategies for optimizing heterologous protein expression in *Escherichia coli. Trends Biotechnol.* **16,** 54–560.
4. Makrides, S. C. (1996) Strategies for achieving high-level expression of genes in *Escherichia coli. Microbiol. Rev.* **60,** 512–38.
5. Marston, F. A. O. (1986) The purification of eukaryotic polypeptides synthesized in *Escherichia coli. Biochem. J.* **240,** 1–12.
6. Cregg, J. M., Vedvick, T. S., and Raschke, W. C. (1993) Recent advances in the expression of foreign genes in *Pichia pastoris. Biotechnology* **11,** 905–910.
7. Faber, K. N., Harder, W., Ab, G., and Veenhuis, M. (1995) Review: methylotropic yeasts as factories for the production of foreign proteins. *Yeast* **11,** 1331–1344.
8. Faber, K. N., Westra, S., Waterham, H. R., Keizer, G. I., Harder, W., and Veenhuis, G. A. (1996) Foreign gene expression in *Hansenula polymorpha.* A system for the synthesis of small functional peptides. *Appl. Microbiol. Biotechnol.* **45,** 72–79.

9. Monsalve, R. I., Lu, G., and King, T. P. (1999) Expression of recombinant venom allergen, antigen 5 of yellojacket (*Vespula vulgaris*) and Paper Wasp (*Polistes annularis*), in bacteria or yeast. *Protein Expr. Purif.* **16,** 410–416.
10. Romanos, M. A., Scorer, C. A., and Clare, J. J. (1992) Foreign gene expression in yeast: a review. *Yeast* **8,** 423–488.
11. Lueking, A., Holz, C., Gotthold, C., Lehrach, H., and Cahill, D. (2000) A system for dual protein expression in *Pichia pastoris* and *Escherichia coli. Protein Expr. Purif.* **20,** 372–378.
12. Cereghino, J. L. and Cregg, J. M. (2000) Heterologous protein expression in the methylotrophic yeast *Pichia pastoris. FEMS Microbiol. Rev.* **24,** 45–66.
13. Tschopp, J. F., Brust, P. F., Cregg, J. M., Stillman, C. A., and Gingeras, T. R. (1987) Expression of the lacZ gene from two methanol-regulated promoters in *Pichia pastoris. Nucleic Acids Res.* **15,** 3859–3876.
14. Macreadie, I. G., Horaitis, O., Verkuylen, A. J., and Savin, K. W. (1991) Improved shuttle vectors for cloning and high-level Cu(2+)-mediated expression of foreign genes in yeast. *Gene* **104,** 107–111.
15. Koller, A., Valesco, J., and Subramani, S. (2000) The CUP1 promoter of *Saccharomyces cerevisiae* is inducible by copper in *Pichia pastoris. Yeast* **16,** 651–656.
16. Schatz, P. (1993) Use of peptide libraries to map the substrate specifity of a peptide-modifying enzyme: A 13 residue consensus peptide specifies biotinylation in *Escherichia coli. Biotechnology* **11,** 1138–1143.
17. Sambrook, J., Fritsch, E. F., and Maniatis, T. (1989) *Molecular Cloning: a laboratory manual,* Cold Spring Harbor Laboratory, Cold Spring Harbor, NY.

4

Purification of Recombinant Proteins from *E. coli* by Engineered Inteins

Ming-Qun Xu and Thomas C. Evans, Jr.

1. Introduction: History of IMPACT Vectors

The IMPACT (Intein-Mediated Purification with an Affinity Chitin-Binding Tag) vectors are designed for the isolation of pure, functional, recombinant proteins by a single affinity chromatography step. The IMPACT technology was developed at New England Biolabs (NEB) by exploiting a novel family of proteins termed inteins (recently reviewed in **ref. *1***). An intein is an internal protein segment responsible for catalyzing an extraordinary post-translational processing event termed protein splicing. Protein splicing results in the precise excision of the intein polypeptide from a protein precursor with the concomitant ligation of the flanking protein sequences, termed exteins. This process requires neither auxiliary proteins nor exogenous energy sources such as ATP (for more information on the requirements and mechanism of protein splicing, *see* **ref. *2***). Once the mechanism of protein splicing was elucidated it was realized that a self-splicing intein could be used for protein purification, because the catalytic steps involved in the fission of the peptide bond at either splice junction could be modulated by mutation of amino acid residues at the splice junctions, as described in detail in the following sections.

1.1. Thiol Inducible N-Terminal Cleavage System

The first series of IMPACT vectors (pTYB1 and its derivatives) were created by engineering the 454-residue intein from the *Saccharomyces cerevisiae* VMA1 gene *(3,4)*. The replacement of the last intein residue, Asn454, with an alanine residue, yielded a mutant which exhibited no splicing or cleavage at

From: *Methods in Molecular Biology, vol. 205, E. coli Gene Expression Protocols*
Edited by: P. E. Vaillancourt © Humana Press Inc., Totowa, NJ

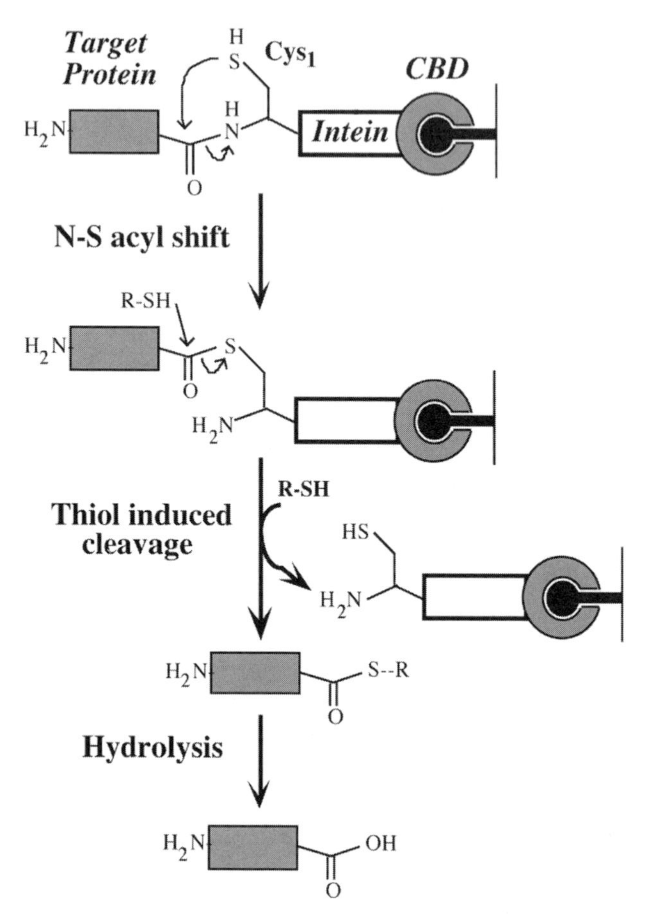

Fig. 1. Purification of a recombinant protein expressed from a C-terminal fusion vector by cleavage at the N-terminus of a modified intein. The C-terminus of a target protein is fused in-frame to the N-terminus of an engineered intein and expressed in *E. coli* as a fusion protein consisting of the target protein-intein-chitin binding domain. The fusion protein is purified by binding to chitin resin. The target protein is released when a thioester bond formed at the intein N-terminal residue (Cys^1) is attacked by a thiol compound (R-SH) such as DTT. The product is eluted following the on-column cleavage reaction while the intein-CBD tag remains bound to chitin.

the C-terminal splice junction but allowed the formation of a thioester linkage between the intein N-terminal cysteine residue and the N-extein (the target protein, *see* **Fig. 1**). Incubation of the *Sce* VMA intein (N454A) mutant protein with thiol reagents such as dithiothreitol (DTT), β-mercaptoethanol, or cysteine led to cleavage of the thioester bond by nucleophilic attack. As a result, the target protein is separated from the intein fusion partner. This intein mutant was then tested for expression and purification of recombinant proteins in conjunction with a small chitin binding domain (CBD) from *Bacillus circulans* *(4,5)*. A target gene was cloned in-frame to the N-terminus of the modified *Sce* VMA intein linked at its C-terminus to the coding region of CBD (*see* **Fig. 1**). A tripartite fusion protein was isolated from *E. coli* cell extract by its binding to chitin resin. The immobilized fusion protein was then induced to undergo intein-mediated cleavage by overnight incubation at 4°C in the presence of DTT. The protein of interest was eluted from the chitin resin while the intein-CBD fusion tag remained bound on the resin. In comparison to other fusion-based affinity purifications, the intein based method allows separation of a protein of interest from the fusion partner without the use of a protease.

1.2. The Use of Mini-Inteins and Intein-Mediated Protein Ligation

The discovery of the thiol-induced cleavage reaction also allowed the expansion of a peptide fusion method described as native chemical ligation *(6)*. The chemistry requires one peptide possessing a C-terminal thioester and another possessing an N-terminal cysteine. The carbonyl of the thioester on the former peptide is attacked by the sulfhydryl group of the N-terminal cysteine in the latter peptide, yielding a thioester linkage between the reacting peptides. A spontaneous S-N acyl rearrangement leads to the formation of a native peptide bond between the two peptide species. The utility of this method was primarily limited by the size of the peptides that could be chemically synthesized.

Several groups pursued the possibility of using the IMPACT system to isolate large thioester tagged recombinant proteins. Muir and coworkers successfully expressed and isolated the protein tyrosine kinase C-terminal Src kinase (CSK) and σ^{70} subunit of *E. coli* RNA polymerase using the commercially available IMPACT vector containing the *Sce* VMA intein (Asn454Ala) mutant *(7,8)*. Following absorption of the *E. coli* expressed fusion proteins onto a chitin resin, the intein cleavage was induced with thiophenol, releasing the protein fragment for ligation with a synthetic peptide. At NEB, we focused on the comparative study of different inteins for their ability to cleave in response to thiol compounds that form a reactive thioester for proficient ligation. At the same time, efforts were made to develop new IMPACT vectors based on mini-inteins of less than 200 residues in size *(9)*. The 198-residue intein from the *Mycobacterium xenopi gyr*A gene (*Mxe* GyrA intein), engineered by replacement of its last residue, Asn198 with Ala, was found to cleave efficiently with DTT. However, DTT-tagged proteins did not permit an efficient ligation reaction. After an extensive screen, a thiol compound, 2-mercaptoethanesulfonic acid (MESNA) was discovered to form a stable, active thioester that allows for ligation at greater than 90% efficiency. Furthermore, the *Mxe* GyrA intein (Asn198Ala) mutant cleaved more efficiently with MESNA than the *Sce* VMA intein (Asn454Ala) mutant, and therefore was suitable for intein-mediated protein ligation (IPL). The *Mxe* GyrA intein, now commercially available from NEB as the pTXB1, pTXB3 or pTWIN1 vector, was subsequently used for the synthesis of two cytotoxic proteins ***(9)***.

Splicing of the 134-residue intein found in the ribonucleoside diphosphate reductase gene of *Methanobacterium thermoautotrophicum* (*Mth* RIR1 intein) was found to be inefficient in *E. coli* and under in vitro conditions with the naturally occurring proline at the position preceding the intein N-terminal cysteine residue ***(10)***. A lucky break came when the –1 proline residue was replaced with glycine, increasing splicing activity substantially. The *Mth* RIR1 intein, supplied by NEB in the pTWIN2 vector, was modified for N-terminal cleavage by the introduction of an Asn134Ala substitution. The mutation blocked C-terminal cleavage and splicing, and resulted in a mutant which underwent efficient cleavage at the N-terminal splice junction at 4°C in the presence of thiol reagents such as DTT or MESNA.

1.3. Thiol-Inducible C-Terminal Cleavage System

One of the properties of the thiol-inducible N-terminal system was that protein expression depended heavily on the expression properties of the target protein. This is desirable if the target protein expresses well, but undesirable if it does not. In addition, an N-terminal methionine is typically required to initiate translation, so expressed proteins start with a methionine, even if they naturally begin with another amino acid residue (due to post-translational modification in the native host). To circumvent these properties a new vector was created that permitted the fusion of a target protein to the C-terminus of an intein. In this way, the expression of the intein-tag-target protein fusion is less dependent on the expression characteristics of the target protein. Furthermore, the target protein need not begin with a methionine residue. The *Sce* VMA intein was again used, but this time its splicing activity was modulated by replacement of the first C-extein residue (cysteine) to prevent splicing and the penultimate His453 residue with Gln to attenuate cleavage at its C-terminal junction ***(11,12)***. The double mutation allowed for the isolation of full length fusion precursors from *E. coli* cells (*see* **Fig. 2**). Interestingly, cleavage at the

---

Fig. 2. Purification of recombinant proteins expressed from an N-terminal fusion vector by cleavage at the C-terminus of a modified intein. The target protein is fused at its N-terminus to the C-terminus of the intein and expressed in *E. coli* as a fusion protein consisting of chitin binding domain, intein, and target protein. The fusion protein is purified by binding to chitin resin. **(A)** fission of the peptide bond at the C-terminus (Asn^{454}) of the modified *Sce* VMA intein in pTYB11 (or pTYB12) triggered by thiol-induced cleavage at the N-terminus of the intein. The CBD is inserted

A **C-terminal cleavage induced by thiol**

Chitin
Cys_1 H S
CBD
Target Protein
O H N
H H_2N
CO_2^-
Intein
O
NH_2
N-S acyl shift
O Asn_{454}
R-SH Cys_1
H_2N S
O
O H N
H_2N
CO_2^-
NH_2
O Asn_{454}
R-SH
Thiol induced cleavage at pH8.0
Cys_1
O
NH
Asn
H_2N S-R H_2N CO_2^-
O
O

B **C-terminal cleavage induced by pH and temperature shift**

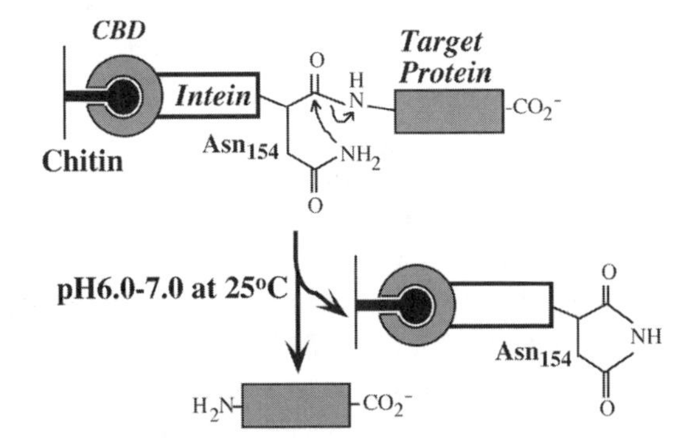

(Fig. 2. *continued from opposite page*) within the intein sequence. The target protein and a 15-residue peptide are eluted from the chitin column. **(B)** cleavage at the C-terminus (Asn^{154}) of the modified *Ssp* DnaB intein (intein 1) in a pTWIN vector induced by a pH and temperature shift on chitin resin. The CBD is fused to the N-terminus of the intein. The product is eluted following the on-column cleavage reaction. The amino acid residues that participate directly in the reactions are shown.

C-terminal splice site appeared to be dependent on thiol-induced cleavage at its N-terminal splice junction (*see* **Fig. 2**). Furthermore, the *B. circulans* CBD was inserted into the homing endonuclease domain, a region not required for splicing, of the *Sce* VMA intein to facilitate the purification of fusion proteins and the cleavage reaction on chitin resin. This *Sce* VMA intein variant was commercialized as the pTYB11 and pTYB12 vectors in the IMPACT-CN kit for isolation of recombinant proteins from *E. coli* cells.

1.4. pH-Inducible C-Terminal Cleavage System

There are cases in which the protein to be purified is sensitive to DTT (a reducing agent) and so its use as a cleavage reagent should be avoided. Furthermore, proteins with an N-terminal cysteine can not be isolated with the pTYB11 and pTYB12 vectors. Proteins with an N-terminal cysteine are required for intein-mediated protein ligation and also can occur naturally. Modification of a new intein, supplied as the pTWIN vectors, resulted in an intein-tag that cleaved in a pH and temperature dependent manner (*see* **Fig. 2**).

This intein was based on the protein splicing element from the *dnaB* gene from *Synechocystis* sp. PCC6803. It was reduced to a 154-residue mini-intein from the full-length 429 amino acid residues (*Ssp* DnaB mini-intein) ***(13)***. A key modulation was performed by substitution of the N-terminal cysteine residue of the *Ssp* DnaB mini-intein with an alanine in order to block its protein splicing and N-terminal cleavage activity ***(14)***. This modified intein mutant underwent cleavage of the peptide bond between the intein and a C-terminal target protein in a pH-dependent manner. The C-terminal cleavage reaction displayed by the *Ssp* DnaB mini-intein was found to be most favored at pH 6.0–7.5, but was essentially blocked below pH 5.5 or above pH 8.0. Taking advantage of this pH response, a purification strategy was developed by fusing a target gene to the C-terminus of the intein with its N-terminus linked to the CBD, and purifying the fusion proteins at pH 8.5 followed by cleavage on chitin resin at pH 6.0–7.0 and 4–25°C. It was also found that the *Mth* RIR1 intein carrying the Cys1Ala mutation underwent cleavage at its C-terminus in a pH- and temperature-dependent manner. Based on these findings, recombinant proteins fused to the C-terminus of either the *Ssp* DnaB or the *Mth* RIR1 mini-inteins were expressed and purified by a single chitin column. The placement of a cysteine residue as the first C-extein residue (the N-terminal residue of a target protein) yielded a protein fragment suitable for ligation with a protein containing a C-terminal thioester. Coincident with the completion of this work, Dr. Marlene Belfort and colleagues studied an intein from the *Mycobacterium tuberculosis recA* gene and successfully used a genetic screen to select for intein mutants that underwent pH-dependent cleavage at its C-terminal splice junction for use in protein purification ***(15)***.

1.5. Protein Cyclization by the Two Intein System (TWIN)

To further capitalize on these scientific discoveries, a new approach was pursued to generate circular peptides or proteins by sandwiching a target protein sequence between two inteins ***(16)***. One intein was engineered for C-terminal cleavage, and the second intein, placed downstream, was modified for thiol-inducible N-terminal cleavage. This approach, termed the two intein (TWIN) system, allowed for generation of both an N-terminal cysteine and a C-terminal thioester on the same bacterially expressed protein. Intramolecular head-to-tail ligation would produce a circular molecule. The intermolecular reaction, which was shown to be concentration dependent, generated multimeric protein species. The IMPACT-TWIN system, from NEB supplied with pTWIN1 and pTWIN2 vectors, allows fusion of a target protein to one or both inteins. Translational fusion to intein 1, the *Ssp* DnaB mini-intein of either vector, allows the target protein to be released by pH inducible cleavage at the C-terminus of the intein. The pH inducible cleavage system avoids exposure to thiols during purification of thiol-sensitive proteins. On the other hand, fusion to the N-terminus of intein 2, the *Mxe* GyrA intein in pTWIN1 or the *Mth* RIR1 intein in pTWIN2, allows the separation of a target protein and the intein-CBD tag by a thiol induced cleavage reaction. Cleavage with MESNA generates a C-terminal thioester for ligation to either a synthetic peptide or a recombinant protein possessing an N-terminal cysteine. Thus, the engineered inteins are not only useful tools for expression and purification of recombinant proteins but also offer many novel approaches for the modification of proteins.

2. Materials

2.1. Maintenance of E. coli *Strains*

1. Luria-Bertani (LB) medium per liter: 10 g tryptone, 5 g yeast extract, 10 g NaCl. Adjust to pH 7.0 with NaOH. Add 15 g agar for plates. Autoclave.
2. 100 mg/mL Ampicillin stock.
3. LB medium supplemented with 100 μg/mL ampicillin.
4. LB agar plates and LB agar plates supplemented with 100 μg/mL ampicillin.

2.2. IMPACT Vectors and Cloning

1. C-terminal fusion vectors; *see* **Subheading 3.2.** and **Note 1**.
2. N-terminal fusion vectors; *see* **Subheading 3.3.** and **Note 1**.
3. Restriction enzymes, agarose, 1X TAE buffer, T4 DNA ligase.
4. *E. coli* competent cells (*see* **Note 2**).

2.3. E. coli *Transformation and Expression*

1. *E. coli* strains ER2566 and BL21(DE3) (or other derivatives) carrying the T7 RNA polymerase gene (*see* **Note 2** for the genotypes).

2. LB broth supplemented with 100 μg/mL ampicillin.
3. 100 m*M* Isopropyl β-D-thiogalactopyranoside (IPTG). Sterilize with a 0.45-μm filter.

2.4. Purification of the Intein-Target Protein Fusion

2.4.1. Preparation of Chitin Resin

1. Chitin beads (NEB, cat. no. S6651S); *see* **Note 3**.
2. Column (Bio-Rad, cat. no. 737–2512, 2.5 cm in diameter and 10 cm in length).
3. Sterile water.
4. Cell lysis/column buffer: 20 m*M* Tris-HCl, pH 8.0, 0.5 *M* NaCl (*see* **Note 4**).

2.4.2. Cell Lysis and Thiol-Inducible Cleavage

1. Cell lysis/column buffer: 20 m*M* Tris-HCl, pH 8.0, 0.5 *M* NaCl (*see* **Note 4**).
2. Dithiothreitol (DTT) (Sigma, cat. no. D6052 or D9163); *see* **Note 5**. 1 *M* DTT stock: Dissolve 3.09 g of DTT in 20 mL of 0.01 *M* sodium acetate, pH 5.2. Sterilize by filtration and store at –20°C.
3. Cleavage buffer: 20 m*M* Tris-HCl, pH 8.0, 0.5 *M* NaCl, 30 m*M* DTT. Dissolve 0.46 g of DTT in 100 mL of column buffer (*see* **Note 6**).
4. 2-Mercaptoethanesulfonic acid (MESNA) (Sigma, cat. no. M-1511; *see* **Note 7**).

2.4.3. Cell Lysis and pH-Inducible Cleavage

1. Cell lysis/column buffer: 20 m*M* Tris-HCl, pH 8.5, 0.5 *M* NaCl (*see* **Note 4**).
2. Cleavage buffer: 20 m*M* Tris-HCl, pH 6.0–7.0, 0.5 *M* NaCl (*see* **Note 8**).

2.4.4. Regeneration of Chitin Resin

1. 0.3 *M* NaOH (*see* **Note 9**).
2. Sterile water.

2.5. Target Protein Detection

2.5.1. SDS-PAGE and Protein Quantitation

1. 3X SDS sample buffer: 187.5 m*M* Tris-HCl, pH 6.8 at 25°C, 6% SDS, 0.03% bromphenol blue, 30% glycerol. DTT should be added to the 3X SDS sample buffer to a final concentration of 40 m*M* DTT (*see* **Note 10**).
2. SDS-PAGE (12–20%), 1X SDS-PAGE running buffer.
3. Bradford solution (Bio-Rad Protein Assay cat. no. 500-0006).

2.5.2. Western Blot Analysis

1. SDS-PAGE and SDS-PAGE running buffer.
2. Nitrocellulose (Schleicher & Schuell, cat. no. 10402599).
3. Transfer buffer.
4. Antichitin binding domain rabbit serum (NEB, cat. no. S6654S).
5. Western blot detection system (Cell Signaling Technology, cat. no. 7071 or 7072).

3. Methods

Molecular Cloning: A Laboratory Manual edited by Sambrook et al. *(17)* is a good source of information for individuals unfamiliar with cloning or expression of foreign proteins in *E. coli* and provides additional information beyond the scope of this article.

3.1. Selection of the Appropriate IMPACT Vectors

Four different modified inteins are now commercially available for cloning a target gene into the polylinker region in-frame to the intein-CBD coding region (*see* **Table 1**). The general features of these vectors are given in **Note 1**. The compatible cloning sites in the polyliner regions (MCS) of the IMPACT vectors allows fusion of a target gene to different inteins and comparison of these constructs for efficiency of expression and purification. The choice of different inteins as fusion partners allows optimization of expression and cleavage efficiency for a specific fusion protein (*see* **Note 11**). A target protein can affect both in vivo and in vitro cleavage of the fusion protein, thereby significantly influencing the yield of purified protein. Both in vivo and in vitro cleavage can be affected by the residue flanking the intein, making determination of whether the target protein places a favorable amino acid residue next to the intein important (*see* **Table 1** and **Note 12**).

Researchers may choose different inteins to fit a specific need in protein purification and modification. If you simply wish to purify a protein and are not concerned with other functions (i.e., DTT sensitivity), we recommend initially cloning the target gene into pTXB1 and/or pTYB1 (or their derivatives) (*see* **Fig. 3**). Some recommendations and considerations are provided below for choosing an N-terminal or C-terminal fusion system or a specific IMPACT vector.

1. An N-terminal vector (pTYB11 or pTYB12, pTWIN1 or pTWIN2) should be chosen for isolation of a protein without an N-terminal methionine residue or with a C-terminal Pro or Asp. The N-terminus of the target protein should be fused to the *Sce* VMA intein in pTYB11 (or its derivative pTYB12) or the *Ssp* DnaB mini-intein (intein 1) in a pTWIN vector (*see* **Fig. 4**).
2. A pTWIN vector should be used for purification of a thiol-sensitive protein or isolation of a protein with an N-terminal cysteine for protein ligation. A target protein possessing an N-terminal Ser, Thr, or Cys should be fused to intein 1 in a pTWIN vector.
3. A C-terminal fusion vector (pTYB1–4, pTXB1 and 3, pTWIN 1 and 2) must be used if the purpose is to generate an activated C-terminal thioester tag for protein ligation or labeling. The *Mxe* GyrA intein in pTXB and pTWIN1 vectors and the *Mth* RIR1 intein in pTWIN2 are preferred because these inteins cleave more efficiently with MESNA than the *Sce* VMA intein in pTYB vectors (*see* **Note 7**).
4. To generate a circular protein or peptide, a target gene is inserted into the polylinker of a pTWIN vector in frame to both inteins, as described in **Note 13**. A

Table 1
IMPACT Vectors

Vectors[a]	Site of target protein fusion	Intein length (amino acids)	Intein-CBD tag (kDa)	Recommended cloning sites	Preferred residues at cleavage site	Method of cleavage
pTYB1	C-terminus	454	56	*Nde*I-*Sap*I	G, LEG	Dithiothreitol
pTYB2				*Nde*I-*Sma*I (or *Xho*I)		(or 2-mercaptoethane-
pTYB3				*Nco*I-*Sap*I		sulfonic acid)[b]
pTYB4				*Nco*I-*Sma*I (or *Xho*I)		pH 8–8.5 at 4°C
pTXB1	C-terminus	198	28	*Nde*I-*Sap*I (or *Spe*I)	M,Y,F, LEM	
pTXB3				*Nco*I-*Sap*I (or *Spe*I)		
pTWIN1		198	28	*Nde*I-*Sap*I (or *Spe*I)	M, Y, F, LEM	
pTWIN2		134	22	*Nde*I-*Sap*I (or *Spe*I)	G, A, LEG	
pTWIN1[c]	N-terminus	154	27	*Sap*I-*Sap*I	S, C, A, G	pH 6–7 and
				*Sap*I-*Pst*I (or *Bam*HI)	CRAM	25°C
				*Bsr*GI-*Pst*I (or *Bam*HI)		
pTYB11	N-terminus	454	56	*Sap*I-*Pst*I (or *Bam*HI)	A, Q, M, G, L,	Dithiothreitol pH
pTYB12				*Bsm*I (or *Nde*I)-*Not*I	N, W, F, Y	8–8.5 at 25°C
pTWIN1[d]	N-terminus and	154	27	*Sap*I-*Sap*I	C, CRAM	Step 1: pH 6–7
	C-terminus	198	28		M, Y, F, LEM	25°C
pTWIN2	N-terminus and	154	27	*Sap*I-*Sap*I	C, CRAM	Step 2: MESNA
	C-terminus	134	22		G, A, LEG	pH 8–8.5 4°C

[a]These vectors are commercially available from New England Biolabs (Beverly, MA); *see* **Note 1**.

[b]2-mercaptoethanesulfonic acid (MESNA) is used for isolation of proteins possessing a C-terminal thioester for ligation, labeling and cyclization (*see* **Notes 7** and **13**).

[c]The CBD-intein 1 tag exhibits an apparent molecular mass of 27 kDa on SDS-PAGE.

[d]The expected molecular mass of a fusion protein is 55 kDa plus the mass of the target protein (*see* **Note 13**).

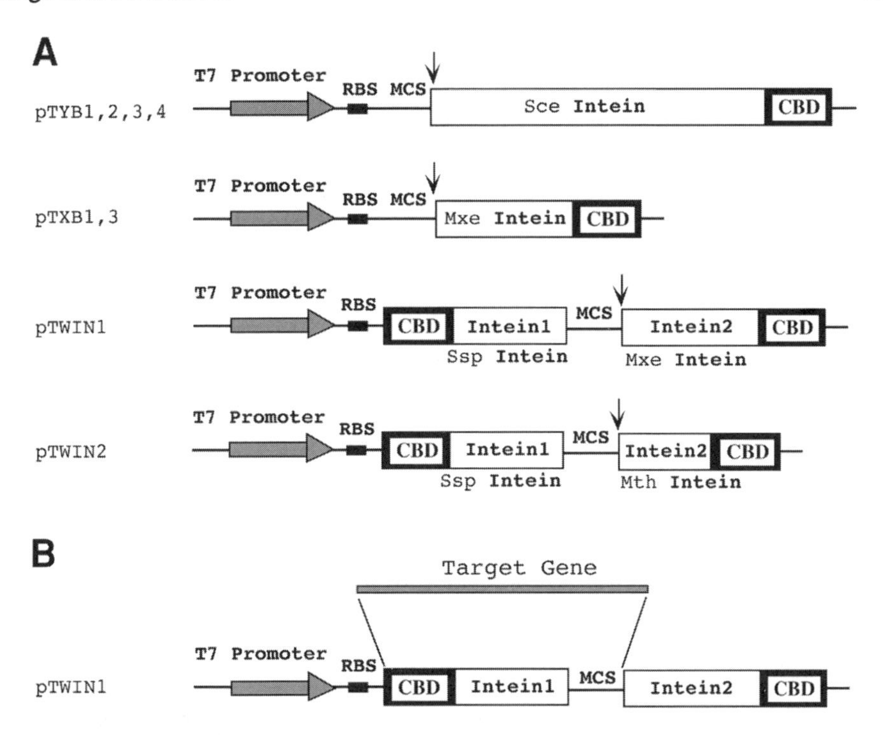

Fig. 3. IMPACT C-terminal fusion vectors. **(A)** schematic representation of vectors. pTYB1-4 series contain the modified *Sce* VMA intein (Sce intein) while pTXB vectors carry the modified *Mxe* GyrA intein (Mxe intein). Intein 2, the Mxe intein in pTWIN1 or *Mth* RIR1 intein (Mth intein) in pTWIN2, has been engineered as a C-terminal fusion tag. pTYB2-4 differ from pTYB1 and pTXB3 differs from pTXB1 only in the multiple cloning site (MCS). The diagram is not to scale and the arrows indicate the sites of thiol-induced cleavage (at the N-terminus of the modified intein). In order to isolate recombinant proteins possessing an active thioester for protein labeling and ligation, 2-mercaptoethanesulfonic acid (MESNA) is used in place of DTT. **(B)** diagram illustrating the fusion of a target gene to the N-terminus of a modified intein (intein 2) in a pTWIN vector.

codon for a cysteine residue must be placed immediately adjacent to the C-terminus of intein 1.

3.2. Use of IMPACT C-Terminal Fusion Vectors

The IMPACT C-terminal fusion vectors, pTYB1, pTYB3, pTXB1, pTWIN1, and pTWIN2, are designed for the translational fusion of a target protein at its C-terminus to the N-terminus of an engineered intein and the release of the target protein on chitin resin induced by a thiol reagent (*see* **Figs. 1** and **3**).

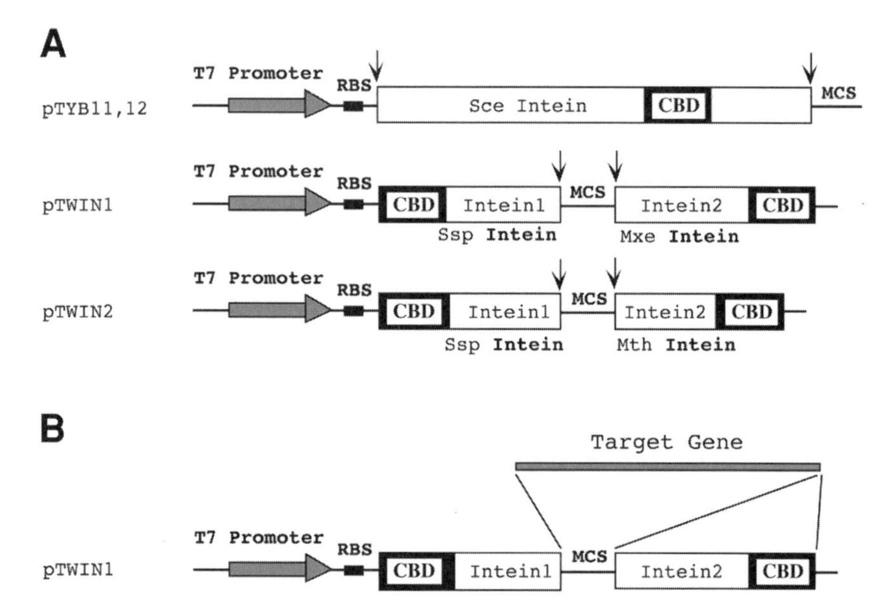

Fig. 4. IMPACT N-terminal fusion vectors. **(A)** schematic representation of vectors. The diagram is not to scale, and the arrows indicate the cleavage sites in an intein fusion protein. **(B)** diagram illustrating the fusion of a target gene to the C-terminus of the modified *Ssp* DnaB intein (Ssp intein or intein 1) in a pTWIN vector. To conduct circularization of a target protein, a target gene fragment is inserted in-frame between intein 1 and intein 2 and must not contain a translation termination codon. To generate polypeptides for ligation or cyclization, a codon for a cysteine residue is placed immediately adjacent to the C-terminus of intein 1.

3.2.1. Cloning Into a C-Terminal Fusion Vector

The *Nde*I or *Nco*I site is usually used for cloning the 5' end of the target gene. When the *Sap*I site is present in the polylinker, it must be used for cloning the 3' end of the target gene; this would yield a target protein without any vector-derived residues at its C-terminus (*see* **Table 1** and **Note 12**). These restriction sites are usually introduced into the ends of the target gene fragment by the polymerase chain reaction (PCR) before it is inserted into the polylinker of an IMPACT vector (*see* **Note 14**).

1. A target gene sequence is amplified by PCR.
2. The PCR fragment or the plasmid carrying the target gene is digested with the appropriate restriction enzymes and the fragment containing the target gene is isolated by agarose gel electrophoresis and ligated to an IMPACT vector digested with appropriate restriction enzymes.

3. *E. coli* cells made competent are transformed with the ligation mixture. Transformants are selected by colony formation on LB agar containing ampicillin.
4. The plasmid DNA samples are extracted from cultures of transformants grown at 37°C overnight in 2 mL LB media supplemented with ampicillin.
5. The structure of recombinant plasmids can be checked by restriction analysis of plasmid DNA. Digestion with enzymes that cut at sites flanking the insert can help to determine the presence of the target gene (*see* **Note 15**). PCR or colony hybridization can be used to screen a large number of transformants for the presence of the insert.
6. DNA sequencing with appropriate primers (available from NEB) should eventually verify the target gene sequence.

Cloning the target gene may be initially performed with a non-expression host strain. If an expression host strain such as ER2566 or BL21(DE3) is used, Coomassie blue-stained SDS-PAGE and Western blot analysis with the anti-CBD antibody can be used to detect fusion proteins in cell lysates as described in the following sections.

3.2.2. Screening for Expression of Fusion Proteins

When ER2566 (or other strain suitable for expression of a target gene under the control of the T7 promoter) is used for cloning, transformants expressing the desired fusion protein can be conveniently identified as follows:

1. Inoculate isolated colonies of potential transformants and of the parental vector (i.e., pTWIN1) into 15-mL tubes containing 2 mL LB broth supplemented with 100 μg/mL ampicillin and grow in a shaking incubator at 37°C until slightly cloudy (3–4 h) or OD_{600nm} of 0.4–0.6.
2. Induce expression by adding IPTG to 0.3 m*M* final concentration and grow for an additional 2–3 h.
3. Mix 40 μL of cell culture with 20 μL 3X SDS sample buffer followed by boiling for 10 min.
4. Analyze 10–15 μL of each sample by a Coomassie blue-stained SDS-PAGE gel.

The presence of a band of expected molecular mass indicates successful expression of the designed polypeptide as a fusion protein. Depending on the target protein and expression conditions, a fusion protein may undergo intein-mediated cleavage in vivo, yielding products corresponding to the intein-CBD species and the target protein (*see* **Note 10** for sample preparation). In addition, proteolysis may result in degradation of the target protein or the fusion protein. Problems of poor transcription, unstable messenger RNA, codon usage, or proteolysis may all contribute to poor expression. In order to determine the optimal expression conditions, different temperatures and host strains should be tested to examine expression level, in vivo cleavage and solubility of the fusion proteins.

3.2.3. Detection of Expressed Proteins by Western Blot Analysis

Some proteins may be expressed poorly so that it is difficult to detect against the background of host proteins using Coomassie staining. A much more sensitive method to detect expression of a fusion protein is to conduct western blot analysis using an antibody against chitin binding domain and/or an antibody against a specific target protein.

1. Induce expression of the fusion protein (*see* **Subheading 3.2.2.**)
2. Prepare samples and perform SDS-PAGE. Normally, 1–2 μL of cell extract or clarified cell extract is loaded.
3. Conduct immunoblotting with anti-CBD antibody.

3.2.4. Determination of Solubility of Fusion Proteins

The expression of heterologous proteins in *E. coli* sometimes results in the formation of inclusion bodies. Samples taken from cell lysate, clarified cell lysate, and the pellet of the cell lysate can be analyzed on SDS-PAGE to determine solubility of an expressed fusion protein (*see* **Note 16**). The presence of the fusion protein in both cell lysate and clarified cell lysate (soluble fraction) suggest that the fusion protein is soluble. However, the presence of the fusion protein in crude cell lysate and pellet of cell lysate, but not in clarified cell lysate (soluble fraction) suggest that the fusion protein is expressed as inclusion bodies. If an intein fusion protein forms inclusion bodies in *E. coli*, various host strains and expression conditions may be tested to improve the solubility. Induction of expression at lower temperature (e.g., 12–15°C) and lower IPTG concentration (less than 0.05 m*M*) should be first examined. If all attempts fail, the protocol described in **Note 17** can be applied to the recovery and refolding of fusion proteins from inclusion bodies.

3.2.5. Expression and Purification Procedures

Once the target protein is cloned into the appropriate intein vector, a scale-up experiment can be conducted. Protein purification and on-column cleavage should be performed at 4°C, and buffers should always be kept at 4°C in order to minimize the cleavage of the target protein-intein-CBD fusion protein due to hydrolysis of the thioester bond. Samples can be taken throughout the experiment to monitor each step.

Day 1: Solution and Strain Preparation

1. Transformation: Competent cells prepared from *E. coli* strain ER2566 are transformed with an IMPACT plasmid containing the desired target gene ***(17)***. Cells are incubated overnight at 30–37°C on an LB agar plate supplemented with 100 μg/mL ampicillin. Alternatively, isolated colonies can be obtained by streaking ER2566 cells bearing the appropriate plasmid on an LB plate supplemented with 100 μg/mL of ampicillin.

Day 2: Protein Expression

2. Cell culture: LB medium (1 L) supplemented with 100 µg/mL ampicillin is inoculated in a 2-L flask with a freshly grown colony or 10 mL of a fresh starter culture and incubated on a rotary shaker at 37°C. It is not recommended to use an overnight culture for inoculation.
3. Induction of gene expression: When culture density reaches OD_{600nm} of 0.5, add 3 mL of 100 m*M* IPTG to the 1-L culture (final concentration of 0.3 m*M*) and transfer to a rotary shaker at 37°C and incubate for 3 h. Before the addition of IPTG, 5 mL of cell culture is transferred to a sterile test tube or flask for continuous incubation and used as a control (cells without IPTG induction). For expression at 15°C overnight, IPTG (to 0.3 m*M*) is added when the culture density (OD_{600nm}) is 0.6–0.7. The conditions for induction and expression may need to be optimized for the specific target protein. Other typical conditions are 30°C for 3 h or 20–25°C for 6 h.
4. Column preparation: Transfer 30 mL of chitin bead slurry into a column (*see* **Note 3**). Wash the column by passage of 200 mL of sterile water and 50 mL of cell lysis/column buffer (*see* **Subheading 2.4.**).

Day 3: Chitin Column Chromatography and On-Column Cleavage

5. Harvesting cells: The cells are collected by centrifugation at 5000*g* for 10 min at 4°C and the cell pellet can be used directly for purification or stored at –20°C. Resuspend the pellet of uninduced cells in 0.5 mL of cell lysis buffer and mix 40 µL with 20 µL of 3X SDS sample buffer (Sample 1: uninduced cells).
6. Cell lysis: Cells from a 1-L culture are resuspended in 50 or 100 mL ice-cold cell lysis/column buffer, and disrupted either by sonication in an ice-chilled water bath or with a French press. Mix 40 µL of the crude cell lysate with 20 µL of 3X SDS sample buffer (sample 2: crude cell extract). Clarified cell extract is prepared by centrifugation at 12,000*g* for 30 min. Mix 40 µL of the supernatant with 20 µL of 3X SDS sample buffer (Sample 3: clarified cell extract; *see* **Subheading 3.2.4.**).
7. Loading: The clarified extract is loaded onto a chitin column at a flow rate not exceeding 0.5–1.0 mL/min. A sample from the cell extract after passing through the column is also analyzed by SDS-PAGE. Comparison of the samples from the flow-through sample and the clarified cell extract provides an indication of the binding efficiency of the fusion protein to the chitin column.
8. Wash: Wash the column with 500–750 mL of cell lysis/column buffer at a higher flow rate (about 2–4 mL/min). All traces of the cell extract should be washed off the sides of the column.
9. On-column cleavage: The target protein is released from the chitin column by inducing the intein to undergo self-cleavage in the presence of DTT. Freshly prepare cleavage buffer containing 30 m*M* DTT (pH 8.0 or 8.5). The column is quickly flushed with 3 bed volumes of cleavage buffer and the flow is stopped. The column is left at 4°C for 12–16 h. If cleavage is observed during equilibra-

tion of the chitin resin with cleavage buffer, 1 *M* DTT stock solution can be added directly to the column to a 30–50 m*M* final concentration, followed by gently mixing the resin. To generate an active thioester at the C-terminus of the target protein for protein ligation and labeling, the column should be flushed with 3 bed volumes of cleavage buffer composed of 50 m*M* Tris-HCl, pH 8.5, 500 m*M* NaCl and 50 m*M* MESNA (*see* **Note 7**).

Day 4: Elution

10. Elution: The target protein is eluted from the column using the cell lysis/column buffer or a specific storage buffer.
11. Analysis by SDS-PAGE: To determine the cleavage efficiency and to examine the residual proteins on the chitin resin, 40 μL of the resin is gently removed from the column, mixed with 20 μL of 3X SDS sample buffer, and boiled for 10 min. The resin is pelleted by centrifugation, and a sample (5 μL) of the supernatant is directly used for SDS-PAGE analysis. In some cases, a target protein is not eluted after on-column cleavage, but is found in the SDS elution fractions, suggesting that the target protein becomes insoluble after cleavage. The bound proteins can be eluted from the column with a column buffer containing 2% SDS at room temperature.

3.3. Use of IMPACT N-Terminal Fusion Vectors

The N-terminal fusion vectors are designed for the fusion of a target protein at its N-terminus to an engineered intein and the release of the target protein is mediated by fission of the peptide bond at the intein C-terminus (*see* **Figs. 2** and **4**). The system provides alternative approaches for intein-mediated purification and thus enhances the probability of successful expression and purification of a target protein and also presents new methods for protein manipulation such as protein ligation, labeling and cyclization. As illustrated in **Fig. 4**, two types of N-terminal fusion vectors are commercially available from New England Biolabs. The intein in pTYB11 and pTYB12 undergoes cleavage of the peptide bonds at both the N-terminal and the C-terminal splice junctions in the presence of a thiol compound such as DTT or cysteine *(11,12)*. In contrast, the intein in a pTWIN vector cleaves only at its C-terminus by incubation at an optimal pH and temperature without the use of a thiol reagent *(14)*. Cleavage without the use of a thiol reagent is beneficial in the purification of thiol-sensitive proteins. A major advantage of the N-terminal fusion system is that proteins can be purified without an N-terminal methionine residue, which, in many cases, may not be present in the mature form of a native protein sequence. Furthermore, multiple applications of pTWIN vectors include generation of recombinant proteins or peptides possessing an N-terminal cysteine for protein ligation, labeling, and cyclization of a target protein or peptide cloned in frame to both modified inteins *(16)*.

3.3.1. Use of pTYB11 and pTYB12

3.3.1.1. Cloning the Target Gene

To purify a target protein with no vector-derived residues, a *Sap*I site in pTYB11 should be used for cloning the 5' end of a target gene, resulting in the fusion of the target protein immediately adjacent to the C-terminus (or the cleavage site) of the intein. Use of pTYB12 yields a target protein with extra residue(s) added to its N-terminus (*see* **Note 18**).

3.3.1.2. Expression and Purification Procedures

Like the C-terminal fusion vectors, a fusion gene in pTYB11 and pTYB12 is under control of the IPTG-inducible T7/*lac* promoter. The protocol for expressing and purifying a target protein in the pTYB11 or pTYB12 vector is essentially the same as those described for the C-terminal fusion vectors except that the cleavage reaction may need higher temperatures (16–25°C) up to 40 h. The N-terminus of a target protein is fused to the C-terminus of the modified *Sce* VMA intein. A 15-residue N-extein sequence is present at the N-terminus of the intein to provide a favorable translational start for protein expression in *E. coli*. After binding of the fusion protein to chitin, the intein is then induced to undergo on-column cleavage by incubation with thiol reagents such as DTT, β-mercaptoethanol, or cysteine at pH 8.0–8.5 (*see* **Note 19**). The target protein is eluted along with the short 15-residue peptide. A dialysis step can be performed to separate the target protein from the N-extein peptide and thiol compounds.

3.3.2. Use of pTWIN1 as an N-Terminal Fusion Vector

Unlike the pTYB11 and pTYB12 vectors, the pTWIN vectors allow purification of recombinant proteins without the use of a thiol reagent (*see* **Figs. 2** and **4**). The C-terminal cleavage activity of intein 1 in pTWIN1 (or pTWIN2) provides a novel approach to isolate recombinant proteins possessing a wide range of N-terminal residues other than a methionine residue without the use of a protease (*see* **Note 20**). The target gene should be cloned in-frame to the C-terminus of the first intein. The cell growth and induction of expression of the intein fusion gene were essentially the same as the protocols described for the C-terminal fusion vectors (*see* **Subheading 3.2.**). The cells are resuspended and lysed in cell lysis/column buffer at pH 8.5, which inhibits cleavage during the purification process. The presence of the *B. circulans* chitin binding domain allows isolation of a CBD-intein-target protein from induced *E. coli* cell extracts by binding to chitin resin. The intein is induced to undergo cleavage on chitin by incubation at 25°C and pH 6.0–7.0 for 16–24 h (*see* **Note 8**). The target protein is eluted while the CBD-intein partner remains bound to chitin.

3.3.2.1. Cloning the Target Gene Into pTWIN1

The steps described in **Subheading 3.2.1.** should be followed for cloning the target gene. The use of the *Sap*I site (adjacent to the 3' end of the coding region of the *Ssp* DnaB mini-intein, intein 1) for cloning the 5' end of the target gene allows purification and isolation of a protein without extra N-terminal amino acids. If the *Sap*I site in a pTWIN vector is chosen for cloning, other sites within the polylinker cannot be used. The second *Sap*I site at the 3' end of the polylinker or one of the unique sites such as *Spe*I, *Pst*I or *Bam*HI downstream of the polylinker should be used as the 3' cloning site. The insert should include a translation termination codon. If the target protein can accommodate non-native sequence, the *Nco*I or *Not*I sites may be used as a 5' cloning site resulting in the inclusion of several vector-derived residues, which are favorable for controlled cleavage, attached to the purified target protein after cleavage.

3.3.2.2. Screening for Expression of Fusion Proteins

If an expression strain such as ER2566 or BL21(DE3) is used, Coomassie Blue stained SDS-PAGE and western blot analysis with the anti-CBD antibody can be used to detect fusion proteins as described in **Subheadings 3.2.2.** and **3.2.3.** In vivo and in vitro cleavage activity of the *Ssp* DnaB mini-intein is dependent on the amino acids adjacent to the intein (*see* **Note 20**). It is advisable to check the solubility of the fusion protein expressed under different conditions (*see* **Subheading 3.2.4.**) before conducting a scale-up expression and purification.

3.3.2.3. Expression and Purification Procedures

It is crucial to perform a rapid purification of the fusion protein under nonpermissive conditions (pH 8.5 and 4°C) before on-column cleavage is conducted at pH 6–7 and 4–25°C. The following is a protocol at a preparative scale for purification of a target protein fused to intein 1 in a pTWIN1 (or pTWIN2) vector from a 1-L *E. coli* culture.

Day 1

1. Transform the appropriate plasmid into an *E. coli* host strain carrying the T7 RNA polymerase gene. Isolated colonies can also be obtained by streaking cells bearing expression plasmids onto LB agar plates containing 100 µg/mL ampicillin and incubating the plate at 37°C overnight.

Day 2

2. Inoculate a freshly grown colony at 37°C into 1 L of LB broth containing 100 µg/mL ampicillin. It is not recommended to use an overnight cell culture for subculturing.

3. For expression at 30–37°C, the cell culture is grown to an OD_{600nm} of 0.5, followed by induction with IPTG (at 0.3 m*M* final concentration); for low temperature induction, the cell culture is grown to an OD_{600nm} of 0.6–0.7, followed by induction with IPTG (0.3 m*M*) overnight at 12–15°C.
4. Fill a column with 30 mL of chitin beads (*see* **Note 3**). Equilibrate the column with 200 mL of distilled water and 100 mL of column buffer.
5. The cells are harvested and the pellet can be stored at –20°C or –70°C until use.

Day 3

6. The pellet is resuspended in 100 mL of column buffer containing 20 m*M* Tris-HCl, pH 8.5, and 0.5 *M* NaCl. Cells are disrupted by sonication at 4°C.
7. The clarified cell extract is prepared by centrifugation and slowly loaded onto the column.
8. The column is washed with 500–750 mL of column buffer in 2–3 h at 4°C
9. Flush the resin with 100 mL (or 5 column volumes) of cleavage buffer (*see* **Note 8**).
10. The column is incubated at 25°C (or room temperature) for 16–24 h. If the target protein is unstable, the cleavage reaction may be carried out at 4°C with longer incubation times (e.g., 3–5 d).

Day 4

11. Elution of the protein product is conducted in 5-mL fractions in the same cleavage buffer. The fractions are then examined by the Bradford assay to determine the protein concentration.
12. Analyze both the eluate and a chitin fraction on SDS-PAGE for cleavage efficiency and protein solubility. A sample of chitin beads can be prepared by mixing 40 µL of beads with 20 µL 3X SDS sample buffer and boiling for 10 min. Samples taken from the cell lysate, clarified cell extract, and cell extract after passage over the chitin column can also be analyzed on SDS-PAGE.

4. Notes

1. The important features of the IMPACT plasmids (**Figs. 3** and **4**) are listed as follows:
   a. All IMPACT vectors use an IPTG-inducible T7/*lac* promoter to provide stringent control of the fusion gene expression in *E. coli* ***(18)***;
   b. The vectors carry their own copy of the *lac*I gene encoding the *lac* repressor. Binding of the *lac* repressor to the *lac* operator sequence immediately downstream of the T7 promoter suppresses basal expression of the intein fusion gene in the absence of IPTG induction;
   c. The presence of the *bla* gene (the Amp^r marker) conveys ampicillin resistance to the host strain. All vectors carry the origin of replication from pBR322;
   d. A T7 transcription terminator is placed downstream of the intein-CBD coding region;
   e. Five tandem transcription terminators (*rrnB* T1) placed upstream of the promoter minimize background transcription;

   f. The vectors possess the origin of DNA replication from bacteriophage M13 which allows for the production of single-stranded DNA by superinfection of cells bearing the plasmid with helper phage, M13KO7.

   The pTWIN vectors allow cloning a target gene in-frame with the C-terminus of the engineered *Ssp* DnaB mini-intein (intein 1), which is fused at its N-terminus to CBD. pTWIN1 differs from pTWIN2 by the replacement of the *Mth* RIR1 intein by the *Mxe* GyrA intein. The vector sequences and technical information are available at the NEB website (www.neb.com)

2. *E. coli* strains ER2566 and BL21(DE3) (or other derivatives), carrying the T7 RNA polymerase gene, must be used as host for expressing a target gene cloned in an IMPACT vector. ER2566 (provided by New England Biolabs with the IMPACT vectors) carries a chromosomal copy of the T7 RNA polymerase gene under control of the IPTG-inducible *lac* promoter, and is deficient in both *lon* and *omp*T proteases. The genotype of ER2566: $F^-\lambda^-$*fhuA2 [lon] ompT lacZ::T7 gene1 gal sulA11 Δ(mcrC-mrr)114::IS10 R(mcr-73::miniTn10-TetS)2 R(zgb-210::Tn10)(TetS) endA1 [dcm]*. The genotype of BL21(DE3): F^- *ompT hsdSB* ($r_B^-m_B^-$) *gal dcm* (DE3). A non-expression *E. coli* strain may be initially used for cloning.
3. The chitin beads have a binding capacity of about 2 mg/mL. Use 20–30 mL chitin slurry to prepare a 10–20 mL bed volume column for a 1 L culture.
4. 20 m*M* Tris-HCl or Na-HEPES (pH 8.0 or 8.5) and 0.5 *M* NaCl. Column buffer or cleavage buffer may be adjusted on the basis of solubility and activity of a target protein. The following conditions are compatible for chitin binding and intein-mediated cleavage: 50 m*M*–2 *M* NaCl; 0–1 m*M* EDTA; 0.1% Triton X-100 or Tween-20. 0.1% Triton X-100 or Tween-20 can be used to reduce non-specific adsorption to the chitin beads, unless the target protein is known to be inactivated by these nonionic detergents. Protease inhibitors such as phenyl-methylsulfonyl fluoride can be added into cell lysis/column buffer. Column buffer containing 1 *M* NaCl, 5 m*M* ATP and 5 m*M* $MgCl_2$ may be used to wash GroEL or DnaK from the chitin resin; these chaperonins may co-purify by binding to the target protein.
5. The presence of a thiol reagent in the column buffer may cause cleavage of the fusion protein expressed from a C-terminal IMPACT vector. TCEP [tris-(2-carboxyethyl)phosphine] (Pierce, cat. no. 20490) and TCCP [tris-(2-cyanoethyl)phosphine] can be used at 0.1–1 m*M* final concentration in the column buffer to stabilize oxidation-sensitive proteins during purification. These compounds specifically reduce disulfide bonds without affecting the intein-mediated cleavage reaction and thus can be used to stabilize proteins with essential thiols.
6. Since DTT is not particularly stable after dilution, the cleavage buffer for proteins expressed from a C-terminal fusion vector should be freshly prepared before use. DTT stock solutions should be aliquoted (to minimize freeze/thaw cycles) and stored at –20°C.
7. It is recommended to use 50 m*M* MESNA in place of DTT if the purpose is to isolate proteins possessing a C-terminal thioester for ligation or cyclization. The

cleavage reaction induced with MESNA yields a relatively stable thioester at the carboxy terminus of a target protein. The thioester-tagged protein products isolated after a 16–24 h cleavage reaction can be ligated with extents typically greater than 90% to polypeptides possessing an N-terminal cysteine. Purified thioester tagged proteins stored at –70°C for more than 6 mo exhibited greater than 80% ligation extents. Efficient ligation requires that the concentration of one of the reactants (e.g., synthetic peptide) is present at no less than 0.5–1.0 m*M* and the reaction proceeds in the pH range of 8.0–9.0. Ligation usually proceeds overnight at 4°C, but can also be carried out at 20–25°C for 2–6 h. The reaction samples plus a sample prior to ligation can be examined by SDS-PAGE. Ligation is indicated by a shift to higher molecular weight in comparison to the starting protein band. If the reaction does not appear to be complete, the sample may be supplemented with one-tenth volume of 1 *M* Tris-HCl, pH 8.5 to a 100 m*M* final concentration and incubated overnight at 4°C. Fusion of two recombinant proteins requires at least one of the protein substrates to be present at a concentration of >500 µ*M*. Following the isolation of the thioester-tagged protein and protein segment possessing an unprotected N-terminal cysteine, both reactants can be mixed and concentrated by a Centriprep or Centricon apparatus (Millipore, Bedford, MA) prior to an overnight incubation at 4°C.

8. Cleavage of intein 1 expressed from a pTWIN vector should be conducted at pH 6–7 at 20–25°C. Cleavage buffer 1: 20 m*M* bis-tris, pH 7.0 at 25°C, 0.5 *M* NaCl; Cleavage buffer 2: 20 m*M* bis-tris, pH 6.0 at 25°C, 0.5 *M* NaCl.
9. The chitin resin can be regenerated by the following procedure: (a) Wash the resin with 3 bed volumes of 0.3 *M* NaOH (stripping solution) to remove residual proteins from the column; (b) Soak the column for 30 min and then wash with an additional 7 bed volumes of 0.3 *M* NaOH; (c) Wash with 20 bed volumes of water followed by 5 bed volumes of column buffer; (d) The column can be stored at 4°C. Sodium azide (0.02%) should be added to the column buffer for long term storage.
10. The presence of thiol reagents such as DTT or 2-mercaptoethanol in the SDS sample buffer can cause cleavage of the thioester intermediates present in the cell extract, especially when relatively large amounts of thioester intermediate accumulates during expression; this would result in an overestimation of in vivo cleavage activity. It is recommended to analyze in vivo cleavage activity using a cell extract sample prepared with SDS sample buffer without DTT. In vivo accumulation of thioester intermediates depend on various factors including intein, expression conditions and amino acids or protein sequences adjacent to the cleavage site.
11. The decision of whether to fuse the C- or the N-terminus of the target protein to the intein may affect expression level and yield. Since each intein has evolved in the context of its own host protein, different inteins may favor different amino acid residues or sequences adjacent to the scissile peptide bond for cleavage. Inducible cleavage is a balancing act in that in vivo cleavage reduces the yield of fusion protein while proficient cleavage on the chitin resin is essential for releas-

ing the target protein from the intein tag. It is possible and recommended to insert a target gene into several different vectors to choose the best fusion partner.

12. The amino acid residue adjacent to the cleavage site has a profound effect on cleavage efficiency ***(10–12,15,16,19)***. Some recommendations are given in **Table 1**, although the profile may be different in a different protein context. The use of a *Sap*I site permits insertion of additional amino acids between a target protein and a modified intein when a favorable residue or sequence is needed to improve cleavage efficiency. The presence of compatible cloning sites in the polylinker region (MCS) of the IMPACT vectors allows an insert to be fused to different inteins in order to determine the optimal construct to express and purify a target protein. For instance, a target gene can be cloned using the *Nde*I and *Sap*I sites in the pTYB1, pTXB1, pTWIN1, and pTWIN2 vectors while the *Nco*I and *Sap*I sites can be used for cloning into pTYB3 and pTXB3. The *Sap*I and *Pst*I sites can be used for cloning a target gene into pTYB11 and pTWIN1 (or pTWIN2).
13. To perform the circularization of the peptide backbone, a target gene is inserted into the polylinker of a pTWIN vector in frame to both inteins using the *Sap*I sites in the polylinker. A codon for a cysteine residue must be incorporated into the forward primer for amplification of the target gene and placed immediately adjacent to the C-terminus of intein 1 after insertion. The following procedure is designed for purification of a circular protein by fusion to both inteins in a pTWIN1 vector.
    a. Clone the target gene into a pTWIN vector using the *Sap*I sites.
    b. Screen the recombinant clones (*see* **Subheading 3.2.1.**).
    c. Transform ER2566 cells with the plasmid expressing the target gene.
    d. Grow the culture in LB media supplemented with ampicillin at 37°C until OD_{600nm} reaches 0.5 (for expression at 37°C) or 0.7 (for expression at 15°C).
    e. Induce expression with 0.3 m*M* IPTG.
    f. Equilibrate chitin resin in 20 m*M* Tris-HCl, pH 8.5, 500 m*M* NaCl.
    g. Lyse the cells in the column buffer and slowly load (ca. 0.5 mL/min) the clarified extract to the resin at 4°C.
    h. Wash the resin with the 10 column volumes of column buffer to remove unbound proteins.
    i. Induce on-column cleavage of Intein 1 in 20 m*M* Tris-HCl, pH 6.0–7.0, and 500 m*M* NaCl.
    j. Leave the column overnight at room temperature.
    k. Wash the column with the intein 1 cleavage buffer at 4°C.
    l. Induce cleavage of Intein 2 by equilibrating the column with 20 m*M* Tris-HCl, pH 8.5, 500 m*M* NaCl, 50 m*M* MESNA.
    m. Incubate the column overnight at 4°C.
    n. Elute the target protein with the intein 2 cleavage buffer, pH 8.5.
    o. Analyze the eluate and an aliquot of the resin by SDS-PAGE to determine cleavage/cyclization efficiency and protein solubility.
14. Appropriate restriction sites, absent in the target gene, should be incorporated into synthetic oligonuceotides (the forward and reverse primers for amplification

of a target gene). The choice of restriction sites determines the amino acid residues that may be attached to the target protein after cleavage of a fusion protein. The reverse primer can be designed such that a favorable residue(s) is adjacent to the scissle peptide bond of the expressed fusion protein. Furthermore, each primer should include six extra nucleotides at the 5' end for efficient digestion of the PCR fragment with restriction enzymes. The PCR fragment can be treated with appropriate restriction enzymes and ligated with a desired IMPACT vector. Alternatively, the DNA fragment generated by PCR can be initially cloned into a blunt-end site of a vector if a PCR product is generated by a proofreading DNA polymerase or a T-vector if a PCR product is produced by a non-proofreading DNA polymerase. The target gene fragment can then be isolated following digestion of the recombinant plasmid with appropriate restriction enzymes and ligated to an IMPACT vector.

15. In general, the plasmid DNA samples isolated from cultures of transformants should be digested with the same restriction enzymes used for cloning the target gene. However, when *Sap*I is used for cloning, other sites adjacent to the insert (e.g., *Spe*I site in pTXB or pTWIN vectors or the *Kpn*I site in pTYB vectors) should be used for analysis. This is because the *Sap*I site is not regenerated after ligation of the insert to the vector.
16. The following is a quick protocol to examine solubility of an expressed fusion protein.
    a. Inoculate isolated colonies of potential transformants into 15-ml tubes containing 3 mL LB broth supplemented with 100 µg/mL ampicillin and grow in a shaking incubator at 37°C until slightly cloudy (3–4 h) or OD_{600nm} of 0.4–0.6.
    b. Induce expression by adding IPTG to 0.3 m*M* final concentration. Incubate 1.5 mL at 30°C in a shaker for an additional 2–3 h and the remaining 1.5 mL culture at 12–15°C for 12–16 h.
    c. Sonicate the cell culture in an ice-chilled water bath or lyse cells by cycles of freeze/thaw. Mix 40 µL of the total cell lysate with 20 µL 3X SDS sample buffer.
    d. Centrifuge cell lysate at 10,000*g* for 5 min at 4°C. Mix 40 µL of clarified cell lysate (the supernatant) with 20 µL 3X SDS sample buffer. A sample may also be prepared from the pellet after washing with LB broth or column buffer and resuspended in 8 *M* urea.
    e. Boil samples for 10 min and load 10–15 µL of each sample on a 10% or 12% SDS-polyacrylamide gel followed by Coomassie blue staining and/or Western blotting.

    If the fusion precursor is detected in the total cell extract but not in the supernatant, the fusion protein is probably expressed in inclusion bodies rather than in soluble form. But, be aware that excessive sonication can also lead to insolubility. Expression at low temperatures may reduce the formation of inclusion bodies, improve the folding and solubility of the fusion protein, and increase the cleavage efficiency of the intein. Low IPTG concentration may reduce expression thereby relieving the solubility problem.

17. The following protocol has been successfully applied to the recovery of proteins fused to intein 1 or intein 2 of the pTWIN1 vector from inclusion bodies. For C-terminal fusion vectors, DTT should not be included in the solutions.
    a. Resuspend the cell pellet from 1 L of *E. coli* culture in 100 mL cell lysis buffer.
    b. Break cells by sonification.
    c. Spin down cell debris containing the inclusion bodies at 15,000*g* and 4°C for 30 min.
    d. Pour out supernatant and resuspend pellet in 100 mL breaking buffer.
    e. Stir solution for 1 h at 4°C.
    f. Spin remaining cell debris down at 15,000*g* and 4°C for 30 min.
    g. Load supernatant into dialysis bag and dialyze against renaturation buffers A, B, C, D, and 2× E. Each step is against 1 L of a renaturation buffer and should take at least 3 h at 4°C. During dialysis the buffer should be stirred by a stir bar.
    h. Centrifuge the dialyzed solution containing the renatured protein at 15,000*g* and 4°C for 30 min to remove any remaining impurities or incorrectly folded protein which is again aggregated.
    i. Use a standard protocol for chitin chromatography and cleavage reaction designed for a specific intein. Elute the protein product and analyze both the eluate and chitin beads for cleavage efficiency and protein solubility.
    j. Solutions:
       Cell lysis buffer: 20 m*M* Tris-HCl, pH 8.0 and 0.5 *M* NaCl.
       Breaking buffer: 20 m*M* Tris-HCl, pH 8.0, 0.5 *M* NaCl, 7 *M* Guanidine-HCl, and 10 m*M* DTT.
       Renaturation buffer A: 20 m*M* Tris-HCl, pH 8.0, 0.5 *M* NaCl, 8 *M* urea, 10 m*M* DTT.
       Renaturation buffer B: 20 m*M* Tris-HCl, pH 8.0, 0.5 *M* NaCl, 6 *M* urea, 1 m*M* DTT.
       Renaturation buffer C: 20 m*M* Tris-HCl, pH 8.0, 0.5 *M* NaCl, 4 *M* urea, 1 m*M* DTT.
       Renaturation buffer D: 20 m*M* Tris-HCl, pH 8.0, 0.5 *M* NaCl, 2 *M* urea, 0.1 m*M* oxidized glutathione, 1 m*M* reduced glutathione.
       Renaturation buffer E: 20 m*M* Tris-HCl, pH 8.0, 0.5 *M* NaCl, 0.1 m*M* oxidized glutathione, 1 m*M* reduced glutathione.
18. Use of *Bsm*I for cloning the 5' end of a target gene adds an alanine residue to the N-terminus of the target protein. A favorable residue (Ala, Gln, Met, Gly) can be added immediately adjacent to the cleavage site by using the *Sap*I site in pTYB11 or the *Bsm*I or *Nde*I site in pTYB12 if an unfavorable residue such as Pro, which completely blocks cleavage, is present at the N-terminus of a protein of interest (*see* the IMPACT-CN manual). Furthermore, Cys, Ser or Thr should not be placed adjacent to the cleavage site (at the +1 position) because they may yield protein splicing (the joining of the target protein and the 15-residue peptide sequence fused to the N-terminus of the intein).

19. Column buffer for proteins expressed from pTYB11 (or pTYB12): 20 m*M* Tris-HCl or Na-HEPES, pH 6.0–8.5, and 0.5 *M* NaCl. Cleavage buffer: 20 m*M* Tris-HCl or Na-HEPES, pH 8.0–8.5, 0.5 *M* NaCl and 50 m*M* DTT.
20. Studies of the effect of the N-terminal residue of a target protein fused to the C-terminus of intein 1 showed that Ser, Cys, Ala, or Gly in the +1 position resulted in the most rapid C-terminal cleavage, and His, Met, Glu, Asp, Trp, Phe, Tyr, Val, and Thr also displayed proficient cleavage of the fusion protein ***(14)***. However, the presence of Gln, Asn, Leu, Ile, Arg, Lys, and Pro in the +1 position resulted in poor cleavage efficiency. This profile, obtained from the studies of a target protein (MBP), may change substantially with a different target protein. The C-extein residues at positions +2 to +5 can also affect cleavage efficiency ***(16)***. Observation of 10–50% in vivo cleavage suggests that the intein is active in this fusion context.

References

1. Xu, M.-Q., Paulus, H., and Chong, S. (2000) Fusions to self-splicing inteins for protein purification. *Methods Enzymol.* **326,** 376–418.
2. Paulus, H. (2000) Protein splicing and related forms of protein autoprocessing. *Annu. Rev. Biochem.* **69,** 447–495.
3. Chong, S., Shao Y., Paulus, H, Benner, J., Perler, F. B., and Xu, M.-Q. (1996) Protein splicing involving the *Saccharomyces cerevisiae* VMA intein: the steps in the splicing pathway, side reactions leading to protein cleavage, and establishment of an in vitro splicing system. *J. Biol. Chem.* **271,** 22,159–22,168.
4. Chong, S., Mersha, F. B., Comb, D. G., et al. (1997) Single-column purification of free recombinant proteins using a self-cleavable affinity tag derived from a protein splicing element. *Gene* **192,** 271–281.
5. Watanabe, T., Ito, Y., Yamada, T., Hashimoto, M., Sekine, S., and Tanaka, H. (1994) The role of the C-terminal domain and type III domains of chitinase A1 from *Bacillus circulans* WL-12 in chitin degradation. *J. Bacteriol.* **176,** 4465–4472.
6. Dawson, P. E., Muir, T. W., Clark-Lewis, I., and Kent, S. B. (1994) Synthesis of proteins by native chemical ligation. *Science* **266,** 776–779.
7. Muir, T. W., Sondhi, D., and Cole, P. A. (1998) Expressed protein ligation: a general method for protein engineering. *Proc. Natl. Acad. Sci. USA* **95,** 6705–6710.
8. Severinov, K. and Muir, T. W. (1998) Expressed protein ligation, a novel method for studying protein-protein interactions in transcription. *J. Biol. Chem.* **273,** 16,205–16,209.
9. Evans, T. C., Jr., Benner, J., and Xu, M.-Q. (1998) Semisynthesis of cytotoxic proteins using a modified protein splicing element. *Protein Sci.* **7,** 2256–2264.
10. Evans, T. C., Jr., Benner, J., and Xu, M.-Q. (1999) The in vitro ligation of bacterially expressed proteins using an intein from *Methanobacterium thermoautotrophicum. J. Biol. Chem.* **274,** 3923–3926.
11. Chong, S., Williams, K. S., Wotkowicz, C., and Xu, M.-Q. (1998) Modulation of protein splicing of the *Saccharomyces cerevisiae* vacuolar membrane ATPase intein. *J. Biol. Chem* **273,** 10,567–10,577.

12. Chong, S., Montello, G. E., Zhang, A., et al. (1998) Utilizing the C-terminal cleavage activity of a protein splicing element to purify recombinant proteins in a single chromatographic step. *Nucleic Acids Res.* **26,** 5109–5115.
13. Wu, H., Xu, M.-Q., and Liu, X.-Q. (1998) Protein trans-splicing and functional mini-inteins of a cyanobacterial DnaB intein. *Biochim. Biophys. Acta* **1387,** 422–432.
14. Mathys, S., Evans, T. C., Jr., Chute, I. C., et al. (1999) Characterization of a self-splicing mini-intein and its conversion into autocatalytic N- and C-terminal cleavage elements: facile production of protein building blocks for protein ligation. *Gene* **231,** 1–13.
15. Wood, D. W., Wu, W., Belfort, G., Derbyshire. V., and Belfort, M. (1997) A genetic system yields self-splicing inteins for bioseparation. *Nat. Biotech.* **17,** 889–892.
16. Evans, T. C., Jr., Benner, J., and Xu, M.-Q. (1999) The cyclization and polymerization of bacterially-expressed proteins using modified self-splicing inteins. *J. Biol. Chem.* **274,** 18,359–18,363.
17. Sambrook, J., Frisch, E. F., and Maniatis, T. (1989) *Molecular Cloning: A Laboratory Manual.* Cold Spring Harbor Laboratory, Cold Spring Harbor, NY.
18. Dubendorff, J. W. and Studier, F. W. (1991) Controlling basal expression in an inducible T7 expression system by blocking the target T7 promoter with lac repressor. *J. Mol. Biol.* **219,** 45–59.
19. Southworth, M. W., Amaya, K., Evans, T. C., Xu, M.-Q., and Perler, F. B. (1999) Purification of proteins fused to either the amino or carboxy terminus of the *Mycobacterium xenopi* Gyrase A intein. *Biotechniques* **27,** 110–120.

5

Calmodulin as an Affinity Purification Tag

Samu Melkko and Dario Neri

1. Introduction

Calmodulin is a small (148 amino acids, 17 kDa) ubiquitous protein in eukaryotes, and is considered the primary intracellular calcium sensor, making it a key regulator of intracellular signal transduction. Upon calcium binding, calmodulin can interact with a variety of proteins *(1)*, mediating effects on gene regulation, DNA synthesis, cell cycle progression, mitosis, cytokinesis, cytoskeletal organization, muscle contraction, and metabolic regulation. Calmodulin is remarkably conserved throughout evolution. Amino acid sequences in multicellular organisms are nearly identical (>90% among mammals, insects and plants). Moreover, the three existing gene copies of calmodulin in humans code for proteins of identical amino acid sequence.

Calmodulin binds 4 calcium ions with 4 recurring motifs called EF-hand domains *(2)* and is an unusually acidic protein, with a net charge of –24 at neutral pH, before Ca^{2+} binding. The interaction of calmodulin to target proteins is mediated by binding to peptidic moieties on these proteins. Such peptides do not show a consensus sequence, but can be classified as positively charged amphiphilic alpha helices *(3)*. Typically, the affinities of calmodulin to its natural target peptides are in the nanomolar range, albeit some synthetic peptides have been identified with even superior binding properties *(4)* ($k_{on} = 9.8 \times 10^5/s\ M$, $k_{off} = 2.2 \times 10^{-6}/s$, $K_D = 2.2 \times 10^{-12}\ M$). Calmodulin is one of the few examples of a small protein capable of binding to peptides with very high affinity, and is therefore an interesting candidate for biotechnological applications. Binding of target peptides generally requires prior binding of calcium ions. The high affinity binding can be abolished under mild conditions by addition of calcium chelators like EDTA or EGTA, making the calmodulin/ligand system an interesting alternative to the avidin-biotin system *(5)*. Structural stud-

From: *Methods in Molecular Biology, vol. 205, E. coli Gene Expression Protocols*
Edited by: P. E. Vaillancourt © Humana Press Inc., Totowa, NJ

ies have shown the sequence of conformational changes of calmodulin interacting with its calcium and target peptides *(6)*. Upon calcium binding, the lobes containing the pairs of EF hands change their position, and hydrophobic patches, mainly consisting of methionine residues, become solvent exposed. Subsequently, calmodulin wraps around the target peptide.

In this methods paper we give a detailed protocol for using calmodulin as an affinity purification tag. We have mainly used calmodulin as a tag for antibodies and recombinant enzymes. Here, we describe the expression and detection of calmodulin fusion proteins, the purification using affinity chromatography, and ion exchange chromatography.

Furthermore, calmodulin can be used as a tag for filamentous phage, allowing their capture using calmodulin-binding peptide. This property represents the basis for a methodology allowing the isolation of novel catalytic activities associated with enzymes displayed on phage *(7)*.

2. Materials

2.1. Expression and Purification of CaM-Fusion Proteins

1. For expression of antibody-calmodulin fusion proteins, we cloned *Xenopus laevis* calmodulin (PCR amplified with primers 5'-AGT TCC GCC ATA GCG GCC GCT GAC CAA CTG ACA GAA GAG CAG-3' and 5'-ATC CAT CGA GAA TTC TTA TCA CTT TGA TGT CAT CAT TTG-3' 1 min at 94°C, 1 min 60°C, 2 min 72°C, 25 cycles, proofreading *Taq* polymerase) into *Not*I and *Eco*RI sites in pHEN1 *(8)*, a vector derived from pUC19. The gene coding for antibody fragment in single chain Fv format was inserted in the *Sfi*I/*Not*I sites of the resulting vector *(5)*. The vector contains a secretion sequence and a *lac* operon for induction of expression. Medium for overnight culture: 2X TY, 100 mg/mL ampicilin, 1% glucose.
2. Fresh medium: 2X TY, 100 mg/mL ampicilin, 0.1% glucose.
3. IPTG: isopropyl-beta-D-thiogalactopyranoside.
4. BBS-EDTA: 0.2 *M* Na-borate, 0.1 *M* NaCl, 1 m*M* EDTA, pH 8.0.

2.1.1. Detection of Functional Fusion Proteins by ELISA

1. Phosphate buffered saline (PBS): 20 m*M* NaH_2PO_4, 30 m*M* Na_2HPO_4, 100 m*M* NaCl, pH 7.4.
2. 96-Well microtiter plates for ELISA (FALCON, Becton Dickson Labware, Oxnard, USA).
3. Powdered fat-skimmed milk.
4. Calmodulin binding peptide: H-CAAARWKKAFIAVSAANRFKKIS-OH *(4)* ($k_{on} = 9.8 \times 10^5$/s *M*, $k_{off} = 2.2 \times 10^{-6}$/s, $K_D = 2.2 \times 10^{-12}$ *M*). The thiol group of the N-terminal cysteine was conjugated to biotin by coupling with iodoacetyl-LC-biotin (Pierce, Rockford, USA). The reaction was performed in a 1.5 mL microtube: 300 µL of 10^{-3} *M* peptide (in TBSC) + 300 µL of 2.5×10^{-3} *M*

iodoacetyl-LC-biotin (in DMSO), 30 min at RT in the dark. Purification was performed with a C-18 reversed phase HPLC column (diameter 2.5 cm, length 10 cm), with the following profile:% acetonitrile at timepoints given: 0/0 min, 0/3 min, 40/20 min, 40/25 min 100/30 min, 100/37 min, 0/40 min, second solvent: H_2O/0.1% trifluoro acetic acid, flow rate: 2 mL/min The peptide elutes after 22 min. As the peptide contains a tryptophan residue, absorbance at 280 nm allows detection of the peptide. In our hands, we did not obtain quantitative coupling of the peptide, but the educt and product of the reaction could be separated by HPLC with the solvent profile shown above. Streptavidin-HRP (ResGen, Huntsville, USA).

5. PBS/Tween: PBS with 0.1% Tween20 (polyoxyethylene sorbitan monolaurate, SIGMA, St. Louis, USA).
6. As colorimetric substrate for ELISA, BM we use blue POD substrate (Roche, Basel, Switzerland), 1 *M* H_2SO_4 was used as stopping reagent.
7. ELISA plate reader with filters for 450 nm and 650 nm. For other reagents that the one we used, check the manual for detection conditions.

2.1.2. Detection of CaM-Fusion Protein by Band Shift Assay

1. We use 10 × 8 cm gels with glass plates and side spacers. The down side of the gel is sealed with 0.4% agarose.
2. Acrylamide/bisacrylamide (37.5/1) 30% solution (e.g., Sigma, St. Louis, USA) Store in fridge.
3. 3 *M* Tris-HCl, pH 8.8.
4. 1 *M* $CaCl_2$.
5. 25% Ammonium persulfate should be freshly prepared on the same day. However, kept at RT, it may be stored for up to one week after preparation., TEMED (Sigma) is stored at 4°C.
6. Cy5 - Cy5-bis-OSU, N,N'-biscarboxypentyl-5,5'-disulfonatoindodicarbocyanine. Labeling of the calmodulin binding peptide with iodoacetamide-CY5 (Amersham Pharmacia, Uppsala, Sweden) is performed in a similar fashion as the coupling with biotin described in **Subheading 3.1.1., item 4**. Pipet 600 µL of 50 µ*M* solution (in TBSC: 10 m*M* Tris-HCl, pH 7.4, 100 m*M* NaCl, 1 m*M* $CaCl_2$) of the peptide + 200 µL 5 m*M* solution of iodoacetamide-CY5 (in DMSO) in a 1.5 mL reaction tube, and let the reaction proceed for 30 min at RT. The separation of product from educt can either be performed with HPLC (as in **Subheading 3.1.1., item 4**, detection at excitation wavelength of CY5 [650 nm] possible) or with a cation exchange chromatography column.
7. 6X Gel loading solution: 50 m*M* Tris-HCl, pH 7.4, 40% glycerol, 25 mg/10 mL bromophenol blue.
8. 5X Gel running buffer: 30.28 g Tris base, 144 g glycine, add 1 L H_2O.
9. Power supply for gel electrophoresis.
10. Fluorescence imager with CY5 fluorescence filters e.g., DIANAII (Raytest, Straubenhardt, Germany).

2.1.3. Affinity Chromatography

1. 1 *M* $CaCl_2$.
2. N-(6-aminohexyl)-5-chloro-1-naphthalenesulfonamide-agarose (Sigma, St. Louis, USA).
3. Tris buffered saline with 1 m*M* $CaCl_2$ (TBSC): 10 m*M* Tris-HCl pH 7.4, 100 m*M* NaCl, 1 m*M* $CaCl_2$.
4. TBSC containing 0.5 *M* NaCl.

2.1.4. Ion Exchange Chromatography

1. For ion exchange chromatography, we use an FPLC system (ÄKTAFPLC from Amersham Pharmacia, Uppsala, Sweden) with an anion exchange column (1 mL Resource Q, Amersham Pharmacia).

2.2. Capture of CaM Displaying Phage

1. Filamentous phage are frequently used as tools for tethering a displayed protein (e.g., an antibody) with the corresponding gene coding for it. In phage display technology *(9)* proteins with desired properties (e.g., binding affinities) are isolated from macromolecular libraries. In our laboratory we use phage vector fd-tet-dog (tet^R) *(8)* or phagemid vector pHEN1 (amp^r) *(8)* for production of phage particles displaying recombinant proteins.
2. Biotinylated CaM-binding peptide described in **Subheading 2.1.2.** of **item 6**.
3. Polyethyleneglycol (PEG): 20% PEG 6000.
4. 2.5 *M* NaCl.
5. Streptavidin-dynabeads (Dynal, Oslo, Norway), preblocking: take 50 µL resuspended dynabeads (3.35×10^7 dynabeads), aspirate solvent with pasteur pipet utilizing Dynal magnetic particle concentrator for microtubes, and resuspend dynabeads in ~500 µL 3% MTBSC (3% w/v skim milk powder in TBSC). Mix at shaking table for 20 min at RT.
6. TBSC/Tween: TBSC with 0.1% Tween20.
7. TBSE: TBS with 20 m*M* EDTA.
8. The *E. coli* strain TG1 (K12, D(*lac-pro*), *supE*, *thi*, *hsdD5*/*F'traD36*, *proA+B+*, *lacIq*, *lacZDM15*) is frequently used when working with filamentous phage display.
9. Most phagemid vectors contain an ampicillin resistance gene, whereas most phage vectors mediate tetracycline resistance. Depending on the system utilized, choose the appropriate antibiotic.

3. Methods

3.1. Expression and Purification of CaM-Fusion Proteins (*see* **Note 1**)

1. Cultures of *E.coli* harboring the CaM-fusion protein expression vector are grown at 37°C overnight.
2. The overnight culture is diluted 1/100 in fresh medium.

3. At OD(600) = 0.8 the bacteria are induced at 22°C with 1 m*M* IPTG and grown overnight.
4. Periplasmic extracts are prepared as follows ***(10)***: centrifuge the bacterial culture at 11000*g* for 20 min at 4°C, resuspend the cells in BBS-EDTA (2 mL/g of cells) and allow stirring for 30 min. After centrifugation at 30000*g* for 30 min at 4°C, the supernatant represents the periplasmic extract (*see* **Note 2**).

3.1.1. Detection of Functional Fusion Proteins by ELISA (*see* Note 3)

1. Wells of microtiter plates are coated with 10 µg/mL antigen in PBS overnight, 100 µL/well total volume. Discard solution after incubation (*see* **Note 4**).
2. The wells are blocked with 2% milk in PBS (300 µL per well) for 2 h at RT.
3. Add 100 µL of the calmodulin fusion protein (periplasmic extract or purified protein) to each well and incubate for 30 min at RT.
4. Pipet 100 µL 2% milk in PBS with 10^{-6} *M* biotinylated calmodulin binding peptide and 10^{-6} *M* streptavidin-HRP into the wells and incubate for 30 min at RT.
5. Wash the wells 4× with PBS/Tween followed by 4 washes with PBS.
6. Add 100 µL of developing reagent to each well, and allow color development. Stop the development by adding 50 µL of 1 *M* H_2SO_4 (*see* **Note 5**).
7. Read the absorbances of the plates at 450 nm and 650 nm. The signal is the substraction product of $OD_{450} - OD_{650}$.

3.1.2. Detection of CaM-Fusion Protein by Band Shift Assay

1. Preparation of native gel (*see* **Note 6**). In a 50 mL Falcon tube, pour sequentially (for 1 gel) 8 mL 30% acrylamide/bisacrylamide (37.5/1) solution, 6.25 mL dd water, 750 µL of 3 *M* Tris-HCl, pH 8.8, and 1.5 µL of 1 *M* $CaCl_2$. Mix the reagents and then sequentially add 45 µL 25% ammonium persulfate and 27 µL TEMED. Pour the mixture immediately into the gel apparatus with a 10 mL pipet, then put a comb on top of the gel. Let the gel polymerize for at least 30 min (*see* **Note 7**).
2. Incubate 10 µL of sample (periplasmic extract or purified protein) with 1 µL of CY5-labeled calmodulin binding peptide for 2 min at RT.
3. Add 2 µL of 6X gel loading solution to samples, mix, and apply samples in sample slots (*see* **Note 8**).
4. Run gel at 150 V until the blue front has migrated 3/4 of the length of the gel.
5. Observe fluorescent bands at emission wavelength 680 nm with an excitation wavelength of 633 nm on a fluorescent imager.

3.1.3. Affinity Chromatography (*see* Note 9)

1. To the periplasmic extract (about 50 mL from 1 L culture broth), add 1 *M* $CaCl_2$ to a final concentration of 20 m*M*.
2. Fill empty 3-mL column with N-(6-aminohexyl)-5-chloro-1-naphtalenesulfonamide-agarose. Preequillibrate in TBSC. Apply the sample to the column.
3. Wash the column with TBSC containing 0.5 *M* NaCl until a baseline in absorption at 280 nm is reached.

4. Elute sample with 20 m*M* ethylene glycol-bis(2-aminoethylether)-*N,N,N',N'*-tetraacetic acid (EGTA).
5. Add 1 *M* $CaCl_2$ to the eluate to a final concentration of 50 m*M*.

3.1.4. Ion Exchange Chromatography (see **Note 10**)

1. Equilibrate anion exchange chromatography column connected to the FPLC with start buffer.
2. Inject sample.
3. Wash column with TBS until a baseline in absorption at 280 nm is reached.
4. Elute sample with a 100 m*M*–1 *M* NaCl gradient in Tris-HCl buffer.

3.2. Capture of CaM-Displaying Phage

1. For each capture experiment or negative control, an equivalent of 10^9 phages (infective particles) are blocked in a final volume of 100 μL TBSC containing 3% BSA for 20 min at RT (*see* **Note 11**).
2. In the binding assay, 100 μL of blocked phage is mixed with 2 μL CaM-binding peptide (10^{-6} *M*) and 48 μL TBSC in a 1.5 mL microtube. For the negative control, no peptide is added. Let the binding proceed for 3 min at RT (*see* **Note 12**).
3. Phage are precipitated by addition of 40 μL PEG and incubation for 30 min on ice.
4. Centrifuge phage 2 min, 15,000*g* at 4°C. A faint pellet may sometimes, but not always, be detected. Resuspend the phage with 200 μL of 2% MTBSC, add 50 μL of preblocked streptavidin-dynabeads, and agitate microtubes 30 min at RT on shaking table (*see* **Note 13**). For removal of unspecifically binding phage, wash dynabeads 6× with TBSC/Tween, and 3× with TBSC (*see* **Note 14**). Elute the phage with 200 μL TBSE, and transfer the supernatant of dynabeads in a fresh 1.5-mL microtube. Add 50 μL of 1 *M* $CaCl_2$ (*see* **Note 15**).
5. Infect appropriate amount of TG1 ($OD_{600} = 0.4 – 0.5$) with phage (*see* **Note 16**). Let the infection of bacteria proceed for 30 min at 37°C without shaking of the bacteria.
6. Plate bacteria on agar plate with appropriate antibiotic. For titer determination, make dilution series. Typical phage titers obtained with this procedure with an input of 10^9 phage particles are: about 10^8 for the binding reaction, and 10^3 in the negative control.

4. Notes

1. We used antibody-CaM fusion-proteins secreted into the periplasmic space, as antibodies require an oxidative environment for proper folding. However, as calmodulin is an intracellular protein, fusions with intracellular proteins can be expressed intracellularly. Moreover, more optimized expression vectors than pUC19-derived ones can be certainly used. Expression times and induction conditions have to be adjusted accordingly.
2. During the stirring of the bacteria in BBS-EDTA, the outer membrane permeabilizes. The periplasmic extract can be further analyzed for functional calmodulin

fusion proteins by ELISA or band-shift assay, and purified via affinity chromatography or ion exchange chromatography. For intracellular fusion-proteins, pelleted bacteria are lysed with ultrasonification or other standard methods.

3. This detection method requires an anti-calmodulin reagent (anti-calmodulin antibody or better: a calmodulin binding peptide) and a reagent able to bind the protein fused to CaM (e.g., if it is an antibody, the corresponding antigen). If the latter reagent is not available, detection is still possible with the band-shift assay in native gel described in **Subheading 4.1.2.** In **Subheading 3.1.1.**, an ELISA protocol for an antibody-CaM fusion protein is described. Horseradish peroxidase covalently linked to a calmodulin binding peptide can be used as well.
4. When washing ELISA plates, carefully rinse the wells and discard liquid with a vigorous shake.
5. The developing time can vary from a few seconds to several minutes. It is therefore advisable to observe the developing color and stop the reaction when a good positive to negative ratio is reached.
6. The gel described is a 15% acrylamide/bisacrylamide gel that is optimal for running native calmodulin. For larger proteins (fusion proteins with calmodulin), run lower percentage gel (less acrylamide/bisacrylamide, more water) without changing anything else.
7. The polymerization process can be monitored using the solution kept in the 50-mL Falcon tube as a reference.
8. Too intense a color from the bromphenolblue in the loading buffer can interfere with the CY5 detection, especially if smears from the loading buffer extend into the lanes.
9. In affinity chromatography, the exposed hydrophobic patches of calcium-loaded calmodulin are exploited. Affinity chromatography of calmodulin with phenyl-sepharose resin has been also successfully performed and may be used as an alternative to the procedure described here ***(11)***.
10. As an alternative (or in addition) to affinity chromatography, calmodulin and calmodulin fusion-proteins can be purified by ion exchange chromatography, as calmodulin is an unusually acidic protein with a pI of 3.9–4.3 ***(1)***. It is possible to perform ion-exchange chromatography with calmodulin either in the presence of calcium or in the presence of a calcium-chelator. The elution profiles in both cases will be different owing to differences in the calmodulin net-charge.
11. Blocking of phage with BSA prevents nonspecific stickiness and recovery of false positives.
12. Each reaction is carried out in a 1.5-mL microtube, and negative controls can either be carried out without addition of a peptide, or with a CaM binding peptide that is not biotinylated.
13. To prevent precipitation of the dynabeads, the microtubes can be turned every few minutes.
14. Before washing, the caps of the microtubes can be cut and discarded, as some contaminant droplets of phage solution on the tips of the caps may not be washed

away, giving a false titer. The magnetic bead concentrator is used throughout the washing and elution procedure. For washing, let the dynabeads attach to the walls of the cups, which takes 1–2 min, aspirate the supernatant carefully with a pasteur pipet that is consequently discarded, and resuspend the dynabeads in the washing solution, before the next round of washing starts by letting the dynabeads precipitate against the microtube wall again.

15. Instead of resuspending the dynabeads in washing solution, the dynabeads are resuspended in 200 μL EDTA. After the dynabeads have attached to the wall, the supernatant with the eluted phages is transferred to a new microtube.
16. TG1 bacteria at $OD_{600} = 0.4$ correspond to 4×10^8 bacteria/mL. For quantitative rescue of eluted phage, the number of bacteria should be greater than the number of infecting phage particles, by at least a factor 5. For preparative phage selections, we therefore take large volumes of bacterial culture (e.g., 50 mL), whereas for titer determination, dilutions of eluted phage can be used (infecting e.g., 1 mL of bacterial culture).

Acknowledgments

S. Melkko receives a bursary from the Boehringer Ingelheim Fonds.

References

1. Rogers, M. S. and Strehler, E. E. (1996) Calmodulin, in *Guidebook to the Calcium-Binding Proteins* (Celio, M. R., Pauls, T., Schwaller, B. eds.), Oxford University Press, Oxford, pp. 34–40.
2. Chattopadhyaya, R., Meador, W. E., Means, A. R., and Quiocho, F. A. (1992) Calmodulin structure refined at 1.7 A resolution. *J. Mol. Biol.* **228,** 1177–1192.
3. O'Neil, K. T. and DeGrado, W. F. (1990) How calmodulin binds its targets: sequence independent recognition of amphiphilic alpha-helices. *Trends Biochem. Sci.* **15,** 59–64.
4. Montigiani, S., Neri, G., Neri, P., and Neri, D. (1996) Alanine substitutions in calmodulin-binding peptides result in unexpected affinity enhancement. *J. Mol. Biol.* **258,** 6–13.
5. Neri, D., de Lalla, C., Petrul, H., Neri, P., and Winter, G. (1995) Calmodulin as a versatile tag for antibody fragments. *Biotechnology (NY)* **13,** 373–377.
6. Weinstein, H. and Mehler, E. L. (1994) Ca(2+)-binding and structural dynamics in the functions of calmodulin. *Annu. Rev. Physiol.* **56,** 213–236.
7. Demartis, S., Huber, A., Viti, F., et al. (1999) A strategy for the isolation of catalytic activities from repertoires of enzymes displayed on phage. *J. Mol. Biol.* **286,** 617–633.
8. Hoogenboom, H. R., Griffiths, A. D., Johnson, K. S., Chiswell, D. J., Hudson, P., and Winter, G. (1991) Multi-subunit proteins on the surface of filamentous phage: methodologies for displaying antibody (Fab) heavy and light chains. *Nucleic Acids Res.* **19,** 4133–4137.

9. Winter, G., Griffiths, A. D., Hawkins, R. E., and Hoogenboom, H. R. (1994) Making antibodies by phage display technology. *Annu. Rev. Immunol.* **12,** 433–455.
10. Skerra, A., Pfitzinger, I., and Pluckthun, A. (1991) The functional expression of antibody Fv fragments in *Escherichia coli*: improved vectors and a generally applicable purification technique. *Biotechnology (NY)* **9,** 273–278.
11. Gopalakrishna, R. and Anderson, W. B. (982) Ca^{2+}-induced hydrophobic site on calmodulin: application for purification of calmodulin by phenyl-Sepharose affinity chromatography. *Biochem. Biophys. Res. Commun.* **104,** 830–836.

6

Calmodulin-Binding Peptide as a Removable Affinity Tag for Protein Purification

Wolfgang Klein

1. Introduction

Protein purification is an important tool for investigations on protein function, structure analysis, and biotechnological use. Therefore a number of different techniques have been developed for fast, reliable, and reproducible overexpression and purification of relevant proteins. Affinity systems have been employed frequently due to speed, yield, and reduction of chromatographic steps necessary in order to get a highly purified protein. Over the years, different tags and matrices have been introduced to the scientific community, each providing a combination of advantages and disadvantages in the light of the protein of interest.

The charm of the calmodulin affinity system described in this chapter is that binding and elution buffers are identical with only the replacement of Ca^{2+} ions by ethylene glycol bis(2-aminoethylether)-*N*,*N*,*N'*,*N'*-tetraacetic acid (EGTA) (*see* **Fig. 1A**), providing extremely gentle and mild conditions throughout the purification procedure. The calmodulin-binding peptide (CBP) fusion technique comprises a complete expression and purification system for proteins that are a genetically engineered hybrid of the removable CBP affinity tag and the protein of interest (POI) *(1*, **Fig. 2**). The system has been designed to create N- or C-terminal fusions with protease-specific sites to remove the affinity tag and gain the pure protein in a second step (*see* **Fig. 1B**) *(2,3)*. Plasmid directed expression in an *E. coli* system is based on the use of the T7 promoter that is repressed under conditions in which expression is undesirable, e.g., for which the protein might exert toxic effects. High-level expression can be achieved by induction of phage T7 RNA polymerase that triggers exclusive transcription of the genetically constructed hybrid protein *(4)*. Calmodulin-binding

From: *Methods in Molecular Biology, vol. 205, E. coli Gene Expression Protocols*
Edited by: P. E. Vaillancourt © Humana Press Inc., Totowa, NJ

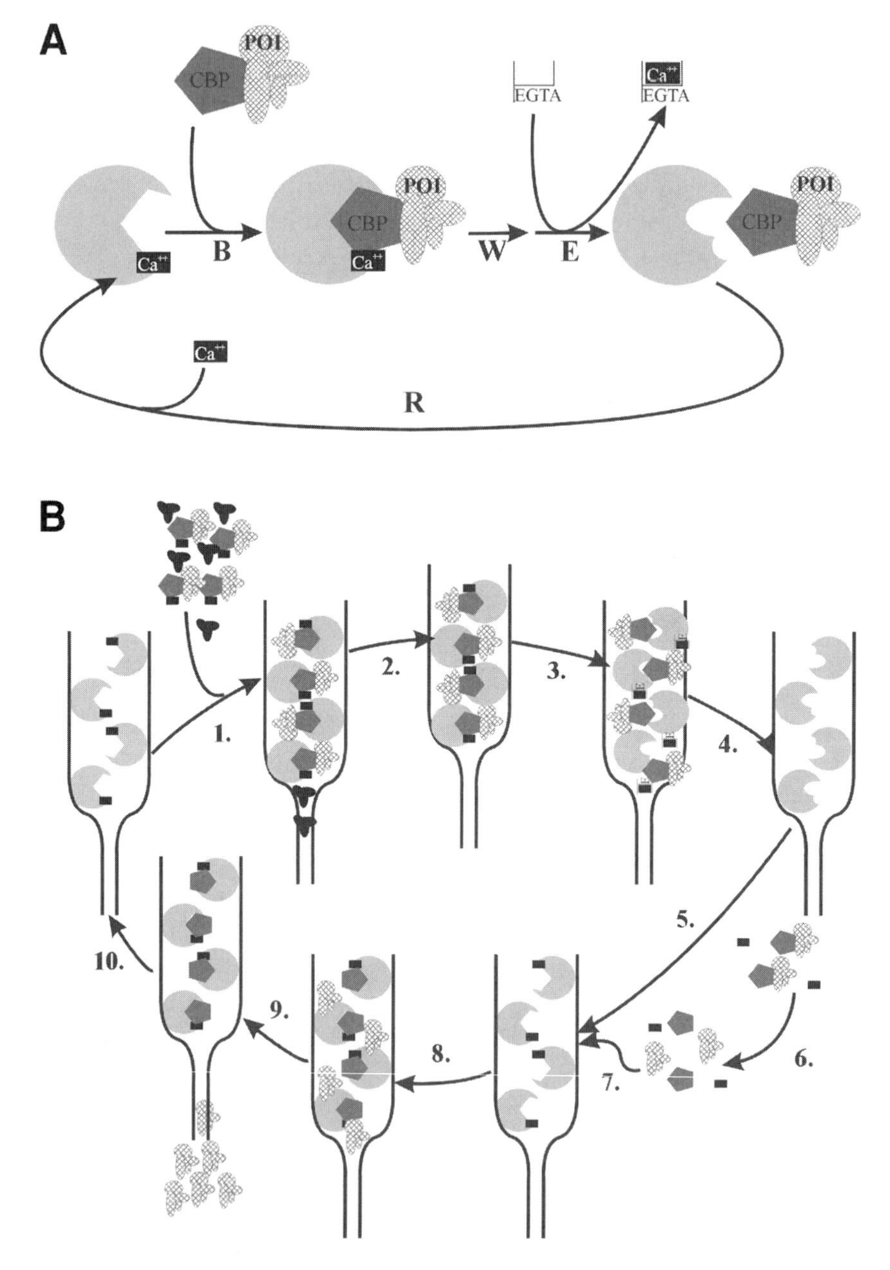

Fig. 1. Binding of calmodulin binding peptide (CBP) tagged hybrid protein to immobilized calmodulin in the presence of calcium ions. **(A)** The calmodulin affinity matrix (represented by grey balls) is equilibrated with $CaCl_2$ containing buffer and the cell extract is loaded for selective binding of the tagged protein (B). Unbound proteins

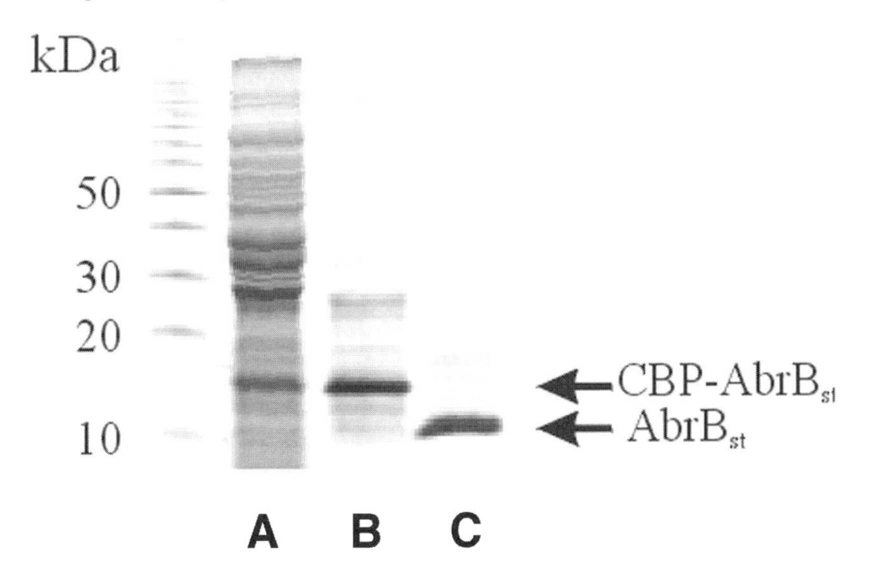

Fig. 2. Example of a protein purification following the two-step method as described in **Fig. 1B**. A CBP-protein of interest hybrid was purified from raw cellular extract (lane A) by a first affinity chromatography (lane B). After addition of calcium ions and thrombin, the affinity tag was split off and the mixture re-applied onto the regenerated column. The protein of interest is now in the flow through of the column (lane C), whereas non-split hybrid protein and affinity tags are retained. The arrowheads on the right side point towards the CBP-fusion protein (upper) and the processed, regular length protein of interest (lower). The mass of molecular weight markers are indicated on the left. This picture is adapted from a preliminary purification of the CBP-AbrB$_{st}$ protein of *Bacillus stearothermophilus* ***(8)***.

---

are removed by extensive washing (W). Elution is initiated by the addition of EGTA that chelates the calcium ions, upon which the calmodulin undergoes a conformational change and releases the tagged protein (E). The matrix is regenerated by several wash buffers that reequilibrate with $CaCl_2$ (R). **(B)** Schematic representation of the two-step method for purifying proteins to near homogeneity. 1. The cleared cell extract is loaded onto the equilibrated affinity matrix. 2. Extensive washing eliminates unbound protein whereas the tagged fusion protein is retained. 3. Chelation of Ca^{2+} by EGTA triggers a conformational change of the immobilized calmodulin and releases the bound CBP-POI fusion protein. 4. The tagged protein elutes in the presence of EGTA. 5. Regeneration of the affinity matrix. 6. Addition of excess $CaCl_2$ and thrombin or enterokinase allow the cleavage of the affinity tag. 7. The protease treated hybrid protein is reapplied onto the affinity column. 8. While removed tags and nonspecifically binding proteins are retained, the processed protein of interest is eluting in the flowthrough fraction (9.). 10. The removed tags are eluted and the column is reequilibrated.

peptide fusions can be purified from crude cell extract to almost homogeneity by one pass through a column containing the calmodulin-affinity resin under moderate conditions and neutral pH, conditions that avoid denaturation of the protein (*see* **Fig. 1**).

The calmodulin-binding peptide (CBP), a 26 amino-acid fragment from the C-terminus of muscle myosin light-chain kinase, serves as protein-affinity tag in this system. This peptide shows a relatively high affinity (K_d of approx 10^{-9} *M*) for calmodulin, strictly dependent on the presence of low concentrations of calcium ions *(5,6)*. Upon the removal of calcium by weak chelators, calmodulin undergoes a conformational change that results in the release of the ligand (*see* **Fig. 1**). For the affinity resin, calmodulin is covalently coupled onto a well characterized chromatography matrix, thereby allowing the selective binding of calmodulin-binding peptide fusion proteins from crude cellular extracts while providing gentle elution conditions. After cleavage of the affinity tag in the presence of calcium, adsorption of the eluate with resin yields highly purified protein, while the affinity tag is absorbed to the resin, allowing fast and quantitative separation of tag and protein of interest (POI) (*see* **Figs. 1B** and **2**). The limited size of the CBP-affinity tag (approximately 4 kDa) is unlikely to affect the functional properties of the POI, in contrast to larger tags like the maltose-binding protein *(7)*. However, effects on the organization of protein multimer formation or topology can occur *(8)*.

Removal of the CBP-affinity tag can be achieved by using the thrombin and/or enterokinase target sequences, as defined by the vector used (*see* **Fig. 3**). Depending on the restriction sites used to genetically construct the hybrid gene, changes in the amino acid sequence and/or the introduction of additional amino acids at the fusion joint are likely. This fact has to be included in setting out the cloning strategy. Some commercially available vectors include additional options such as ligation-independent cloning (LIC) overhang sequences for seamless cloning *(9)*, or internal target sequences for efficient radiolabeling of the fusion protein by protein kinase A. Such features allow the generation of highly specific probes for interaction cloning protocols and sequential blot overlay analysis *(10)*.

2. Materials

2.1. E. coli *Strains for Cloning and Overproduction*

For construction of the hybrid gene on the plasmid, the use of well characterized cloning strains, such as the XL or JM series, is recommended. An inactive T7 RNA polymerase (or deficiency of this gene) and a *recA* background may facilitate cloning. A *lacI* allele can further inhibit basal expression during the subcloning when using a strain with an IPTG inducible T7 polymerase

pCAL-n:

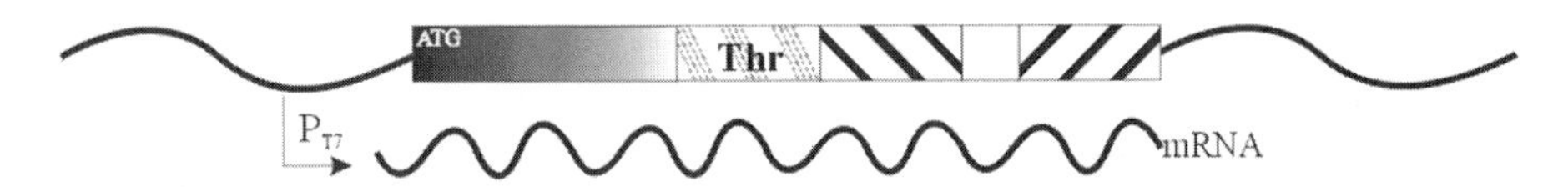

pCAL-n-EK:

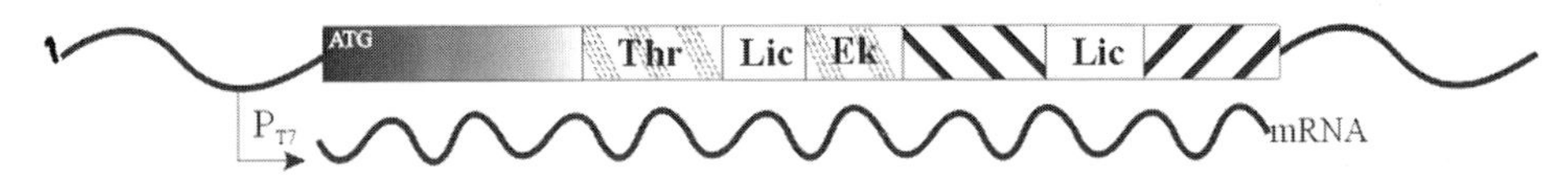

pCAL-c:

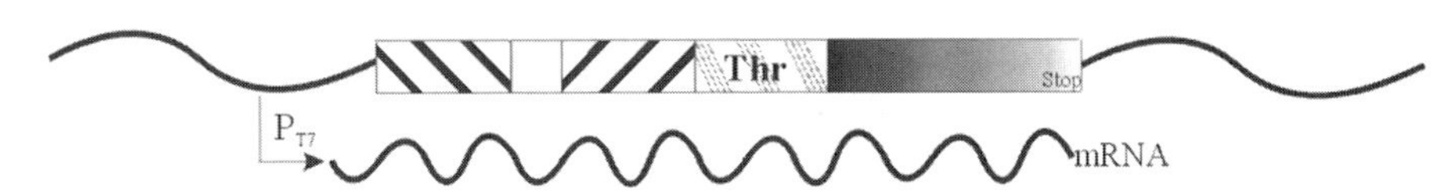

pCAL-kc:

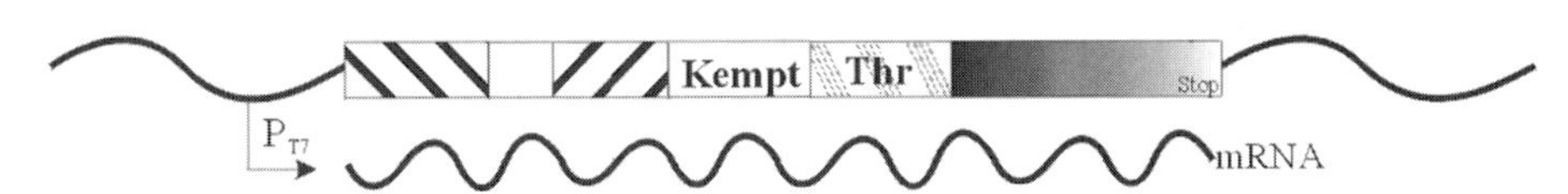

Legend:

ATG Stop Calmodulin-binding peptide sequence including stop or start codon;

upper and lower multiple cloning sites;

Thr/Ek protease recognition sequences for thrombin or enterokinase;

Lic LIC overhang cloning sequences, Kempt kemptide protein labeling sequence.

Fig. 3. Calmodulin-Binding-peptide (CBP) fusion vectors. Schematic drawing of vectors for the construction of N- and C-terminal CBP-fusion and expression of the hybrid proteins. The vectors are ColE1 based with the β-lactamase resistance gene for selection. Upstream of the cloning sites, a P_{T7} *lac* promoter triggers expression. A T7 termination site is positioned downstream of the expression cassette.

system. As all commercially available vectors for the CBP-affinity system are based on tightly regulated T7 RNA polymerase-dependent promoters, the use of transgenic *E. coli* strains bearing the gene encoding this DNA-dependent RNA polymerase under an inducible promoter is necessary for efficient expression of the cloned hybrid. Two different types of strains are commonly used. One type is represented by *E. coli* K38 hosting the kanamycin-selectable pGP1-2 plasmid, encoding the phage T7 RNA polymerase. Polymerase expression by this plasmid is temperature regulated with silent expression at temperatures below 28°C, and induction of the polymerase (and therefore of the CBP-fusion protein) by a short temperature shock to 42°C and prolonged expression at 37°C. Therefore, transformation and selection of transformants within this strain background have to be performed at temperatures below 28°C to ensure non-inducing conditions.

The other type of strain is represented by *E. coli* BL21 and its derivatives that carry the λDE3 lysogen with its immunity region, a cassette with the *lacI* gene, and a *lacUV5*-promoted T7 RNA polymerase construct. Upon the addition of IPTG to the medium, expression of T7 RNA polymerase is induced that results in expression of T7-promoted genes. Strain BL21 is generally regarded as a good protein expression strain due to deficiency in *lon* and *ompT* proteases that can degrade proteins during overexpression and purification ***(4,11,12)***. Some BL21 strains host the pLysS vector, mediating low level expression of T7 lysozyme. T7 lysozyme binds to T7 RNA polymerase and inhibits transcription, an effect that is overcome upon the induction of T7 RNA polymerase by isopropyl-thio-β-D-galactopyranoside (IPTG) ***(13,14)***.

2.2. Expression Vectors

A series of suitable plasmids for cloning and expression of CBP tagged proteins is available from Stratagene (La Jolla, USA). These plasmids are based on the pET-11 vectors, providing the features of a well characterized T7 gene promoter and leader sequence, showing an outstanding selectivity for the T7 RNA polymerase, tight repression of T7 RNA polymerase in the uninduced state due to a copy of the *lacI* gene, and high-level expression in the induced state. The plasmids pCAL-c and pCAL-kc create fusion proteins with the CBP affinity tag at the C-terminus by having a thrombin target-*cbp* sequence 3' to the multiple cloning site (MCS, *see* **Fig. 3**). Cloning of inserts occurs between the *Nco*I site containing the ATG codon in optimized spacing to the ribosomal binding site, and the *Bam*HI site. The pCAL-kc vector contains an additional kemptide sequence located between the thrombin cutting and *Bam*HI site. Thrombin digest of proteins of this type results in a C-terminal addition of four amino acids (Met-Tyr-Pro-Arg) originating from the thrombin recognition sequence.

The plasmid pCAL-n has the CBP-coding sequence upstream of its multiple cloning site, placing the CBP-tag at the N-terminus of the cloned insert ***(15)***. The thrombin recognition sequence is located between the *cbp*-sequence and the MCS, resulting in the N-terminal modification of the cleaved protein. This modification is a combination of the addition of a glycine residue from the thrombin target sequence and the exchange of several amino acids based on the formation of the cloning site. The pCAL-n-EK vector contains an enterokinase splitting site in addition to the thrombin target sequence (*see* **Fig. 3**). The efficient translation of the CBP-tag in *E. coli* ensures that the whole N-terminally tagged hybrid protein will be efficiently expressed, albeit the expression of some genes containing rarely used codons in *E. coli* might be enhanced by using a BL21 derivative containing additional tRNA genes for these codons.

2.3. Cloning

Luria-Bertani (LB)-medium is generally used for growing strains, albeit various other media are possible ***(16)***. The preparation of competent cells and transformation procedure described here is based on the convenient $CaCl_2$ technique, but other techniques like electroporation, the use of TSS solution ***(17)***, or $RbCl_2$ will work fine. The user might refer to some very comprehensive laboratory manuals for general protocols on isolation and cloning of the gene of interest as this topic is out of the scope of this chapter ***(18,19)***.

1. LB medium per liter: dissolve 10 g tryptone, 5 g yeast extract, and 5 g NaCl and sterilize by autoclaving at 121°C for 20 min. Antibiotics and IPTG are added after the autoclaved medium has cooled to 55°C. For plating of bacteria, 1.6 g agar is added to the medium prior to sterilization, and the medium with additions is poured into Petri dishes.
2. IPTG stock solution: IPTG at 1 *M* is dissolved in distilled water and sterilized by use of 0.2 µm sterile filters. This solution is kept frozen at –20°C.
3. Antibiotics are prepared as 1000X stock solutions with ampicillin (50 mg/mL) and kanamycin (70 mg/mL) dissolved in distilled water, filter sterilized and stored at 4°C. Chloramphenicol (34 mg/mL) is dissolved in 70% ethanol and stored at –20°C.
4. For transformation of DNA, prepare a 50 m*M* solution of $CaCl_2$ and sterilize by autoclaving. Store at 4°C.
5. Distilled water, sterilized by autoclaving.
6. Dimethyl sulfoxide (DMSO).
7. Have enough sterile culture tubes, microcentrifuge cups, cryo tubes, pipet tips, and the like ready for use.
8. The use of terrific broth (TB) medium is recommended for cloning and transformation as it allows a tighter regulation and therefore a reduced leakiness in case of gene products that exert growth-hampering effects on the *E. coli* host cells. TB consists of 15 g/l Bacto tryptone and 8 g/l NaCl.

2.4. The Calmodulin Affinity Matrix

The calmodulin affinity matrix consists of calmodulin that has been covalently linked to a beaded matrix. This allows its use in column and batch purification procedures (*see* **Subheadings 3.4.** and **3.5.**). The matrix is relatively stable towards a number of commonly used reagents like sodium chloride (up to 1 *M*), potassium chloride (up to 1 *M*), dithiothreitol (up to 5 m*M*), mercapto ethanol (up to 10 m*M*), ammonium sulfate (up to 1 *M*), detergents like Triton X-100 or Nonidet P-40 up to 0.1% (v/v), and imidazole up to some m*M*. As calmodulin is an immobilized protein, it is irreversibly damaged by proteases. Therefore, for long-term storage the addition of minor amounts of a commercial protease inhibitor like phenylmethylsulfonyl fluoride (PMSF) or 4-(2-aminoethyl)-benzenesulfonyl fluoride (AEBSF) in a buffered 20% (v/v) ethanol solution is recommended. As a consequence, the matrix has to be carefully equilibrated before use. For equilibration, prepare approx 10 vol of your selected binding buffer (*see* **Subheading 2.5.**).

2.5. Buffers and Additional Material

1. Binding buffer: this buffer has to be designed according to the protein of interest. A number of different buffer salts like Tris-HCl, $NaHPO_4$, and $KHPO_4$ have been tested by the author and were found to be compatible with the matrix. The pH is best around the neutral, and salt concentrations should be in the 50–300 m*M* range for effective reduction of nonspecific binding, whereas higher concentrations can affect the interaction of the fusion protein with the affinity matrix. A first test might be done using a 50 m*M* Tris-HCl or phosphate buffer, pH 7.0–8.0, with 2 m*M* $CaCl_2$ and 150 m*M* NaCl.
2. Elution buffer: this buffer might be similar to the binding buffer, differing only as the $CaCl_2$ is replaced by 2–5 m*M* EGTA for chelating the Ca^{2+} ions, thereby releasing protein (*see* **Fig. 1A**). Variations and a complete change of the binding buffer system are possible.

2.6. Induction of Overexpression

1. Prepare LB medium in Erlenmeyer flasks and sterilize by autoclaving. To test the purification strategy by the small scale batch analysis, use 250 mL of LB medium in a 500–1000-mL Erlenmeyer flask. For large scale expression and purification, use several 400-mL aliquots of LB medium in 2-L flasks as these conditions will ensure good aeration during growth.
2. Prepare an ampicillin (50 mg/mL) as well as an IPTG (1 *M*) stock solution and sterilize by use of 0.2 μm sterile filters.
3. Fresh plate of LB including ampicillin.
4. Clean centrifugation bottles (Sorvall GSA or equivalent)
5. French pressure cell for effective cell rupture. In case you can not access a French press, prepare a cell lysis buffer according to current standard protocols ***(18)***.

2.7. Small Scale Batch Analysis

1. Prepare batches of about 100 mL of each binding-, elution-, and wash buffers as well as a number of microcentrifuge tubes for handling and spinning. Use 0.2 µm sterile filters to filtrate the buffers, degassing is not necessary for the small scale batch method.
2. The use of clear tubes and a fixed-angle rotor in a benchtop centrifuge is recommended for best visualization of the glassy matrix pellet.

2.8. Large Scale Affinity Column Chromatography

1. Prepare approx 12–15 column vol of your selected binding buffer per fast performance liquid chromatography (FPLC) assisted purification run.
2. Prepare 3–5 column vol of your elution buffer per column run.
3. Prepare 5 column volumes of each wash buffer (*see* **Subheading 2.9.**) for each regeneration cycle.

All buffers should be filtered and degassed. Check FPLC equipment, column, and tubing material for the specifications desired.

2.9. Affinity Matrix Regeneration

1. Wash buffer 1: 0.1 *M* $NaHCO_3$, pH 8.5, including 2 m*M* EDTA.
2. Wash buffer 2: 1 *M* NaCl containing 2 m*M* $CaCl_2$.
3. Wash buffer 3: 100 m*M* acetate, pH 4.4, containing 2 m*M* $CaCl_2$.
4. Wash buffer 4: your selected binding buffer containing 2 m*M* $CaCl_2$.

Approx 5 column vol of each buffer are necessary per regeneration cycle between each column run when using a FPLC system. Prepare 100 mL of each wash buffer for the small-scale batch method. Buffers should be filtered, and degassed for the use in the FPLC system.

2.10. Removing the Calmodulin-Binding Peptide Tag

1. Prepare a small amount of 1 *M* $CaCl_2$ stock solution.
2. Thrombin cleavage buffer: 20 m*M* Tris-HCl, pH 8.4, 150 m*M* NaCl, 2.5 m*M* $CaCl_2$. Dissolve thrombin in thrombin cleavage buffer in a concentration of approx 5 U/µL. This stock solution can be stored frozen for several weeks but repeated freeze-thawing should be avoided.
3. Enterokinase splitting buffer: 50 m*M* Tris-HCl, pH 8.0, 50 m*M* NaCl, 2.0 m*M* $CaCl_2$, 0.1% (v/v) Tween-20. Enterokinase stock solution should be in the range of 2–5 U/µL.

2.11. Analysis of Purified Proteins

To analyze the protein expression and purification pattern, sodium dodecylsulfate-polyacrylamide gel electrophoresis (SDS-PAGE) is employed. The Lämmli system is the most convenient in terms of speed, ease of use, and general performance ***(20)***.

1. 4X Separating gel buffer: 1.5 *M* Tris-HCl, pH 8.8; 50 mL.
2. 4X Stacking gel buffer: 0.5 *M* Tris-HCl, pH 6.8; 50 mL.
3. 10% (w/v) SDS in water; 50 mL.
4. 10% (w/v) ammonium peroxodisulfate in water, stored frozen in aliquots of 500 μL.
5. Electrophoresis buffer: 25 m*M* Tris base, 192 m*M* glycine, 0.1% (w/v) SDS.
6. Get a ready to use acrylamide/bis-acrylamide stock solution (40%; 38:2 mixture), tetramethylethylenediamine (TEMED), and a commercial molecular weight ladder to calibrate your gel.
7. 4X Standard protein loading mix of 500 m*M* Tris-HCl, pH 6.8, 8% (w/v) SDS, 40% (v/v) glycerol, 20% (v/v) β-mercaptoethanol, and 5 mg/mL bromphenol blue.
8. Coomassie staining solution of 50% ethanol, 10% acetic acid, and 250 mg/L Coomassie brilliant blue G250; 250 mL.
9. Destaining solution of 20% ethanol with 10% acetic acid; 1 L.

3. Methods

3.1. Maintenance of Strains

To select for and maintain *E. coli* transformants, strains are grown on LB-agar plates supplemented with the appropriate antibiotic for selection. Media and antibiotic concentrations are according to known standard procedures ***(16)***. The term ‘antibiotic as appropriate’ will refer to the use of either carbenicillin, ampicillin, or penicillin G/K (50 μg/mL final concentration) to select for the pCAL vectors, and the use of additional antibiotics for supplemental vectors like chloramphenicol (34 μg/mL) for the pLysS plasmid and kanamycin (80 μg/mL) for the K38-pGP1-2 strain. For long term storage, an aliquot of a fresh cell culture is brought to 7% DMSO (final concentration) and stored in a –80°C freezer. With the exception of *E. coli* K38 (grown at the non-inducing temperature of 28°C), all strains are usually grown aerobically at 37°C.

3.2. Cloning into the Expression Vectors, Transformation, and Screening for Positive Clones

For cloning of the CBP-POI fusion protein, the DNA sequence of the gene of interest has to be known. To allow proper in-frame cloning, restriction sites are generated at both ends of the gene of interest by PCR using appropriate primers. To create fusions with an N-terminal CBP tag, the ATG specifying the original start codon of the POI can be retained or changed (*see* **Subheading 2.2.**). The primer defining the C-terminal cloning site must include a stop codon. To construct fusions with a C-terminal tag, the 5' primer should include the ATG codon to initiate translation of the fusion construct, and the spacing of this feature relative to the ribosomal binding site should be optimal for efficient translation. The C-terminal primer creating the fusion junction to the tag should not contain a stop codon, in order that the gene of interest may be fused in frame with the CBP tag sequence provided by the vector. At the fusion point between CBP and POI, the cloning will result in the change and/or addition of some amino acids encoded by the restriction site. Careful planning and *in silico* cloning for selection of the restriction sites and maintenance of the reading frames is therefore essential.

The coding sequence of the gene of interest is amplified by PCR, and the resulting DNA fragment is purified from the PCR reaction mixture. The ends of the fragment are trimmed using the appropriate restriction endonucleases and ligated into the chosen vector using compatible sites. Aliquots of this ligation reaction are transformed into *E. coli* cells by standard methods ***(19)***. In the our lab, highly competent cells of efficient cloning strains like the XL or JM series were routinely used for this step. After selection on LB-antibiotic agar plates, plasmid DNA is prepared from colonies and analyzed for successful cloning by restriction analysis according to standard procedures ***(19)***. Candidate plasmids are introduced into overexpression strains (*see* **Subheading 2.1.**) and selected under non-inducing conditions. To test for expression of the fusion protein, transformants are inoculated in a small volume of LB medium and grown under inducing conditions. The following protocol is given for use of BL21 competent cells which allows some faster testing for positive clones (*see* **Note 1**).

1. Streak BL21 cells from the freezer stock onto fresh LB plates to obtain single colonies. Incubate at 37°C overnight.
2. Inoculate a single colony of BL21 cells into 3 mL of LB medium using a sterile inoculation loop and grow overnight at 37°C. In the morning, dilute the culture 1:100 into 20 mL fresh LB medium. Grow cells at 37°C until cell density reaches an OD_{600} of 0.3. Chill the whole culture on ice for 15 min.
3. Harvest cells in a pre-chilled centrifuge at 2000*g* for 5 min, discard the supernatant, and resuspend cells in 10 mL of sterile, ice-cold 50 m*M* $CaCl_2$ solution. Keep on ice for 10 min.
4. Collect cells by spinning again and resuspend the cell pellet in 1 mL of ice-cold 50 m*M* $CaCl_2$. Keep cells on ice for at least 90 min. Overnight storage on ice is possible.
5. In a chilled microcentrifuge tube, mix 100 µL of competent cells with 10–150 ng of plasmid DNA of your restriction analysis verified clones. Choose several of the CBP-POI candidates as well as the parental vector for control purpose. Incubate on ice for 20 min.
6. Perform a heat shock by incubating 2 min at 42°C.
7. Add 400 µL of 37°C-prewarmed LB medium and allow a 30–60 min phenotype expression at 37°C.
8. Plate the transformation mixture onto selective LB-ampicillin agar plates and incubate overnight at 37°C. Colonies should appear within 2 d. To avoid the appearance of satellite colony formation, Petri disks should be wrapped by parafilm and stored at 4°C.
9. Using a sterile inoculation loop, cells from well-isolated colonies are inoculated in 3 mL LB medium supplemented with the appropriate antibiotic and 0.5 m*M* IPTG (final concentration). Allow growth at 37°C overnight. A culture of a transformant bearing the parental plasmid vector serves as control.

10. Two mL of the cell culture are harvested using microcentrifuge cups by spinning 14,000*g* for 10 min. The supernatant is discarded and cells are resuspended in 40 µL distilled water.
11. Forty µL of SDS-PAGE loading mix are added and mixed by pipetting up and down several times. The sample is heated to 95°C for 10 min, cooled to room temperature, and centrifuged 10 min at full speed in a benchtop centrifuge. The supernatant is transferred into a new microcentrifuge tube and stored for analysis.
12. Pour an analytical SDS-PAGE according to the Lämmli system ***(20)***. For selection of the mesh grade, correlate the calculated size of your CBP-POI fusion protein with the resolution range of the acrylamide systems (*see* also **Subheading 3.8.**). Load 10–20 µL of the cell extract (**step 11**) into the wells, place a suitable commercial calibration marker in another well, and run the SDS-PAGE according to the manufacturer's instructions. After completing the run, stain your gel with Coomassie brilliant blue ***(21)***.
13. By comparing the lane of the parental vector with the candidate plasmids, the appearance of a discrete band of approximately the size of the hybrid protein allows the selection of effectively overproducing strains.
14. The rest of the overnight cultures from effective overproducers identified should be brought to 7% DMSO (final concentration) and saved as freezer stocks at –80°C.

3.3. Induction of Overexpression of the Target Protein

1. Grow an overnight culture of the candidate overproducer strain (BL21 based) in LB medium with appropriate antibiotic selection at 37°C overnight.
2. In the morning, dilute the fresh overnight culture 1:100 into pre-warmed LB medium supplemented with antibiotics.
3. Grow cells to an OD_{600} of 0.5.
4. Induce expression of the *cbp*::*poi* by adding IPTG at a final concentration of 0.5 m*M* and allow expression to continue for about 3 h.
5. Harvest cells by centrifugation at 5000*g* for 10 min.
6. Cells can be used directly for further purification or stored as frozen cell pellet at –20°C.

Induction conditions can be given only as a rule of thumb, and to maximize expression of your POI, IPTG concentration and expression time may have to be adjusted for best results. To optimize these conditions for maximum yield, a series of 5 mL cultures with IPTG concentrations ranging from 0.05–5 m*M* and 2–24 h incubation, followed by yield analysis according to the small batch scale, is recommended. The culture volume can be chosen over a wide range. Only the need for good aeration by use of notched flasks seemed to be critical in the authors lab (*see* **Note 2**).

Using *E. coli* K38, overnight culture and pre-induced growth are performed at 28°C. Induction is mediated by a sudden temperature shift to 42°C for 2 min and prolonged expression at 37°C for 5 h.

3.4. Preparation of the Calmodulin Affinity Matrix (see *Note 3*)

Commercially available calmodulin matrix is typically stored in buffer containing 20% (v/v) ethanol. Before using the matrix, the resin is equilibrated with the selected binding buffer, and loaded into a column. For the small batch technique, all equilibration steps are performed in microcentrifuge cups (*see* **Subheading 3.4.8.**).

1. Decant the ethanol-containing storage buffer from the storage-settled calmodulin affinity matrix. Add 5 vol of your selected $CaCl_2$ binding buffer, resuspend well and allow the slurry to settle. Carefully decant any fines and perform a second equilibration step.
2. Decant the supernatant from the resin and resuspend in 3–5 vol of $CaCl_2$ binding buffer.
3. Allow the resin to settle again, decant the supernatant, and add one vol of $CaCl_2$ binding buffer. Resuspend well and degas the equilibrated affinity resin.
4. Fill the selected column with approx 10% column vol of $CaCl_2$ binding buffer to eliminate air bubbles in the lower connection fitting that might impair the performance of your calmodulin affinity column. Clamp the lower column adapter.
5. Pour the degassed affinity resin into the column by running down a glass pipet to avoid the entrapment of air bubbles during column packing. Fill the remaining column space with $CaCl_2$ binding buffer and allow the affinity matrix to settle.
6. Fill the column to the top with $CaCl_2$ binding buffer and affix the opened upper column adapter (filled with $CaCl_2$ binding buffer and connected to a reservoir of buffer). Connect the column to a pump.
7. Open the lower connection fitting and set the pump on run at a flow rate of about 2 mL/min to create a proper bed of resin. Take care not to exceed flow and pressure limitations given by the resin and the column as specified by the manufacturers. When the resin bed is stable, complete packing by lowering the top adaptor to meet the top of the resin bed.
8. For use in the small batch method, the equilibration steps of aliquots (50–100 μL) of calmodulin affinity resin are performed in microcentrifuge tubes. Allow 15 min for each equilibration step, and pellet the matrix by spinning at 1500*g* for 1 min in a benchtop centrifuge. Three to four equilibration steps with a fourfold volume of $CaCl_2$ binding buffer are recommended.

3.5. Small Scale Batch Analysis

This rapid method is suitable for the purification of 15–150 μg CBP fusion protein using a microcentrifuge tube. It is very useful for the optimization of buffer conditions, washing and elution procedures for a subsequent large-scale column purification, and to determine the best expression conditions by variation of induction and expression time.

1. Resuspend the calmodulin matrix in storage buffer by shaking and aliquot 50–100 μL affinity resin into microcentrifuge tubes and equilibrate as described in **Subheading 3.4.8.**

2. Resuspend the equilibrated affinity resin in the initial volume of your selected $CaCl_2$ binding buffer. Mix with up to 200 µL of a crude *E. coli* cell lysate in a total volume of 300 µL (*see* **Note 4**).
3. Incubate the slurry with agitation for 2 h at 4°C to allow binding of the fusion protein.
4. Pellet the matrix beads by centrifugation for 2 min at 1500*g* in a benchtop centrifuge. Remove the supernatant with unbound proteins and save this fraction for further analysis.
5. Wash the beads 4–8× with 300 µL $CaCl_2$ binding buffer. Centrifuge as described above and save the supernatants for analysis. The final wash fraction should contain no protein as determined by SDS-PAGE analysis or spectrophotometric protein determination ***(20,22)***.
6. Elute the bound protein(s) by three or more washing steps using 200 µL of your elution buffer containing 2 m*M* EGTA and centrifugation as described until the fractions do no longer show detectable levels of purified protein.
7. Perform a final elution step using elution buffer supplemented with 1 *M* NaCl (final concentration). This step should remove tightly binding proteins from the affinity resin.
8. Check the performance of the affinity purification by SDS-PAGE analysis (**Subheadings 2.11.** and **3.9.**) and regenerate the affinity resin as described in **Subheading 3.7.**

3.6. Purification of Target Proteins

1. Resuspend the bacterial cell pellet from **Subheading 3.3.** in approx $1/50^{th}$ of the culture vol of $CaCl_2$ binding buffer. Using frozen cell pellets, directly apply the binding buffer onto the frozen cells and thaw on ice.
2. Lyse the cells by a conventional chemical or physical method. The use of a French pressure cell (3 passages at 1000 psi, SIM Aminco 5.1 or equivalent) is recommended due to speed and effectiveness of cell rupture. Spin the crude lysate in a centrifuge at >5000*g* and 4°C for 15 min to generate a clear cell extract.
3. Connect your column to a FPLC system, making sure to have enough binding and elution buffer prepared. Load your sample at a flow rate of 2 mL/min. We recommend approx 1.0 mL of calmodulin affinity resin for every 2.0 mg fusion protein estimated in the extract as judged by SDS-PAGE.
4. Wash the column with $CaCl_2$ binding buffer to remove unbound protein at a flow rate of up to 5 mL/min. Usually, 5 column vol of $CaCl_2$ binding buffer were found to be sufficient, as determined by UV monitoring. However, extensive washing or more stringent conditions (e.g., different ionic strength of the buffer) may be necessary.
5. Proteins are eluted from the column by removal of the calcium ions from the calmodulin resin (*see* **Fig. 1**). This removal is preferably achieved by an elution buffer that is basically identical to the binding buffer but replaces $CaCl_2$ by 2 m*M* EGTA for chelating the calcium ions on the matrix. The elution buffer can tolerate a wide range of chemical constituents, and EGTA may be replaced by EDTA where necessary. In some cases tightly binding proteins may require an addi-

tional step using a buffer with up to 1 *M* NaCl for recovery of the protein (*see* **Note 5**). The same flow rate should be used for this additional step.

6. After each run, regenerate the column as described in **Subheading 3.7.** (*see* **Fig. 1A**).

3.7. Regeneration of the Calmodulin Affinity Matrix

1. Wash the calmodulin affinity column with 3–5 column vol of wash buffer 1 at a flow rate of 3–5 mL/min.
2. Wash with 3 column vol of wash buffer 2.
3. Wash with 3 column vol of wash buffer 3.
4. Equilibrate with 5 column vol of wash buffer 4 or your binding buffer containing 1–2 m*M* $CaCl_2$.
5. Store the calmodulin affinity column at 4°C in the same buffer as step 4. Long term storage should be in binding buffer containing 20% (v/v) ethanol and 0.5 m*M* PMSF protease inhibitor.

In some instances denatured proteins or lipids may hamper regeneration of the column. These can be washed off using up to 0.1% of a non-ionic detergent such as Nonidet P-40 or Triton X-100 for a few minutes, followed by extensive re-equilibration in binding buffer.

3.8. Removing of the Calmodulin-Binding Peptide Tag

For some applications the affinity tag is a useful tool to selectively immobilize the protein (*see* Chapter 5 in this volume). In certain instances, however, the CBP may influence protein activity or structure, and removal of the tag is required. This can be done by using the thrombin or enterokinase target sites located adjacent to the affinity tag.

1. Pool the fractions containing the purified protein and dialyze against thrombin cleavage buffer. As an alternative, a suitable volume of the 1 *M* $CaCl_2$ stock solution can be added directly to the protein eluate to compensate the EGTA of the elution buffer and give a final concentration of 2.5 m*M* free $CaCl_2$.
2. Add approx 5 U thrombin/mg fusion protein and incubate at 37°C. For fusions containing the enterokinase target site, dilute or dialyze the CBP fusion protein into enterokinase cleavage buffer. Add 1 U of enterokinase/100 µg fusion protein.
3. To determine the efficiency of the proteolytic cleavage of the tag, take several aliquots of the reaction at different time points ranging from several minutes up to 24 h and determine the quality of the processing by SDS-PAGE analysis (*see* **Subheadings 2.11.** and **3.9.**). This efficiency varies for different proteins and should be optimized for each fusion protein. In our lab, we routinely achieved complete processing of 5 mg protein by 20 U thrombin in 4 h *(8)*. Higher thrombin to target protein rates may be applied to avoid inconveniently long incubation times.
4. To separate the affinity tag from the processed target protein, the complete cleavage mixture is again applied onto the regenerated calmodulin affinity column

(*see* **Fig. 1B**). Whereas the cleaved affinity tag is retained by the resin, the processed protein will be in the flow through and can be used for further investigations. Thrombin can be removed by anti-thrombin resin (Sigma) or it is inactivated by the addition of 0.5 m*M* PMSF.

5. Elute the CBP affinity tags with elution buffer and regenerate the affinity resin as described in **Subheading 3.7.**

3.9. Analysis of the Purified Proteins

Analysis of protein expression and purification is monitored by SDS-polyacrylamide gel electrophoresis (SDS-PAGE). The polyacrylamide concentration determining the mesh size should be adjusted to the size of the proteins under investigation (*see* **ref.** ***19***). Whereas the Lämmli system might be most convenient ***(20)***, other protocols can improve the resolution of small proteins ***(23)***. Fast and convenient staining uses Coomassie brilliant blue G250 with 30 min of staining and about 2 h of destaining with several changes of destaining solution ***(21)***. Alternatively, a more sensitive rapid silver staining method can be employed, especially to visualize the homogeneity of the purified protein ***(24)***.

1. Pour a SDS-PAGE of appropriate acrylamide mesh using a degassed solution of acrylamide, separation buffer, and 0.1% SDS (w/v, final concentration) in a volume of 10 mL. Start polymerization with 5 µL TEMED and 50 µL APS stock solution.
2. After polymerization, pour an appropriate 3–5% stacking gel (acrylamide, stacking gel buffer, 0.1% SDS in 5 mL vol; start polymerization by 5 µL TEMED and 25 µL APS) and insert comb to form sample wells.
3. Mix your protein samples with a one quarter volume of loading mix, heat for 10 min at 95°C and load a suitable amount (10–20 µL) onto the gel. Electrophorese the sample with constant mA until the bromophenol blue dye reaches the bottom of the gel.
4. Stain with Coomassie brilliant blue staining solution for 30 min with slight agitation.
5. Destain for approx 2 h or until appropriate signal strength is reached with slight agitation. The destaining solution should be changed several times during the course of destaining the gel.

4. Notes

1. Usually, the plasmids of the pCal series seemed very stable in our hands. Problems may arise by the cloning of proteins that exert toxic effects on the host cells. This can be seen by the formation of reduced colony size where inducing conditions should nearly abolish growth, giving extremely small or even unvisible colonies. Notice that some good growing colonies appear that usually are mutants having defects within the coding sequence of your gene of interest or an impaired expression. As the expression of toxic proteins is difficult and does not allow high yields, the addition of rifampicin 5 min after induction of T7 RNA polymerase might be beneficial. Rifampicin is inactivating the *E. coli* RNA polymerase at concentrations of 120–150 µg/mL whereas T7 RNA poly-

merase is not affected. This allows the nearly exclusive expression of T7 promoted genes.

2. The formation of inclusion bodies is another problem hampering protein purification, caused by improper folding (expression rate may be too high) and/or aggregation within the cell. Frequently, such problems can be solved by induction at a temperature of 30°C or less, or by inducing with a lower concentration of IPTG ***(25)***. It has recently been reported that changing the medium osmolarity in combination with the addition of osmoprotective substances may be helpful ***(26)***. Inclusion bodies may be solubilized with urea or guanidinium-HCl and using a higher ionic strength in the buffer systems for the purification ***(27)***. If the protein aggregates during the purification, perform the procedure at room temperature, or try the addition of solubilizing reagents like 0.05% Triton X-100 or Tween-20. If the target protein is naturally secreted, formation of inclusion bodies may occur. In such cases, inclusion body formation may be avoided using a C-terminal affinity tag, thereby allowing secretion. Secretion may be monitored by the analysis of periplasmic shock fluid ***(28)***.
3. Whereas the efficiency of the extraction procedure depends mostly on the correlation between the amount of affinity resin per volume of crude cell lysate, the level of expression of the hybrid protein is another important variable. Typical binding rates are within a level of 1.0–3.0 mg of pure protein/mL of resin in the batch method, with the column purification yielding even higher amounts. However, these values depend on size, overall net charge, and some conformational and physicochemical characteristics of the target protein to be purified.
4. Cell lysis can be achieved by sonification. We have tested 4 pulses of 20 s each and a relative output of 40% with a Branson sonifier. Samples were kept on ice.
5. If the protein eluate contains unwanted contaminants, the ionic strength of the binding and washing buffers have to be increased. Another possibility is that your column is too large. If the POI is an unstable protein when expressed in *E. coli*, add protease inhibitor to the extract and perform all further steps at 4°C. It should however be noted that protease inhibitors should be completely removed by extensive dialysis prior to removal of the tag by use of thrombin or enterokinase.

Acknowledgments

Many thanks to ‘Bimmler’ Martin Hahn for enthusiastic collaboration and proofreading and to Beatrice van Saan-Klein for persistent encouragement.

References

1. Vaillancourt, P., Simcox, T. G., and Zheng, C. F. (1997) Recovery of polypeptides cleaved from purified calmodulin-binding peptide fusion proteins. *Biotechniques* **22,** 451–453.
2. Zheng, C. F., Simcox, T., Xu, L., and Vaillancourt, P. (1997) A new expression vector for high level protein production, one step purification, and direct isotopic labeling of calmodulin-binding peptide fusion proteins. *Gene* **186,** 55–60.

3. Vaillancourt, P., Zheng, C. F., Hoang, D. Q., and Breister, L. (2000) Affinity purification of recombinant proteins fused to calmodulin or to calmodulin-binding peptides. *Methods Enzymol.* **326,** 340–362.
4. Studier, F. W. and Moffatt, B. A. (1986) Use of bacteriophage T7 polymerase to direct selective high-level expression of cloned genes. *J. Mol. Biol.* **189,** 113–130.
5. Carr, D. W., Stofko-Hahn, R. E., Fraser, I. D., et al. (1991) Interaction of the regulatory subunit (RII) of cAMP-dependent protein kinase with RII-anchoring proteins occurs through an amphipathic helix binding motif. *J. Biol. Chem.* **266,** 14,188–14,192.
6. Stofko-Hahn, R. E., Carr, D. W., and Scott, J. D. (1992) A single step purification for recombinant proteins. Characterization of a microtubule associated protein (MAP 2) fragment which associates with the type II cAMP-dependent protein kinase. *FEBS Lett.* **302,** 274–278.
7. Maina, C. V., Riggs, P. D., Grandea, A. G., 3rd, et al. (1988) An *Escherichia coli* vector to express and purify foreign proteins by fusion to and separation from maltose-binding protein. *Gene* **74,** 365–373.
8. Klein, W., Winkelmann, D., Hahn, M., Weber, T., and Marahiel, M. A. (2000) Molecular characterization of the transition state regulator AbrB from *Bacillus stearothermophilus. Biochim. Biophys. Acta* 1**493,** 82–90.
9. Aslanidis, C. and de Jong, P. J. (1990) Ligation-independent cloning of PCR products (LIC-PCR). *Nucleic Acids Res.* **18,** 6069–6074.
10. Blanar, M. A. and Rutter, W. J. (1992) Interaction cloning: identification of a helix-loop-helix zipper protein that interacts with c-Fos. *Science* **256,** 1014–1018.
11. Phillips, T. A., VanBogelen, R. A., and Neidhardt, F. C. (1984) The *lon* gene product of *Escherichia coli* is a heat-shock protein. *J. Bacteriol.* **159,** 283–287.
12. Studier, F. W., Rosenberg, A. H., Dunn, J. J., and Dubendorff, J. W. (1990) Use of T7 RNA polymerase to direct expression of cloned genes. *Methods Enzymol.* **185,** 60–89.
13. Dubendorff, J. W. and Studier, F. W. (1991) Controlling basal expression in an inducible T7 expression system by blocking the target T7 promoter with *lac* repressor. *J. Mol. Biol.* **219,** 45–59.
14. Dubendorff, J. W. and Studier, F. W. (1991) Creation of a T7 autogene. Cloning and expression of the gene for bacteriophage T7 RNA polymerase under control of its cognate promoter. *J. Mol. Biol.* **219,** 61–68.
15. Wyborski, D. L., Bauer, J. C., Zheng, C. F., Felts, K., and Vaillancourt, P. (1999) An *Escherichia coli* expression vector that allows recovery of proteins with native N-termini from purified calmodulin-binding peptide fusions. *Protein Expr. Purif.* **16,** 1–10.
16. Miller, J. H. (1992) A short course in bacterial genetics. *A laboratory manual and handbook for* Escherichia coli *and related bacteria.* Cold Spring Harbor Laboratory, Cold Spring Harbor, New York.
17. Chung, C. T., Niemela, S. L., and Miller, R. H. (1989) One-step preparation of competent *Escherichia coli*: transformation and storage of bacterial cells in the same solution. *Proc. Natl. Acad. Sci. USA* **86,** 2172–2175.

18. Ausubel, F. M., Brent, R., Kingston, R. E., Moore, D. D., Seidman, J. G., Smith, J. A., and Struhl, K. (1992) *Short protocols in molecular biology.* John Wiley & Sons, Harvard Medical School.
19. Sambrook, J., Fritsch, E. F., and Maniatis, T. (1989) *Molecular cloning. A laboratory manual.* Cold Spring Harbor Laboratory, Cold Spring Harbor, New York.
20. Lämmli, U. K. (1970) Cleavage of structural proteins during the assembly of the head of bacteriophage *T4. Nature* **227,** 680–685.
21. Bennett, J. and Scott, K. J. (1971) Quantitative staining of fraction I protein in polyacrylamide gels using Coomassie brillant blue. *Anal. Biochem.* **43,** 173–182.
22. Bradford, M. M. (1976) A rapid and sensitive method for the quantification of microgram quantities of protein utilizing the principle of protein-dye binding. *Anal. Biochem.* **72,** 248–254.
23. Schägger, H. and Jagow, G. V. (1987) Tricine-sodium dodecyl sulfate-polyacrylamide gel electrophoresis for the separation of proteins in the range from 1 to 100 kDa. *Anal. Biochem.* **166,** 368–379.
24. Bloom, H., Beier, H., and Gross, H. S. (1987) Improved silver staining of plant proteins, RNA, and DNA in polyacrylamide gels. *Electrophoresis* **8,** 93–99.
25. Wingfield, P. T. (1995) Preparation of soluble proteins from *Escherichia coli.* in *Current protocols in protein science*, Vol. 1 (Coligan, J. E., Dunn, B. M., Ploegh, H. L., Speicher, D. W., and Wingfield, P. T., eds.), Wiley and Sons, New York.
26. Chong, Y. and Chen, H. (2001) Preparation of functional recombinant protein using a nondetergent sulfobetaine. *BioTechniques* **45,** 24–26.
27. Wingfield, P. T., Palmer, I., and Liang, S.-M. (1995) Folding and purification of insoluble (inclusion-body) proteins from *Escherichia coli.* in *Current protocols in protein science*, Vol. 1 (Coligan, J. E., Dunn, B. M., Ploegh, H. L., Speicher, D. W., and Wingfield, P. T., eds.), Wiley and Sons, New York.
28. Neu, H. C. and Heppel, L. A. (1965) The release of enzymes from *Escherichia coli* by osmotic shock and during the formation of spheroplasts. *J. Biol. Chem.* **240,** 3685–3692.

7

Maltose-Binding Protein as a Solubility Enhancer

Jeffrey D. Fox and David S. Waugh

1. Introduction

A major impediment to the production of recombinant proteins in *Escherichia coli* is their tendency to accumulate in the form of insoluble and biologically inactive aggregates known as inclusion bodies. Although it is sometimes possible to convert aggregated material into native, biologically-active protein, this is a time consuming, labor-intensive, costly, and uncertain undertaking *(1)*. Consequently, many tricks have been employed in an effort to circumvent the formation of inclusion bodies *(2)*. One approach that shows considerable promise is to exploit the innate ability of certain proteins to enhance the solubility of their fusion partners. Although it was originally thought that virtually any highly soluble protein could function as a general solubilizing agent, this has not turned out to be the case. In a direct comparison with glutathione S-transferase (GST) and thioredoxin, maltose-binding protein (MBP) was decidedly superior at solubilizing a diverse collection of aggregation-prone passenger proteins *(3)*. Moreover, some of these proteins were able to fold into their biologically active conformations when fused to MBP. It is not entirely clear why MBP is such a spectacular solubilizing agent, but there is some evidence to suggest that it may be able to function as a general molecular chaperone in the context of a fusion protein by temporarily sequestering aggregation-prone folding intermediates of its fusion partners and preventing their self association *(3–6)*. The ability to promote the solubility of its fusion partners is not an exclusive attribute of MBP (*see* Chapters 8 and 9), but to the best of our knowledge MBP is the only general solubilizing agent that is also a natural affinity tag. Consequently, we consider MBP to be the best "first choice" fusion partner for the production of recombinant proteins in *E. coli.*

From: *Methods in Molecular Biology, vol. 205, E. coli Gene Expression Protocols*
Edited by: P. E. Vaillancourt © Humana Press Inc., Totowa, NJ

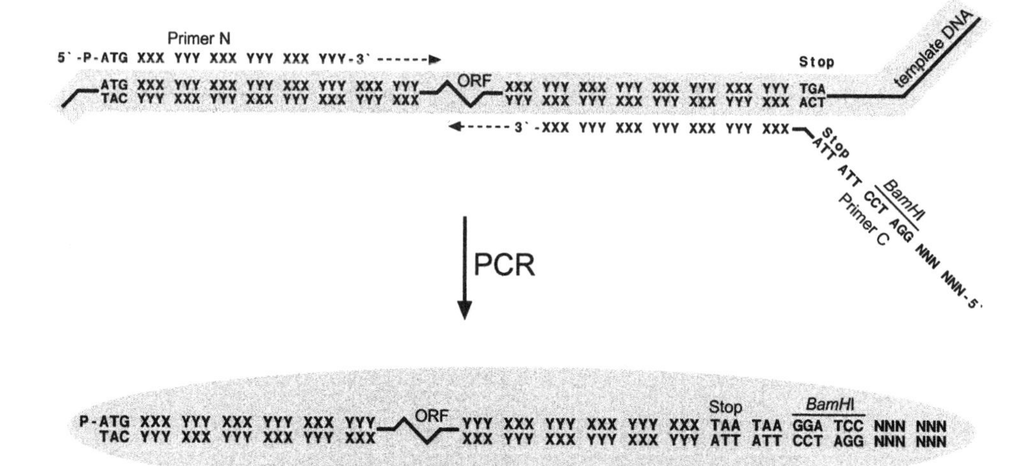

Fig. 1. PCR strategy for conventional cloning into pMAL-c2X. The template DNA is amplified with primers N and C. The primers are designed to base-pair with 20–25 bp of the 5' and 3' ends of the coding region respectively. Primer N is phosphorylated to allow blunt-ligation with the *Xmn*I site of pMAL-c2X. Primer C includes a 5' extension with a *BamH*I site for ligation with the *BamH*I site in pMAL-c2X.

Basic protocols for constructing MBP fusion vectors and for assessing the solubility and folding state of the fusion proteins are described herein. Special attention is given to a rapid and efficient method of generating fusion vectors by recombinational cloning. In addition, a method is described to quickly evaluate the folding state of a passenger protein by intracellular processing of a fusion protein with TEV protease. More detailed descriptions of commercially available MBP fusion vectors and methods for the purification of MBP fusion proteins by amylose affinity chromatography have been presented elsewhere *(7)*.

2. Materials

2.1. Conventional Vector Construction

1. The desired pMAL vector (New England Biolabs, Beverly, MA, USA).
2. Reagents and thermostable DNA polymerase for PCR amplification (*see* **Note 1**).
3. Appropriate synthetic oligodeoxyribonucleotide primers for PCR amplification (*see* **Fig. 1**).
4. Restriction enzymes and matching reaction buffers for screening putative clones.
5. Tris-Acetate-EDTA (TAE)-agarose, ethidium bromide, and an apparatus for submarine gel electrophoresis of DNA (*see* **Note 2**).
6. QIAquick™ gel extraction kit (QIAGEN, Valencia, CA, USA) for the extraction of DNA from agarose gels.

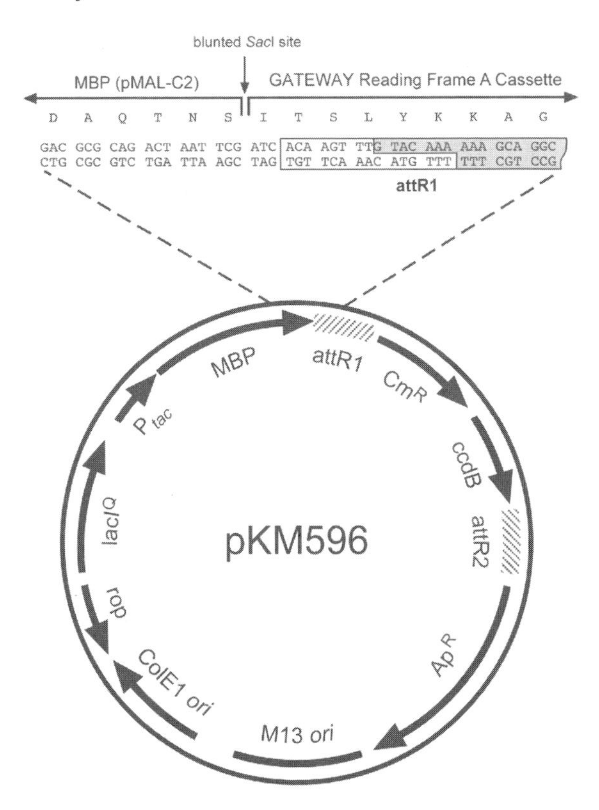

Fig. 2. Schematic representation of the Gateway™ destination vector pKM596. This vector can be recombined with an entry vector that contains an ORF of interest, via the LxR reaction, to generate an MBP fusion protein expression vector.

7. T4 DNA ligase, reaction buffer, and ATP.
8. Competent *E. coli* cells (e.g., DH5α or similar; *see* **Note 3**).
9. Luria-Bertani (LB) medium and LB agar plates containing ampicillin (100 μg/mL). LB medium: Add 10 g bactotryptone, 5 g yeast extract, and 5 g NaCl to 1 L of H_2O and sterilize by autoclaving. For LB agar, also add 12 g of bactoagar before autoclaving. To prepare plates, allow medium to cool until flask or bottle can be held in hands without burning, then add 1 mL ampicillin stock solution (100 mg/mL in H_2O, filter sterilized), mix by gentle swirling, and pour or pipet ca. 30 mL into each sterile Petri dish (100 mm diameter).
10. Reagents for small-scale plasmid DNA isolation (*see* **Note 4**).

2.2. Recombinational Vector Construction

1. The Gateway™ destination vector pKM596 (*see* **Fig. 2**).
2. Reagents and thermostable DNA polymerase for PCR amplification (*see* **Note 1**).
3. Synthetic oligodeoxyribonucleotide primers for PCR amplification (*see* **Fig. 3**).

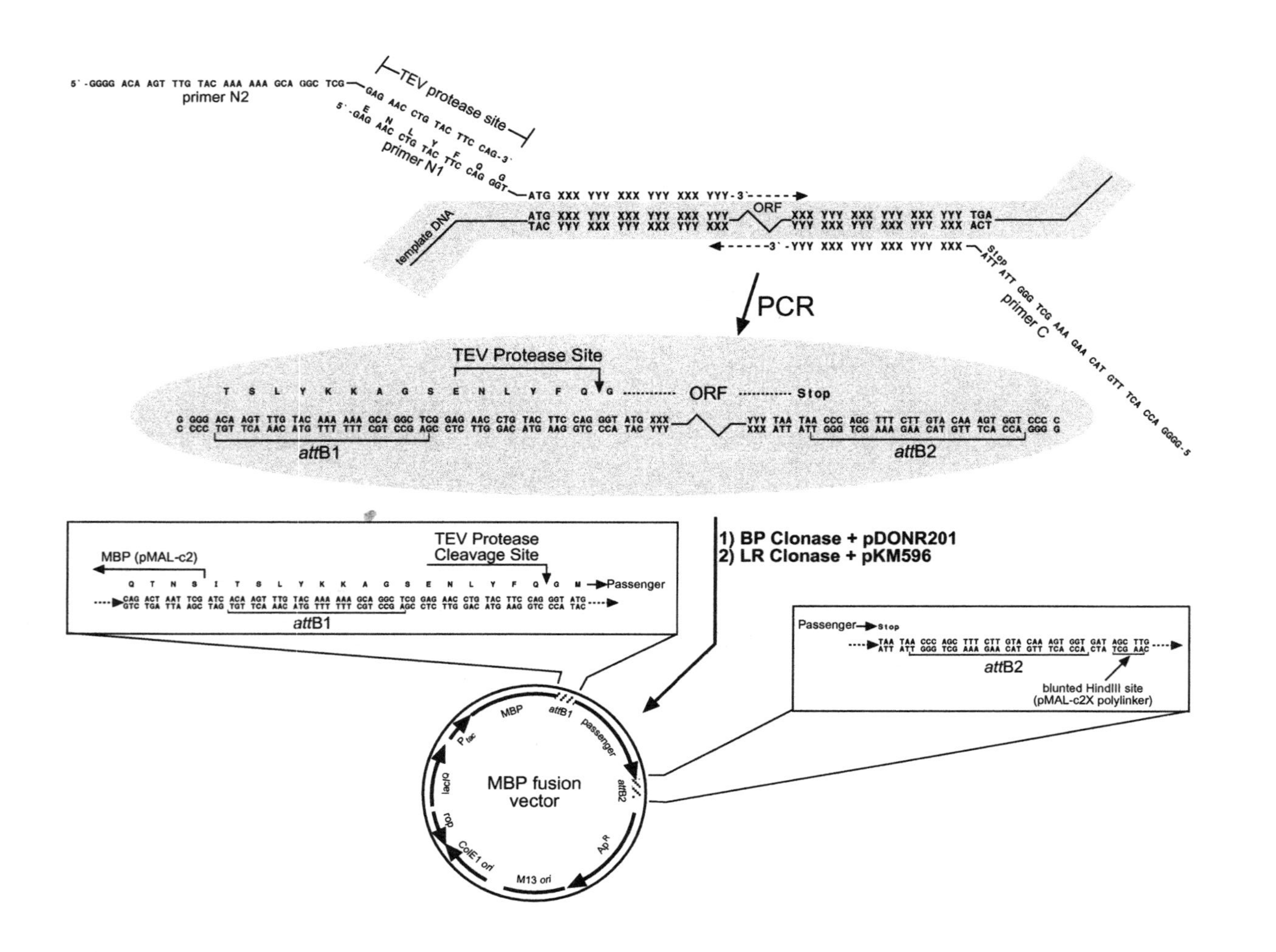
TEV protease site
5`-GGGG ACA AGT TTG TAC AAA AAA GCA GGC TCG
primer N2
primer N1
template DNA
ORF
Stop
primer C
PCR
TEV Protease Site
ORF
Stop
attB1
attB2
1) BP Clonase + pDONR201
2) LR Clonase + pKM596
MBP (pMAL-c2)
TEV Protease
Cleavage Site
Passenger
attB1
Passenger
Stop
attB2
blunted HindIII site
(pMAL-c2X polylinker)
MBP fusion
vector
MBP
attB1
passenger
attB2
Ap-R
M13 ori
ColE1 ori
rop
lacI^q
P tac

4. Tris-EDTA (TE) buffer: 10 m*M* Tris-HCl, pH 8.0, 1 m*M* ethylenediaminetetra-acetic acid (EDTA).
5. TAE-agarose and an apparatus for submarine gel electrophoresis of DNA (*see* **Note 2**).
6. QIAquick™ gel extraction kit (QIAGEN, Valencia, CA, USA) for the extraction of DNA from agarose gels.
7. Competent DB3.1 cells (Invitrogen, Carlsbad, CA, USA) for propagating pKM596 and pDONR201 (*see* **Note 3**).
8. Gateway™ PCR Cloning System (Invitrogen, Carlsbad, CA, USA).
9. LB medium and LB agar plates containing ampicillin (100 µg/mL). *See* **Subheading 2.1.9.** for preparation.
10. Reagents for small-scale plasmid DNA isolation (*see* **Note 4**).

2.3. Assessing the Solubility of MBP Fusion Proteins

1. Competent BL21-CodonPlus™-RIL cells (Stratagene, La Jolla, CA, USA) (*see* **Notes 3** and **5**).
2. LB agar plates and broth containing both ampicillin (100 µg/mL) and chloramphenicol (30 µg/mL). See **Subheading 2.1.**, item 9 for LB broth, LB agar, and ampicillin stock solution recipes. Prepare stock solution of 30 mg/mL chloramphenicol in ethanol and store at 4°C for up to 1 mo. Dilute antibiotics 1000-fold into LB medium or molten LB agar.
3. Isopropyl-thio-β-D-galactopyranoside (IPTG). Prepare a stock solution of 200 m*M* in H_2O and filter sterilize. Store at –20°C.
4. Shaker/incubator set at 37°C.
5. 250-mL baffle-bottom flasks (sterile).
6. Cell lysis buffer: 20 m*M* Tris-HCl, pH 8.0, 1 m*M* EDTA.
7. Sonicator (with microtip).
8. 2X Sodium dodecyl sulfate-polyacrylamide gel electrophoresis (SDS-PAGE) sample buffer: 100 m*M* Tris-HCl, pH 6.8, 200 m*M* dithiothreitol (DTT), 4% SDS, 0.2% bromphenol blue, 20% glycerol.
9. SDS-PAGE gel, electrophoresis apparatus, and running buffer (*see* **Note 6**).
10. Gel stain (e.g., Gelcode® Blue from Pierce, Rockford, IL, USA)

Fig. 3. *(opposite page)* Construction of an MBP fusion vector using PCR and Gateway™ cloning technology. The ORF of interest is amplified from the template DNA by PCR, using primers N1, N2, and C. Primers N1 and C are designed to base-pair to the 5' and 3' ends of the coding region, respectively, and contain unpaired 5' extensions as shown. Primer N2 base-pairs with the sequence that is complementary to the unpaired extension of primer N1. The final PCR product is recombined with the pDONR201 vector to generate an entry clone via the BxP reaction. This entry clone is subsequently recombined with pKM596 and LxR Clonase to yield the final MBP fusion vector.

2.4. Intracellular Processing of MBP Fusion Proteins by TEV Protease

1. Competent DH5αPRO or BL21PRO cells (Clontech, Palo Alto, CA, USA) containing the TEV protease expression vector pRK603 and the tRNA plasmid pKC1 (*see* **Notes 3**, **5**, and **7**).
2. A derivative of pKM596 (*see* **Subheading 3.3.1.**), or a pMAL vector that produces an MBP fusion protein with a TEV protease recognition site in the linker between domains.
3. LB medium and agar plates containing ampicillin (100 µg/mL), kanamycin (25 µg/mL), and chloramphenicol (30 µg/mL). See **Subheadings 2.1.**, **item 9** and **2.3., item 2.** for preparation. Prepare a stock solution of 25 mg/mL kanamycin in H_2O and store at 4°C for up to 1 mo. Dilute antibiotics 1000-fold into LB medium or molten LB agar.
4. Anhydrotetracycline. Prepare a 1000X stock solution by dissolving in ethanol at 100 µg/mL. Store in a foil-covered tube at –20°C.
5. Other materials as in **Subheading 2.3.**

3. Methods

3.1. Construction of MBP Fusion Vectors by Conventional Techniques

Workers are encouraged to consult the instructions and technical literature available from New England Biolabs related to their MBP fusion product line.

3.1.1. Selecting a pMAL Vector

Before constructing an expression vector, the proper plasmid backbone must be selected. A range of choices is currently available from New England Biolabs. These include pMAL-c2X, pMAL-p2X, pMAL-c2E, pMAL-p2E, pMAL-c2G, and pMAL-p2G; where p or c indicates periplasmic or cytoplasmic localization; and X, E, or G denote the identity of the protease cleavage site that is present in the fusion protein linker. (*See* **Subheading 3.2.** for more information about linkers and proteases.) Many labs also still have in their possession the older vectors pMAL-c2 and/or pMAL-p2 (*see* **Note 8**).

The properties of the passenger protein dictate the proper choice between cytoplasmic and periplasmic expression of an MBP fusion protein. This relates mainly to whether disulfide bonds are expected in the passenger, in which case the more oxidizing environment of the periplasm may be desirable. The methods described in this article pertain specifically to the production of MBP fusion proteins in the cytoplasm. In general, the yield of fusion protein is much greater in the cytoplasm, and purification by amylose affinity chromatography usually removes the majority of contaminating cytoplasmic proteins (*see* **Note 9**).

3.1.2. Assembling an Expression Vector

1. Assuming that pMAL-c2X has been selected (cloning strategies are similar for all six pMAL vectors) (*see* **Note 10**), a suitable restriction fragment encompassing the open reading frame (ORF) of interest must be prepared for ligation with the vector DNA. PCR is by far the most efficient means by which to generate this fragment. For general PCR protocols, *see* **ref. *8***. Typically, oligodeoxyribonucleotide primers are used to amplify the ORF while also extending either or both ends to introduce appropriate restriction site(s) for cloning (*see* **Fig. 1**). If the *Xmn*I site in the pMAL-c2X polylinker is to be used for cloning then the 5' PCR primer (Primer N) must either be phosphorylated or include a properly positioned blunt restriction site (*see* **Note 11**). The 3' extension adds a *Bam*HI site immediately after the stop codon (*see* **Note 12**).
2. The PCR product is digested with the appropriate restriction enzyme(s) (e.g., *Bam*HI in the example in **Fig. 1**) and purified by agarose gel electrophoresis (*see* **Note 2**).
3. pMAL-c2X is digested with *Xmn*I and *Bam*HI followed by gel purification of the large fragment (*see* **Note 2**).
4. The PCR fragment and digested vector backbone are combined and incubated with T4 DNA ligase and ATP (*see* **Note 13**).
5. The products of the ligation reaction are transformed into an appropriate *E. coli* strain (e.g., DH5α; *see* **Note 3**) and then spread on LB agar plates containing 100 µg/mL ampicillin. The plates are incubated overnight at 37°C.
6. Plasmid DNA is isolated from saturated cultures that were inoculated with individual ampicillin-resistant colonies and screened by restriction analysis to identify clones with the desired properties.
7. It is advisable to submit putative clones for sequence analysis to verify the proper construction and lack of PCR-induced mutations.

3.2. Protease Cleavage Sites

In almost every case, the investigator would like to obtain the protein of interest free from its fusion partner and with a minimum of extraneous amino acids. New England Biolabs offers vectors with three different options for protease cleavage: factor Xa, enterokinase, and genenase I *(7)*. However, we have found that the tobacco etch virus (TEV) protease, which can be purchased from Invitrogen, is superior to the three alternatives offered by New England Biolabs. The major advantage of this protease is its exceptionally high specificity. In contrast to factor Xa, enterokinase, and thrombin, there have never been any documented reports of cleavage by TEV protease at locations other than the designed site in fusion proteins. However, New England Biolabs does not offer a pMAL vector with a TEV protease cleavage site already in the linker. Therefore, to utilize this protease, a recognition site must be introduced by PCR. For an example of a TEV protease site introduced by PCR, *see* **Fig. 3**.

3.3. Construction of MBP Fusion Vectors by Recombinational Cloning

Recombinational cloning can greatly simplify the construction of MBP fusion vectors. Although several different methods for recombinational cloning have been described, we strongly recommend the Gateway™ Cloning System based on the site specific recombination reactions that mediate the integration and excision of bacteriophage lambda into and from the *E. coli* chromosome, respectively. For detailed information about this system, the investigator is encouraged to consult the technical literature supplied by Invitrogen.

3.3.1. Cloning with Gateway™

To utilize the Gateway™ system for the production of MBP fusion proteins, one must first construct or obtain a suitable "destination vector". Currently there are no commercial sources for such vectors. An example of a destination vector that can be used to produce MBP fusion proteins (pKM596) is shown in **Fig. 2**. pKM596 was constructed by replacing the DNA between the *Sac*I and *Hin*dIII restriction sites in the New England Biolabs vector pMAL-c2 with the RfA Gateway™ Cloning Cassette. The Gateway™ cassette consists of two different recombination sites (*att*B1 and *att*B2) separated by DNA encoding two gene products: chloramphenicol acetyl transferase, which confers resistance to chloramphenicol, and the DNA gyrase poison CcdB. The former marker provides a positive selection for the presence of the cassette, which is useful when one is constructing a destination vector. The latter gene product provides a negative selection against the donor vector and various recombination intermediates so that only the desired recombinant is obtained when the end products of the recombinational cloning reaction are transformed into *E. coli*. pKM596 and other vectors that carry the *ccdB* gene must be propagated in a host strain with a *gyrA* mutation (e.g., *E. coli* DB3.1) that renders the cells immune to the action of CcdB.

The Gateway™ Cloning System has several noteworthy advantages. First, it is much faster and more efficient than conventional cloning techniques that utilize restriction endonucleases and DNA ligase. Second, because it does not rely on restriction endonucleases to generate substrates for ligation, Gateway™ cloning is never complicated by the existence of restriction sites within the ORF of interest that are also used for cloning. In fact, with the exception of the gene-specific primers that are used for PCR amplification, the Gateway protocol is completely generic and therefore readily amenable to automation. Finally, once an ORF has been cloned into a Gateway™ vector, it can easily be transferred by recombinational cloning into a wide variety of destination vectors that are available from Invitrogen. This gives the investigator the flexibility to experiment with various modes of expression (e.g., different fusion tags) and/or hosts. There is even a destination vector for yeast two-hybrid screening.

3.3.2. An Abbreviated Gateway™ Cloning Protocol

The investigator is encouraged to refer to the detailed protocols in the technical literature from Invitrogen. The easiest way to construct an MBP fusion vector by recombinational cloning is to start with a PCR amplicon wherein the ORF of interest is bracketed by *att*B1 and *att*B2 sites on its N- and C-termini, respectively, which can be generated by amplifying the target ORF with PCR primers that include the appropriate *att*B sites as 5' unpaired extensions (*see* **Fig. 3**). The 3' ends of these PCR primers are chosen so that the primer will be able to form 20–25 bp with the template DNA. So that the passenger protein can be separated from MBP, a target site for TEV protease (or an alternative reagent) is also incorporated between the N-terminus of the ORF and the *att*B1 site in this PCR amplicon. Although it is possible to accomplish this by using a single N-terminal PCR primer for each gene, typically on the order of 75 nucleotides long, we have found that it is convenient to perform the PCR amplification with two N-terminal primers instead, as outlined in **Fig. 3**. Two gene-specific primers (N1 and C) are required for each ORF. The C-terminal primer (C) includes the *att*B2 recombination site as a 5' extension. The 5' extension of the N-terminal primer (N1) includes a recognition site for TEV protease. The PCR product generated by these two primers is subsequently amplified by primers N2 and C to yield the final product. Primer N2 anneals to the TEV protease recognition site and includes the *att*B1 recombination site as a 5' extension. This generic PCR primer can be used to add the *att*B1 site to any amplicon that already contains the TEV protease recognition site at its N-terminal end. The PCR reaction is performed in a single step by adding all three primers to the reaction at once (*see* **Note 14**). To favor the accumulation of the desired product, the *att*B-containing primers are used at typical concentrations for PCR but the concentration of the gene-specific N-terminal primer (N1) is 20-fold lower.

1. The PCR reaction mix is prepared as follows (*see* **Note 15**): 1 µL template DNA (~10 ng/µL), 10 µL thermostable DNA polymerase 10X buffer, 16 µL dNTP solution (1.25 m*M* each), 2.5 µL primer N1 (~1 µ*M*, or 13 ng/µL for a 40 mer), 2.5 µL primer N2 (~20 µ*M*, or 260 ng/µL for a 40 mer), 2.5 µL primer C (~20 µ*M*, or 260 ng/µL for a 40 mer), 1 µL thermostable DNA polymerase, 64.5 µL H_2O (to 100 µL total volume).
2. The reaction is placed in the PCR thermal cycler with the following program (*see* **Note 16**): Initial melt: 94°C, 5 min, 25 cycles of 94°C, 30 s (melting); 55°C, 30 s (annealing); 72°C, 60 s (extension), final extension: 72°C, 7 min, hold at 4°C.
3. Purification of the PCR amplicon by agarose gel electrophoresis (*see* **Note 2**) is recommended to remove *att*B primer-dimers.

4. To create the MBP fusion vector, the PCR product is recombined first into pDONR201 to yield an entry clone intermediate (BxP reaction), and then into pKM596 (LxR reaction; *see* **Note 17**).
   a. Add to a microcentrifuge tube on ice: 300 ng of the PCR product in TE, 300 ng of pDONR201 DNA, 4 µL of BxP reaction buffer, and enough Tris-EDTA (TE) or H_2O to bring the total volume to 16 µL. Mix well.
   b. Thaw BP Clonase enzyme mix on ice (2 min) and then vortex briefly (2 s) twice (*see* **Note 18**).
   c. Add 4 µL of BP Clonase enzyme mix to the components in (a.) and vortex briefly twice
   d. Incubate the reaction at room temperature for at least 4 h (*see* **Note 19**).
   e. Add to the reaction: 1 µL of 0.75 *M* NaCl, 3 µL (ca. 450 ng) of the destination vector (pKM596), and 6 µL of LR Clonase enzyme mix (*see* **Note 18**). Mix by vortexing briefly.
   f. Incubate the reaction at room temperature for 3–4 h.
   g. Add 2.5 µL of 10X stop solution and incubate for 10 min at 37°C.
   h. Transform 2 µL of the reaction into 50 µL of competent DH5α cells (*see* **Note 3**).
   i. Pellet the cells by centrifugation, gently resuspend pellet in 100–200 µL of LB broth and spread on an LB agar plate containing ampicillin (100 µg/mL). Incubate the plate at 37°C overnight (*see* **Note 20**).
5. Plasmid DNA is isolated from saturated cultures started from individual ampicillin-resistant colonies, and screened by PCR using the gene-specific primers N1 and C to confirm that the clones have the expected structure. Alternatively, plasmids can be purified and screened by conventional restriction digests using appropriate enzymes. At this stage, we routinely sequence putative clones to ensure that there are no PCR-induced mutations.

3.4. Assessing the Solubility of MBP Fusion Proteins

The fusion protein is overproduced on a small scale to assess its solubility. The amount of fusion protein in the soluble fraction of the crude cell lysate is compared by SDS-PAGE with the total amount of fusion protein in the cells, and the results are analyzed by visual inspection of the stained gel.

3.4.1. Selecting a Host Strain of E. coli

The pMAL vectors and derivatives of pKM596 can be used in virtually any strain of *E. coli*. However, we prefer BL21 *(9)* because of its robust growth characteristics and the fact that it lacks two proteases (Lon and OmpT) present in most *E. coli* K12 strains. To improve the likelihood of obtaining a high yield of MBP fusion protein, we routinely use BL21 cells containing accessory plasmids that overproduce the cognate tRNAs for codons that are rarely used in *E. coli* (e.g., BL21-CodonPlus™-RIL or BL21 cells containing pKC1; *see* **Note 5**).

3.4.2. Pilot Expression Experiment

1. Inoculate 2–5 mL of LB medium containing ampicillin (100 μg/mL) and chloramphenicol (30 μg/mL) in a culture tube or shake-flask with BL21-CodonPlus™-RIL cells harboring an MBP fusion vector. Use a single colony from an LB agar plate containing ampicillin (100 μg/mL) and chloramphenicol (30 μg/mL) as the inoculum. Grow to saturation overnight at 37°C with shaking. *See* **Subheadings 2.1.9.** and **2.3.2.** for the preparation of LB medium and antibiotic stock solutions.
2. The next morning, inoculate 25 mL of the same medium in a 250-mL baffled-bottom flask with 0.25 mL of the saturated overnight culture. Label this flask "+". Also prepare a duplicate culture and label it "–".
3. Grow the cells at 37°C with shaking to mid-log phase (OD_{600nm} ~0.5), and then add IPTG to the "+" flask (1 m*M* final concentration).
4. Continue shaking for 3–4 h at 37°C.
5. Measure the OD_{600nm} of the cultures (dilute cells 1:10 in LB to obtain an accurate reading). An OD_{600nm} of approx 3–3.5 is normal, although lower densities are possible. If the density of either culture is much lower than this, it may be necessary to adjust the volume of the samples that are analyzed by SDS-PAGE (*see* **Subheading 3.4.4.**).
6. Transfer 10 mL of each culture to a 15-mL conical centrifuge tube and pellet the cells by centrifugation.
7. Resuspend the cell pellets in 1 mL of lysis buffer (*see* **Subheading 2.3.6.**) and then transfer the suspensions to a 1.5-mL microcentrifuge tube.
8. Store the cell suspensions at –80°C overnight. Alternatively, the cells can be disrupted immediately by sonication (after freezing and thawing) and the procedure continued without interruption, as described below.

3.4.3. Sonication and Sample Preparation

1. Thaw the cell suspensions at room temperature, then place them on ice.
2. Lyse the cells by sonication (*see* **Note 21**).
3. Prepare samples of the total intracellular proteins from the induced and uninduced cultures (T+ and T–, respectively) for SDS-PAGE by mixing 50 μL of each sonicated cell suspension with 50 μL of 2X SDS-PAGE sample buffer.
4. Pellet the insoluble cell debris (and proteins) by centrifuging the sonicated cell suspension from the "+" culture at maximum speed in a microcentrifuge for 10 min.
5. Prepare a sample of the soluble intracellular proteins (S+) for SDS-PAGE by mixing 50 μL of the supernatant with 50 μL of 2X SDS-PAGE sample buffer.

3.4.4. SDS-PAGE

We typically use precast Tris-glycine SDS-PAGE gels (10–20% gradient) to assess the yield and solubility of MBP fusion proteins (*see* **Note 6**). Of course, the investigator is free to choose any appropriate SDS-PAGE formulation, depending on the protein size and laboratory preference.

1. Heat the T–, T+ and S+ protein samples at 90°C for approx 5 min and then spin them at maximum speed in a microcentrifuge for 5 min.
2. Dilute 10 µL of each sample with enough 1X SDS-PAGE sample buffer to fill the well of the gel.
3. Assemble the gel in the electrophoresis apparatus, fill it with SDS-PAGE running buffer, load the samples, and carry out the electrophoretic separation according to standard lab practices. T+ and S+ samples are loaded in adjacent lanes to allow easy assessment of solubility. Molecular weight standards may also be loaded on the gel, if desired.
4. Stain the proteins in the gel with GelCode® Blue reagent, Coomassie Brilliant Blue, or a suitable alternative.

3.4.5. Interpreting the Results

The MBP fusion protein should be readily identifiable in the T+ sample after the gel is stained since it will normally be the most abundant protein in the cells, whereas there will be very little or no fusion protein in the T– (uninduced) sample. Molecular weight standards can also be used to corroborate the identity of the fusion protein band. If the S+ sample contains a similar amount of the fusion protein, this indicates that it is highly soluble in *E. coli*. On the other hand, if little or no fusion protein is observed in the S+ sample, then it can be concluded that the protein is poorly soluble. Of course, a range of intermediate states is also possible. Yet, even when the solubility of the MBP fusion protein is relatively poor, an adequate amount of soluble material usually can be obtained by scaling up production.

3.4.6. Improving the Solubility of MBP Fusion Proteins

Not every MBP fusion protein will be highly soluble. However, solubility usually can be increased by reducing the temperature of the culture from 37 to 30°C or even lower during the time that the fusion protein is accumulating in the cells (i.e., after the addition of IPTG). In some cases, the improvement can be quite dramatic. It may also be helpful to reduce the IPTG concentration to a level that will result in partial induction of the fusion protein ***(2)***. The appropriate IPTG concentration must be determined empirically, but is generally in the range of 10–20 µ*M*. Under these conditions, longer induction times (18–24 h) are required to obtain a reasonable yield of fusion protein.

3.5. Determining the Folding State of a Passenger Protein

MBP is an excellent solubilizing agent, but some passenger proteins are unable to fold into their native conformations even after they have been rendered soluble by fusing them to MBP. These proteins evidently exist in a soluble but nonnative form that resists aggregation only as long as they remain fused to MBP. Consequently, it is difficult to assess the folding state of the

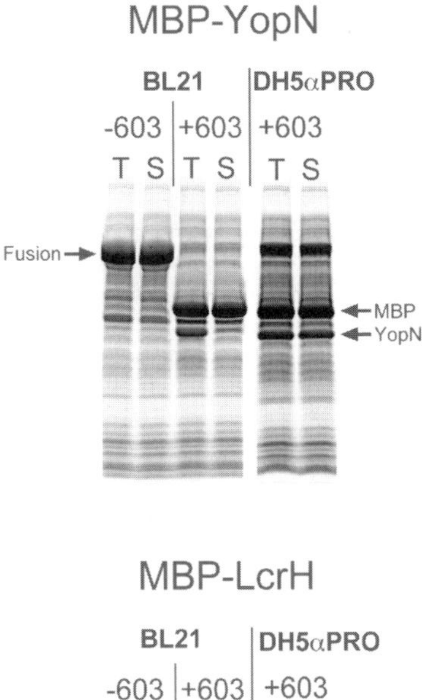

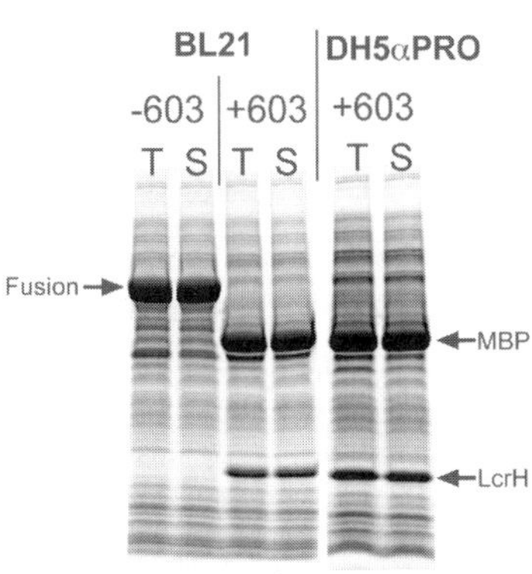

Fig. 4. Intracellular processing of MBP fusion proteins by TEV protease. Two MBP fusion proteins were processed in vivo by TEV protease to illustrate the utility of this method. YopN and LcrH are essential virulence factors from *Yersinia pestis* ***(12)***. They were both expressed from derivatives of pKM596 in *E. coli* strains BL21 and DH5αPRO. In all cases, the cells also contained the tRNA accessory plasmid pKC1 (*see* **Note 5**). TEV protease was produced in vivo by pRK603 (*see* **Note 7**). The production of TEV protease is constitutive in BL21 cells because no Tet repressor is present. However, TEV protease is not produced in DH5αPRO cells until the resident Tet repressor is displaced from the synthetic P_L/tetO promoter/operator by the addition of anhydrotetracycline ***(10)***. Both fusion proteins were processed essentially to completion in BL21 cells. Whereas all of the LcrH was soluble after cleavage, the YopN protein was almost completely insoluble. In DH5αPRO cells, the production of TEV protease was induced 2 h after induction of the fusion proteins with IPTG. Under these circumstances, virtually all of the free YopN protein became soluble. It should be noted, however, that sometimes intracellular processing is less efficient when the induction of TEV protease is delayed for 2 h, as is clearly the case with the MBP-YopN fusion protein.

passenger protein while it is still attached to MBP. In our lab, we have developed a simple method to rapidly ascertain whether a fusion protein will yield a soluble product after cleavage ***(10)***. For this purpose, we use another plasmid vector (pRK603; *see* **Note 7**) to coexpress TEV protease along with the fusion protein substrate. First, IPTG is added to the log phase culture and the fusion protein is allowed to accumulate for a period of time. Then, we stimulate the production of TEV protease by adding anhydrotetracycline to the culture. This protocol must be performed in a strain of *E. coli* that produces the Tet repressor (e.g., DH5αPRO or BL21PRO cells from Clontech); otherwise, the expression of TEV protease will be constitutive. The cells are harvested after the protease has had time to digest the fusion protein, and then samples of the total and soluble protein are prepared and analyzed by SDS-PAGE (*see* **Subheading 3.4.**). If the passenger protein is soluble after intracellular processing, then it is also likely to be soluble after the fusion protein has been purified and processed in vitro. Examples of how this method can be used are illustrated in **Fig. 4**.

3.5.1. Intracellular Processing of MBP Fusion Proteins by TEV Protease

Transform competent DH5αPRO or BL21PRO cells that already contain pRK603 and pKC1 with the MBP fusion protein expression vector (*see* **Note 3**) and spread them on an LB agar plate containing ampicillin (100 μg/mL), chloramphenicol (30 μg/mL), and kanamycin (30 μg/mL). Incubate the plate overnight at 37°C. Inoculate 2–5 mL of LB medium containing ampicillin (100 μg/mL), chloramphenicol (30 μg/mL), and kanamycin (30 μg/mL) in a culture tube or shake-flask with a single colony from the plate. Grow to saturation overnight at 37°C with shaking. *See* **Subheadings 2.1.9.**, **2.2.6.**, and **2.3.2.** for the preparation of LB medium and antibiotic stock solutions. The next morning, inoculate 25 mL of the same medium in a 250-mL baffled-bottom flask with 0.25 mL of the saturated overnight culture. Label this flask "+". Also prepare a duplicate culture and label it "–". Grow at 37°C with shaking to mid-log phase (OD_{600nm} ~0.5), and then add IPTG to the "+" flask (1 m*M* final concentration). After 2 h, add anhydrotetracycline to both flasks (100 ng/mL final concentration), and adjust the shaker temperature to 30°C (the optimum temperature for TEV protease cleavage). After 2 more hours, pellet the cells by centrifugation, prepare T–, T+ and S+ samples for SDS-PAGE, and analyze results as described in **Subheadings 3.4.4.** and **3.4.5.** It is advisable also to include a total protein sample from cells producing the same fusion protein in the absence of TEV protease (i.e., the T+ sample prepared in **Subheading 3.4.3.**) on the gel to facilitate interpretation of the results. Examine the gel to determine approximately what fraction of the fusion protein was cleaved and what fraction of the cleaved passenger protein was soluble.

3.5.3. Checking the Biological Activity of the Passenger Protein

Occasionally, a passenger protein may accumulate in a soluble but biologically inactive form after intracellular processing of an MBP fusion protein. Exactly how and why this occurs is unclear, but we suspect that fusion to MBP somehow enables certain proteins to evolve into kinetically trapped, folding intermediates that are no longer susceptible to aggregation. Therefore, although solubility after intracellular processing is a useful indicator of a passenger protein's folding state in most cases, it is not absolutely trustworthy. For this reason, we strongly recommend that a biological assay be employed (if available) at an early stage to confirm that the passenger protein is in its native conformation.

4. Notes

1. We recommend a proofreading polymerase such as *Pfu* Turbo (Stratagene, La Jolla, CA, USA) or Deep Vent (New England Biolabs, Beverly, MA, USA) to minimize the occurrence of mutations during PCR. This is especially important when attempting to ligate a blunt-ended PCR fragment with a vector fragment produced by digestion with a restriction endonuclease that generates blunt ends, because thermostable polymerases without proofreading activity (e.g., *Taq* polymerase) will add an extra unpaired adenosine residue to the 3' end of the DNA.
2. We typically purify fragments by horizontal electrophoresis in 1% agarose gels run in TAE buffer (40 m*M* Tris-acetate, 1 m*M* EDTA, pH 8). It is advisable to use agarose of the highest possible purity (e.g., Seakem-GTG from FMC BioPolymer, Philadelphia, PA, USA). Equipment for horizontal electrophoresis can be purchased from a wide variety of scientific supply companies. DNA fragments are extracted from slices of the ethidium bromide-stained gel using a QIAquick™ gel extraction kit (QIAGEN, Valencia, CA, USA) in accordance with the instructions supplied with the product.
3. While any method for the preparation of competent cells can be used (e.g., $CaCl_2$), we prefer electroporation because of the high transformation efficiency that can be achieved. Electrocompetent cells can be purchased from various sources (e.g., Stratagene, Invitrogen, Clontech, Bio-Rad, Novagen). In addition, detailed protocols for the preparation of electrocompetent cells and electrotransformation procedures can be obtained from the electroporator manufacturers (e.g., Bio-Rad, BTX, Eppendorf). Briefly, the cells are grown in 1 L of LB medium (with antibiotics, if appropriate) to mid-log phase (OD_{600} ~0.5) and then chilled on ice. The cells are pelleted at 4°C, resuspended in 1 L of ice-cold 10% glycerol, then pelleted again. After several such washes with 10% glycerol, the cells are resuspended in 3–4 mL of 10% glycerol, divided into 50-μL aliquots, and then immediately frozen in a dry ice/ethanol bath. The electrocompetent cells are stored at –80°C. Immediately prior to electrotransformation, the cells are thawed on ice and mixed with 10–100 ng of DNA (e.g., a plasmid vector, a ligation reaction, or a Gateway™ reaction). The mixture is placed into an ice-cold electroporation

cuvet and electroporated according to the manufacturers recommendations (e.g., a 1.8 kV pulse in a cuvet with a 1-mm gap). 1 mL of SOC medium *(8)* is immediately added to the cells, and they are allowed to grow at 37°C with shaking for 1 h. 5–200 μL of the cells are then spread on an LB agar plate containing the appropriate antibiotic(s).

4. We prefer the Wizard® miniprep kit (Promega, Madison, WI, USA) or the QIAprep™ Spin miniprep kit (QIAGEN, Valencia, CA, USA), but similar kits can be obtained from a wide variety of vendors.
5. To circumvent the problem of codon bias in *E. coli*, we routinely express proteins in BL21-CodonPlus™-RIL cells (Stratagene). The RIL plasmid is a derivative of the p15A replicon that carries the *E. coli argU*, *ileY*, and *leuW* genes, which encode the cognate tRNAs for AGG/AGA, AUA, and CUA codons, respectively. These codons are rarely used in *E. coli*, but occur frequently in ORFs from other organisms. Consequently, the yield of some MBP fusion proteins will be significantly greater in cells that harbor RIL, particularly if the target ORF contains tandem runs of rare codons ***(11)***. When this is not the case, RIL can be omitted. RIL is selected for by resistance to chloramphenicol (30 μg/mL). In addition to the tRNA genes for AGG/AGA, AUA, and CUA codons, the accessory plasmid in the recently introduced Rosetta™ host strain (Novagen, Madison, WI, USA) also includes tRNAs for the rarely used CCC and GGA codons. Hence, the Rosetta™ strain may turn out to be even more useful than BL21-CodonPlus™-RIL cells. Like RIL, the Rosetta™ plasmid is a chloramphenicol-resistant derivative of the p15A replicon. For intracellular processing experiments with TEV protease (*see* **Subheading 3.5.1.**), we use pKC1, a derivative of the low copy number plasmid pSC101 that is compatible with the p15A-derived TEV protease expression vector pRK603. pKC1 carries only the *ileX* and *argU* genes and is also selected for with chloramphenicol.
6. We find it convenient to use precast SDS-PAGE gels (e.g., 1.0 mm × 10 well, 10–20% gradient), running buffer, and electrophoresis supplies from Novex, a subsidiary of Invitrogen (Carlsbad, CA, USA).
7. pRK603 is a derivative of the p15A replicon that produces TEV protease when induced by anhydrotetracycline ***(10)***. pRK603 is selected for by its resistance to kanamycin.
8. These older vectors are essentially the same as pMAL-c2X and pMAL-p2X, respectively. However, in the new vectors, an *Nco*I site has been removed from within the MBP coding sequence and an *Nde*I site has been placed immediately at the start of the MBP open reading frame. Also, an *Nde*I site has been removed from another location that existed in the older vectors.
9. Purification of MBP fusion proteins from the periplasm does not rupture the inner membrane and release the contents of the cytosol *(7)*. Consequently, periplasmic expression often results in higher initial purity of the fusion protein prior to amylose affinity chromatography.

10. When using pMAL-c(or p)2E, the *Kpn*I-digested vector must be filled in to yield a blunt end before ligation with a blunt-ended PCR fragment. Cloning in pMAL-c(or p)2G is like the X vectors, except that *Sna*BI is utilized as the blunt site in the vector (*see* **ref. *7***).
11. Alternatively, one of the other restriction sites in the pMAL polylinker can be used to join the N-terminus of the passenger protein to the C-terminus of MBP (e.g., *Eco*RI), but this would have the effect of adding extra nonnative residues to the linker between MBP and the passenger protein. It is possible that an increase in the length of the linker would affect MBP's ability to promote the solubility of the passenger protein. Moreover, if one intends to exploit a protease cleavage site that is already contained in the linker, the additional residues would end up on the N-terminus of the passenger protein after digestion.
12. If a particular ORF happens to contain a *Bam*HI restriction site, then any of the other sites in the pMAL polylinker may be used instead (*e.g., Eco*RI, *Xba*I, *Sal*I, *Pst*I, or *Hin*dIII).
13. *See* **ref. *8*)** for tips on setting up ligation reactions. A typical reaction contains ~300–400 ng of DNA. The two fragments should be present at approximately equimolar concentrations. The two DNA fragments, 2 µL of 10X ligase buffer, ATP (1 m*M* final concentration), 1 µL of T4 DNA ligase, and H_2O are combined in a total volume of 20 µL. The reaction is incubated at room temperature for several hours or at 16°C overnight.
14. Alternatively, the PCR reaction can be performed in two separate steps, using primers N1 and C in the first step and primers N2 and C in the second step. The PCR amplicon from the first step is used as the template for the second PCR reaction. All primers are used at the typical concentrations for PCR in the two-step protocol.
15. The PCR reaction can be modified in numerous ways to optimize results, depending on the nature of the template and primers. *See* **ref. *8*** (Vol. 2, Chapter 8) for more information.
16. PCR cycle conditions can also be varied. For example, the extension time should be increased for especially long genes. A typical rule-of-thumb is to extend for 60 s/kb of DNA.
17. This "one-tube" Gateway™ protocol bypasses the isolation of an "entry clone" intermediate. However, the entry clone may be useful if the investigator intends to experiment with additional Gateway™ destination vectors, in which case the LxR and BxP reactions can be performed sequentially in separate steps; detailed instructions are included with the Gateway™ PCR kit. Alternatively, entry clones can easily be regenerated from expression clones via the BxP reaction, as described in the instruction manual.
18. Clonase enzyme mixes should be thawed quickly on ice and then returned to the –80°C freezer as soon as possible. It is advisable to prepare multiple aliquots of the enzyme mixes the first time that they are thawed in order to avoid repeated freeze-thaw cycles.

19. At this point, we remove a 5 µL aliquot from the reaction and add it to 0.5 µL of 10X stop solution. After 10 min at 37°C, we transform 2 µL into 50 µL of competent DH5α cells (*see* **Note 3**) and spread 100–200 µL on an LB agar plate containing kanamycin (25 µg/mL). From the number of colonies obtained, it is possible to estimate the percent conversion of the PCR product to entry clone in the BxP reaction. Additionally, entry clones can be recovered from these colonies in the event that no transformants are obtained after the subsequent LxR reaction.
20. If very few or no ampicillin-resistant transformants are obtained after the LxR reaction, the efficiency of the process can be improved by incubating the BxP reaction overnight.
21. We routinely break cells with two or three 30 s pulses using a VCX600 sonicator (Sonics & Materials, Newtown, CT, USA) with a microtip at 38% power. The cells are cooled on ice between pulses.

Acknowledgments

We wish to thank Rachel Kapust and Karen Routzahn for their valuable contributions to the development of these methods and Invitrogen/Life Technologies for granting us early access to the Gateway™ cloning technology.

References

1. Lilie, H., Schwarz, E., and Rudolph, R. (1998) Advances in refolding of proteins produced in *E. coli*. *Curr. Opin. Biotechnol.* **9,** 497–501.
2. Baneyx, F. (1999) In vivo folding of recombinant proteins in *Escherichia coli*, in *Manual of Industrial Microbiology and Biotechnology* (Davies, J. E., Demain, A. L., Cohen, G., et al., eds.), American Society for Microbiology, Washington, D. C., pp. 551–565.
3. Kapust, R. B. and Waugh, D. S. (1999) *Escherichia coli* maltose-binding protein is uncommonly effective at promoting the solubility of polypeptides to which it is fused. *Protein Sci.* **8,** 1668–1674.
4. Fox, J. D., Kapust, R. B., and Waugh, D. S. (2001) Single amino acid substitutions on the surface of *Escherichia coli* maltose-binding protein can have a profound impact on the solubility of fusion proteins. *Protein Sci.* **10,** 622–630.
5. Richarme, G. and Caldas, T. D. (1997) Chaperone properties of the bacterial periplasmic substrate-binding proteins. *J. Biol. Chem.* **272,** 15,607–15,612.
6. Sachdev, D. and Chirgwin, J. M. (1998) Solubility of proteins isolated from inclusion bodies is enhanced by fusion to maltose-binding protein or thioredoxin. *Protein Expr. Purif.* **12,** 122–132.
7. Riggs, P. (2000) Expression and purification of recombinant proteins by fusion to maltose-binding protein. *Mol. Biotechnol.* **15,** 51–63.
8. Sambrook, J. and Russell, D. W. (2001) *Molecular Cloning: A Laboratory Manual.* Cold Spring Harbor Laboratory Press, Cold Spring Harbor, NY.

9. Studier, F. W., Rosenberg, A. H., Dunn, J. J., and Dubendorff, J. W. (1990) Use of T7 RNA polymerase to direct expression of cloned genes. *Methods Enzymol.* **185,** 60–89.
10. Kapust, R. B. and Waugh, D. S. (2000) Controlled intracellular processing of fusion proteins by TEV protease. *Protein Expr. Purif.* **19,** 312–318.
11. Kane, J. F. (1995) Effects of rare codon clusters on high-level expression of heterologous proteins in *Escherichia coli. Curr. Opin. Biotechnol.* **6,** 494–500.
12. Cornelis, G. R., Boland, A., Boyd, A. P., et al. (1998) The virulence plasmid of *Yersinia*, an antihost genome. *Microbiol. Mol. Biol. Rev.* **62,** 1315–1352.

8

Thioredoxin and Related Proteins as Multifunctional Fusion Tags for Soluble Expression in *E. coli*

Edward R. LaVallie, Elizabeth A. DiBlasio-Smith, Lisa A. Collins-Racie, Zhijian Lu, and John M. McCoy

1. Introduction

Escherichia coli has traditionally been a popular host for the production of heterologous proteins because of its ease of genetic manipulation and growth. Recombinant proteins produced in *E. coli* have been useful for a variety of purposes, including the study of protein tertiary structure, structure-function experiments, enzymology, and as bio-pharmaceuticals. Despite an impressive body of literature describing the production of numerous non-native proteins in *E. coli,* successful results are in no way assured. The use of *E. coli* as a robust expression system has been hampered by several integral pitfalls. Low or undetectable expression levels can often be caused by inefficient translation initiation of eukaryotic mRNAs on bacterial ribosomes *(1)*. Recombinant proteins produced in *E. coli* sometimes retain the N-terminal initiator methionine residue, as they may be a poor substrate for the host methionine aminopeptidase *(2)*. In addition, individual purification schemes must be devised for each native recombinant protein produced in *E. coli*. But perhaps most importantly, it is very common for recombinant proteins expressed in *E. coli* to form insoluble, misfolded cytoplasmic complexes known as "inclusion bodies" *(3)*. The likelihood of inclusion body formation is unpredictable but appears to increase in proportion to the size and complexity of the protein.

The use of *E. coli* thioredoxin (TrxA) and related proteins as protein fusion partners was devised as a potential solution to these problems *(4–7)*. Thioredoxin has several inherent properties that make it well suited as a protein fusion partner.

From: *Methods in Molecular Biology, vol. 205, E. coli Gene Expression Protocols*
Edited by: P. E. Vaillancourt © Humana Press Inc., Totowa, NJ

When thioredoxin is expressed from plasmid vectors in *E. coli*, it can accumulate to 40% of the total cellular protein while remaining fully soluble ***(8)***. This extraordinary level of soluble expression suggests that thioredoxin is translated very efficiently. This property is often conferred to heterologous proteins fused to thioredoxin, especially when thioredoxin is positioned at the amino-terminus of the fusion where protein translation initiates. In the tertiary structure of thioredoxin ***(9)***, both the N- and C-termini of the molecule are surface accessible, and therefore are possible fusion points to link thioredoxin to other proteins. In addition, thioredoxin has a very compact fold, with >90% of its primary sequence involved in strong elements of secondary structure. This helps to explain its observed high thermal stability (T_m = 85°C) ***(6)***, and its robust folding characteristics no doubt contribute to its success as a fusion partner protein. In fact, nature has previously discovered the utility of thioredoxin fusions. Complete thioredoxin domains are found in a number of naturally occurring multidomain proteins, including *E. coli* DsbA ***(10)***, the mammalian endoplasmic reticulum proteins ERp72 ***(11)***, and protein disulfide isomerase (PDI; *see* **ref. *11***). In addition, thioredoxin is small (11,675 M_r) and therefore constitutes only a minor proportion of the total mass of most protein fusions.

Thioredoxin is distinguished from many other potential fusion partners by its propensity to confer solubility to many proteins that otherwise form inclusion bodies when expressed in *E. coli* ***(4)***. We have proposed that thioredoxin may serve as a covalently joined chaperone protein by keeping folding intermediates of linked heterologous proteins in solution long enough for them to adopt their correct final conformations ***(12)***. This characteristic of thioredoxin could be viewed as analogous to the covalent chaperone role proposed for the N-terminal propeptide regions of a number of protein precursors ***(13,14)***.

Other unusual attributes of thioredoxin have been exploited to extend its utility as a fusion partner. Its active-site comprises a surface-accessible loop naturally flanked by cysteine residues which can serve as a permissive site for internal, constrained peptide insertions. Purification of thioredoxin fusion proteins may be achieved by utilizing the molecule's remarkable ability to be released from the bacterial cytoplasm by simple osmotic-shock ***(4)***, by taking advantage of the molecule's high thermal stability ***(4)***, by using avidin or streptavidin matrices to bind thioredoxin variants (BIOTRX) modified to allow for in vivo biotinylation ***(7)***, or by using engineered forms of thioredoxin (His-patch Trx) with affinity for metal chelate column matrices ***(6)***. A final manifestation of thioredoxin fusion technology utilizes the secreted thioredoxin homolog, *E. coli* DsbA ***(10)***. Fusions to secretory DsbA are localized to the periplasmic space in *E. coli*, which is sometimes beneficial for enhancing protein folding, disulfide bond formation, and activity ***(5)***.

Here we provide detailed protocols for using the thioredoxin gene fusion expression system. These protocols include information on choice and manipulation of expression vectors (**Subheading 3.1.**) and host strains (**Subheading 3.2.**), as well as procedures for expression (**Subheading 3.3.**), characterization (**Subheading 3.4.**), purification (**Subheading 3.5.**), and site-specific proteolytic cleavage (**Subheading 3.6.**), of thioredoxin and DsbA fusion proteins.

2. Materials

2.1. Thioredoxin Gene Fusion Construction

1. A cDNA or PCR product containing the coding sequence for the protein of interest.
2. A thioredoxin fusion vector of choice: pALtrxA-781, pTRXFUS, pHis-patch-TRXFUS, pBIOTRXFUS-BirA or pDsbAsecFUS (*see* **Fig. 1**)

2.2. E. coli *Host Strain Transformation*

Throughout this section, all filter sterilization steps should be preceded by prerinsing the filter with deionized H_2O (*see* **Note 1**).

1. An appropriate *E. coli* host strain, chosen from **Table 1**.
2. 10X M9 salts: Mix 60 g Na_2HPO_4 (0.42 *M* final concentration), 30 g KH_2PO_4 (0.24 *M* final concentration), 5 g NaCl (0.09 *M* final concentration), 10 g NH_4Cl (0.19 *M* final concentration), deionized H_2O to 1 L. Adjust to pH 7.4 with 1 *M* NaOH, autoclave or filter sterilize through a 0.2 µm filter, store ≤ 6 mo at room temperature.
3. 1 *M* KCl: Mix 7.4 g KCl with deionized H_2O to 100 mL. Filter sterilize through a 0.2 µm filter, store ≤ 6 mo at room temperature.
4. 1 *M* NaCl: Mix 5.8 g NaCl with deionized H_2O to 100 mL. Filter sterilize through a 0.2 µm filter, store ≤ 6 mo at room temperature.
5. 2 *M* Mg^{2+}: Mix 20.3 g with $MgCl_2$ hexahydrate and 24.6 g $MgSO_4$ heptahydrate with deionized H_2O to 100 mL. Filter sterilize through a 0.2 µm filter, store ≤ 6 mo at room temperature.
6. 2 *M* Glucose: Add 36 g D-glucose to deionized H_2O to 100 mL. Filter sterilize through a 0.2 µm filter, store ≤ 6 mo at room temperature.
7. SOB medium: Mix 20 g bactotryptone (Difco), 5 g yeast extract (Difco), and 0.5 g NaCl with deionized H_2O to 1 L. Adjust pH to 7.5 + 0.05 with 1 *M* KOH. Filter sterilize through a 0.2 µm filter, store ≤ 6 mo at room temperature.
8. SOC media: Mix 20 g bactotryptone (Difco), 5 g yeast extract (Difco), 10 mL 1 *M* NaCl, and 2.5 mL 1 *M* KCl with 970 mL deionized H_2O. Stir to dissolve, autoclave, cool to room temperature. Add 10 mL 2 *M* Mg^{2+} solution and 10 mL of 2 *M* glucose solution. Filter sterilize through a 0.2 µm, store ≤ 6 mo at room temperature. pH should 7.0 + 0.1.
9. 10% glycerol: Mix 100 mL redistilled glycerol with 900 mL deionized H_2O. Filter sterilize through a 0.2 µm filter. Store ≤ 6 mo at 4°C.
10. IMC plates containing 100 µg/mL ampicillin: Mix 15 g agar (Difco; 1.5% [w/v] final concentration) and 4 g casamino acids (Difco-certified; 0.4% [w/v] final

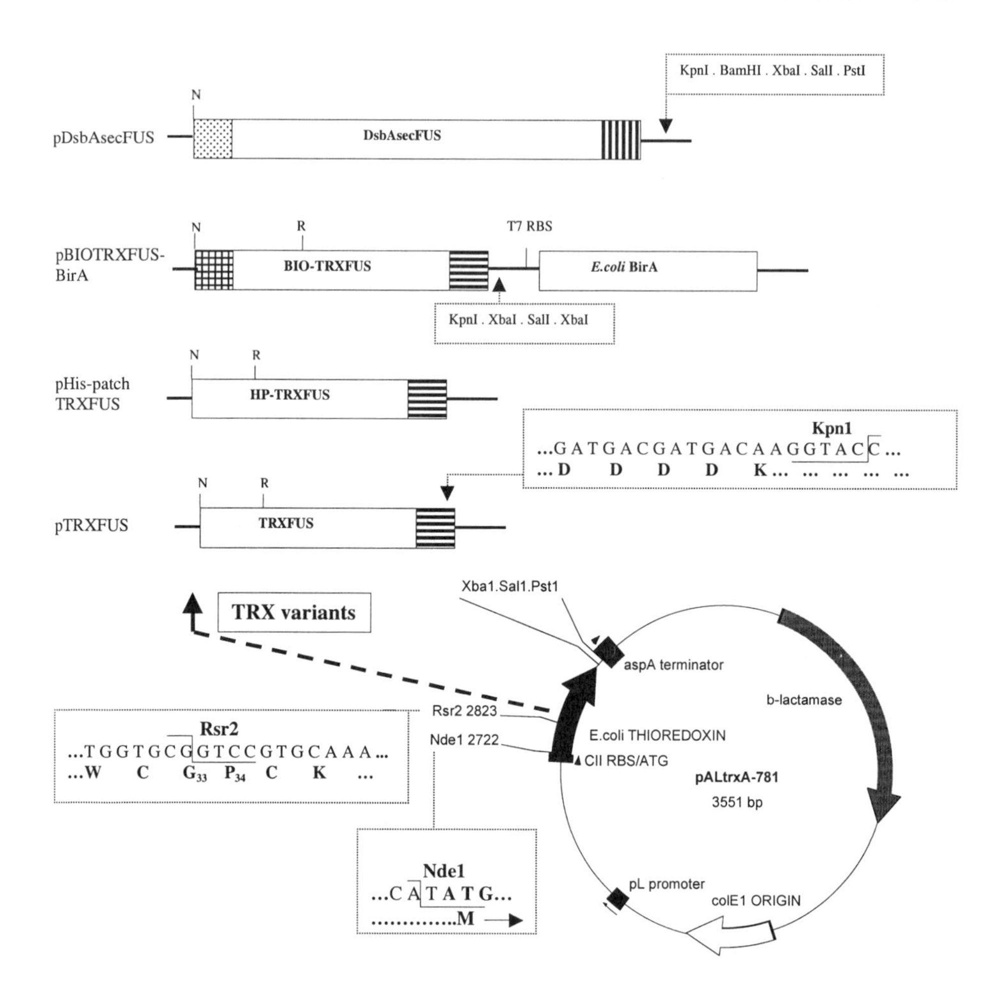

Fig. 1. Expression vectors for creating fusions to thioredoxin and its variants. aspA terminator, *E. coli* aspA transcriptional terminator; N, Nde1 restriction site at initiator methionine codon; pL promoter, bacteriophage λ major leftward promoter; R, Rsr2 restriction site in the TRX active-site loop, suitable for creating internal fusions; TRX, *E. coli* thioredoxin; T7 RBS, bacteriophage T7 gene 10 ribosome-binding site. The spacer peptide "-GSGSGDDDDK-" which includes an enteropeptidase cleavage site, is denoted by horizontal hatch marks. The BIOTRX N-terminal peptide "-MASSLRQILDSQ<u>K</u>MEWRSNAGGS-", which is biotinylated on the underlined lysine residue by BirA, is denoted by a cross-hatched box. The DsbA native signal peptide is depicted as a dotted box. The linker peptide "-GSGSGHHHHHHDDDDK-", containing a hexahistidine purification tag and an enteropeptidase recognition site, is shown as vertical hatch marks. The 3' polylinker providing convenient restriction sites for producing translational fusions is depicted by a black horizontal line at the end of each construct.

Table 1
Recommended *E. coli* Strains and Expression Conditions for Producing Thioredoxin Fusion Proteins

Strain	Preinduction growth temperature (°C)	Expression temperature (°C)	Time for maximal induction (h)
GI723	37	37	4
GI724	30	37	4
		30	6
GI698	25	25	18
		20	10
		15	20

concentration) (*see* **Note 2**) with 858 mL deionized H_2O (sterile). Autoclave for 30 min, cool in a 50°C water bath, then mix with: 100 mL 10X M9 salts (sterile; 1X final concentration), 40 mL 20% (w/v) glucose (sterile; 0.5% final concentration), 1 mL 1 *M* $MgSO_4$ (sterile; 1 m*M* final concentration), 0.1 mL 1 *M* $CaCl_2$ (sterile; 0.1 m*M* final concentration), 1 mL 2% (w/v) vitamin B1 (sterile; 0.002% final concentration), 10 mL 10 mg/mL ampicillin (sterile; optional; 100 µg/mL final concentration). Mix well and pour into Petri plates, store wrapped ≤ 1 mo at 4°C. (*see* **Note 3**).

11. CAA/glycerol/ampicillin 100 medium: Autoclave 800 mL 2% (w/v) Casamino Acids (Difco-Certified, 1.6% final concentration) (*see* **Note 2**). for 30 min, cool to room temperature, then add 100 mL 10X M9 salts (sterile; 1X final concentration), 100 mL 10% (v/v) glycerol (sterile; 1% final concentration), 1 mL 1 *M* $MgSO_4$ (sterile; 1 m*M* final concentration), 0.1 mL 1 *M* $CaCl_2$ (sterile; 0.1 m*M* final concentration), 0.1 mL 1 *M* $CaCl_2$ (sterile; 0.1 m*M* final concentration), 1 mL 2% (w/v) vitamin B1 (sterile; 0.002% final concentration), 10 mL 10 mg/mL ampicillin (sterile; 100 µg/mL final concentration).

2.3. Induction of Thioredoxin Gene Fusion Expression

1. IMC medium containing 100 µg/mL ampicillin: Mix 200 mL 2% (w/v) casaminoacids (Difco-certified, sterile; 0.4% final concentration) (*see* **Note 2**), 100 mL 10X M9 salts (sterile; 1X final concentration), 40 mL 20% (w/v) glucose (sterile; 0.5% final concentration), 1 mL 1 *M* $MgSO_4$ (sterile; 1 m*M* final concentration) 0.1 mL 1 *M* $CaCl_2$ (sterile; 0.1 m*M* final concentration), 1 mL 2% (w/v) vitamin B1 (sterile; 0.002% final concentration), 658 mL deionized H_2O (sterile), and 10 mL 10 mg/mL ampicillin (optional, sterile; 100 µg/mL final concentration). Use fresh.
2. 10 mg/mL tryptophan in deionized H_2O (sterile, *see* **Note 4**).

2.4. Characterization of Protein Production

1. 1X Sodium dodecyl sulfate-polyacrylamide gel electrophoresis (SDS-PAGE) sample buffer (final concentration): 15% (v/v) glycerol, 0.125 *M* Tris-HCl,

pH 6.8, 5 m*M* ethylenediaminetetraacetic acid disodium salt (Na_2EDTA), 2% (w/v) SDS, 0.1% (w/v) bromphenol blue. Store indefinitely at room temperature. Add 2-mercaptoethanol (β-ME) to a final concentration of 1% (v/v) immediately before use.
2. 2X SDS-PAGE sample buffer (final concentration): 30% (v/v) glycerol, 0.25 *M* Tris-HCl, pH 6.8, 10 m*M* Na_2EDTA, 4% (w/v) SDS, 0.2% (w/v) bromphenol blue. Store indefinitely at room temperature. Add 2-mercaptoethanol (β-ME) to a final concentration of 2% (v/v) immediately before use.
3. Coomassie brilliant blue stain (final concentration): 10% glacial acetic acid (v/v), 25% isopropyl alcohol (v/v), 0.05% Coomassie brilliant blue G-250 (w/v).
4. Gel destainV solution (final concentration): 10% glacial acetic acid (v/v), 10% isopropyl alcohol (v/v).

2.5. Post-Induction Protein Release and Fractionation from E. coli *Cells*

1. Cell lysis buffer: 50 m*M* Tris-HCl, pH 7.5, containing 1 m*M* PABA (p-aminobenzamidine) and 1 m*M* PMSF phenylmethylsulfonyl fluoride (PMSF). **Caution:** PMSF is toxic. Read the Material Safety Data Sheet that accompanies it and handle with appropriate caution.
2. Osmotic shock buffer: 20 m*M* Tris-HCl, 2.5 m*M* EDTA, pH 8, 20% sucrose.
4. Osmotic release buffer: 20 m*M* Tris-HCl, 2.5 m*M* EDTA, pH 8.

2.6. Purification of Thioredoxin Fusion Proteins

1. IDA column equilibration buffer A: 25 m*M* Tris-HCl, pH 7.5, 200 m*M* NaCl, 2 m*M* imidazole.
2. IDA column elution buffer B: 25 m*M* Tris-HCl, pH 7.5, 100 m*M* NaCl, 100 m*M* imidazole.

2.7. Site-Specific Cleavage of Fusion Proteins

1. Recombinant bovine enterokinase light chain (commercially available from Invitrogen, New England Biolabs, Promega, Stratagene, and others; *see* **Note 5**).

3. Methods

3.1. Thioredoxin Gene Fusion Expression Vectors

There are several varieties of thioredoxin gene fusion expression vectors that allow creation of protein fusions to native thioredoxin, thioredoxin variants that have enhanced purification properties, or secreted thioredoxin homologs that localize to the bacterial periplasm. The structure of these expression vectors is shown in **Fig. 1**. There are some general features shared by all of the vectors. They are all based on the parent plasmid pUC-18 ***(15)*** and contain its colE1 origin of replication and β-lactamase selectable marker. The *lac* promoter of pUC-18 has been replaced in these vectors by the bacteriophage λ pL promoter *(16)* positioned upstream of the thioredoxin/thioredoxin-related sequence. The fusion vectors are designed to allow joining of various proteins to the carboxyl-terminal end of TrxA. This configuration provides efficient and consistent translation initiation on the TrxA coding sequence. An intervening spacer sequence "-GSGSGDDDDK-" is encoded in place of the thioredoxin's native translation termination codon in all vectors except pALtrxA-781 (not designed for carboxyl-terminal fusions). This spacer sequence permits the two protein domains (Trx and the carboxyl-terminal partner) to fold independently, and also provides a recognition site for the highly specific protease enteropeptidase, allowing for subsequent in vitro separation of the two protein domains. Downstream of the spacer-encoding sequence lies a multiple cloning site polylinker providing convenient restriction sites to facilitate precise translational fusions to the 3' end of *trx*A. The transcriptional terminator sequence from the *E. coli asp*A gene is placed just beyond the polylinker to define the end of the mRNA. Following is a detailed description of each plasmid to aid in choosing the appropriate vector for the application.

3.1.1. Choosing the Appropriate Vector

1. pALtrxA-781

This vector contains the native *E. coli* thioredoxin gene with its intrinsic stop codon at the end of the coding sequence. When transcription from the pL promoter in this vector is induced in the appropriate E. coli strains (*see* **Subheading 3.1.2.**), "wild-type" thioredoxin is produced in high levels in the bacterial cytoplasm. pALtrxA-781 is not intended for creating C-terminal fusions; rather, it is used for constructing internal fusions (peptides) in the active site loop. Nature has provided a perfectly positioned *RsrII* site within the coding sequence of the active-site loop, ideally suited for precise insertion of peptide-encoding DNA sequences (*see* **Fig. 1**). Such internal peptide fusions have been shown to be accessible on the surface of the protein *(4)*, which has expanded their usefulness into the areas of antigen production, epitope mapping, and binding interaction studies (*17–19*, unpublished data). In addition, a unique *NdeI* site has been engineered at the 5' end of the thioredoxin coding sequence, incorporates the initiator methionine, and can be used to create N-terminal fusions toTrxA. While such fusions appear to retain many of the physical attributes of C-terminal thioredoxin fusions (unpublished data), expression levels may suffer because of the uncertainty of efficient translation initiation.

2. pTRXFUS

This is the "original" thioredoxin fusion vector *(4)*, which provides a convenient platform for creating in-frame fusions to the 3' end of wild-type *E. coli*

TrxA. It retains the active-site *RsrII* site and 5' *NdeI* site of pALtrxA-781, so concurrent fusions could also be constructed internally or at the 5' end. This can be advantageous; for instance, a peptide with a particular affinity for a purification matrix can be fused at the Trx 5' end or in the active-site loop to facilitate the physical isolation of a 3' fusion partner protein *(7)*.

A unique attribute of thioredoxin fusions is that they are often produced at high levels in soluble and active form. While solubility is usually highly desirable, it can pose a problem when it comes to purification of the protein from the other *E. coli* cellular proteins. Some other fusion partners like glutathione-S-transferase ***(20)*** or maltose-binding protein ***(21)*** have a natural binding affinity that can be utilized for protein purification. Thioredoxin has certain physical characteristics that can sometimes be exploited to fractionate fusions from other proteins (*see* **Subheadings 3.6.1.** and **3.6.2.**). However, not all protein fusions to thioredoxin retain these characteristics. As a result, we developed modified thioredoxin proteins that have been engineered to provide universal purification affinities.

3. pHis-patch-TRXFUS

This vector contains a modified form of thioredoxin, called "histidine-patch" thioredoxin or hpTRX. Using a rational protein engineering approach we mutated two surface-exposed residues of thioredoxin to histidines (E30H and Q62H). Although these residues lie more than 30 residues apart in the primary sequence of the protein, they are brought together in the tertiary structure of the protein with a naturally occurring histidine at position 6 and can coordinate binding of a metal ion ***(6)***. A third modification (D26A) was made to restore the thermal stability of this modified thioredoxin. These modifications do nothing to degrade thioredoxin's performance as a fusion partner but rather augment it, as the resulting hpTRX is capable of binding to nickel ions on IMAC resins (immobilized metal ion affinity chromatography) ***(22)*** such as nitrilotriacetic acid (NTA)-Sepharose or iminodiacetic acid (IDA)-Sepharose. This vector is otherwise analogous to pTRXFUS, so in-frame fusions to hpTRX can be made to the amino terminus, carboxyl terminus, or within the active-site loop. Such fusions then can be purified from contaminating *E. coli* proteins, often in a single step and in high yield. Once bound to an IMAC resin, carboxyl-terminal fusions proteins can be specifically released from the matrix by incubating the bound fusion proteins with enteropeptidase. This results in cleavage of the fusion within the spacer domain, leaving the hpTRX portion of the fusion bound to the IMAC resin while liberating the carboxyl-terminal partner in highly purified form.

4. pBIOTRXFUS-BirA

This vector produces another modified form of thioredoxin that has an intrinsic binding "handle" incorporated into its sequence. In this instance, a 23 residue peptide joined to the amino terminus of thioredoxin can be quantitatively biotinylated in vivo, providing proteins fused to BIOTRX's carboxyl terminus or active-site loop with strong and specific affinity for biotin-binding proteins such as streptavidin and avidin. This peptide (MASSLRQILDSQKMEWRSNAGGS-) was found in work by Schatz ***(23)*** to serve as a substrate for the intrinsic *E. coli* enzyme biotin holoenzyme synthetase (BirA), which attaches a biotin to the single lysine underlined in the peptide sequence above. Because this vector typically produces high levels of fusion protein, endogenous levels of BirA protein are insufficient for biotinylation of all of the protein produced by the vector without supplementation. The pBIOTRXFUS-BirA vector provides additional BirA enzyme by producing it as part of an operon fusion with the BIOTRX protein. The additional BirA produced by the vector, in conjunction with exogenous biotin (or biotin analog) added to the growth medium, results in quantitative biotinylation of BIOTRX fusion proteins (*see* **Note 6** and **ref.** *7*).

Applications that rely on strong binding benefit from the high affinity of avidin or streptavidin for BIOTRX. For instance, interaction assays for proteins bound to biotinylated BIOTRX can readily utilize surface plasmon resonance instruments (e.g., BIAcore) by using streptavidin-coated chips. BIOTRX fusions are especially amenable to this technology because the fusion protein is tethered to the chip at a single point in the amino terminus of the protein, leaving the active-site and/or carboy-terminal fusions accessible for secondary "sandwich" interactions. In addition, BIOTRX fusions can be readily imaged using avidin or streptavidin conjugates, obviating the need for individual antibody reagents.

5. pDsbAsecFUS

There are instances when it is advantageous or necessary to direct a fusion protein to the periplasmic space in *E. coli*. For instance, some proteins require the oxidizing environment and sequential folding that secretion provides in order to achieve their proper conformation and activity ***(5,24)***. Since thioredoxin is a cytoplasmic protein, protein fusions to thioredoxin also reside in the reducing environment of the cytoplasm. As an alternative to the cytoplasmic thioredoxin expression vectors, the vector pDsbAsecFUS was developed. This vector utilizes the gene for the secreted thioredoxin homolog DsbA ***(10)*** as a platform for carboxyl-terminal fusions. Amino-terminal fusions are not supported by this vector because of the presence of the signal sequence, while active-site loop fusions may be possible but have not been tried. DsbA protein fusions appear to be comparable to thioredoxin fusions in terms of their expression levels and solubilizing capability (*see* **Note 7**).

The interdomain "spacer" sequence of the DsbA fusion vectors ("-GSGSG-HHHHHHDDDDK-") is unique to these vectors in that they encode a hexa-histidine sequence along with the "GS" repeat and enteropeptidase recognition sequence common to the other pTRXFUS-based vectors. The hexa-histidine sequence confers metal-ion binding capability to DsbA fusions proteins and enables their purification on IMAC.

3.1.2. Constructing Gene Fusions to Thioredoxins

Detailed descriptions of the methodology used to create gene fusions at the DNA level are beyond the scope of this chapter, but rely solely on standard molecular biological techniques such as those described in Sambrook et al. ***(25)***.

1. To create carboxyl-terminal fusions to TRX, hp-TRX, BIOTRX, or DsbA, the unique *Kpn*I site in the polylinker can be used to generate a precise translational fusion. This is accomplished by digesting the expression vector with *Kpn*I and trimming back the resulting 3' overhang with the Klenow fragment of *E. coli* DNA polymerase. A cDNA containing the desired coding sequence to be fused to the Trx (or Trx variant) gene can usually be adapted to this blunt end to form an in-frame translational fusion by designing a synthetic oligonucleotide duplex that has a blunt 5' end and reconstitutes the coding sequence of the desired fusion partner to a convenient restriction site close to its 5' end. One of the other polylinker sites in the Trx vectors can then be used to ligate to the 3' end of the desired cDNA (*see* **Note 8**).
2. Insertional fusions into the Trx active-site loop can be created in pALtrxA-781, pTRXFUS, pHis-patch-TRXFUS, and pBIOTRXFUS-BirA . Such fusions are accomplished by taking advantage of the naturally occurring *Rsr*II site in the Trx coding sequence. Typically, oligonucleotide duplexes encoding the desired peptide insertion sequence are designed so that cohesive ends compatible with the *Rsr*II overhangs are generated when the oligonucleotides are annealed. These overhangs should regenerate the glycine and serine codons of the active site loop and, by virtue of the fact that the *Rsr*II cohesive ends are 3 bases long and asymmetric, the synthetic oligonucleotide duplex will orient in only one direction.
3. Amino-terminal fusions can be constructed by utilizing the unique *Nde*I site at the 5' end of the coding sequence in pALtrxA-781, pTRXFUS and pHis-patch-TRXFUS. The initiator methionine codon for Trx is incorporated into the recognition sequence for *Nde*I, so the full *Nde*I site should be reconstituted by the insert so that the methionine is still positioned at the 5' end of the coding sequence. When cloning into the *Nde*I site, it should be noted that the insert can orient both forward and backward, so the desired orientation must be determined by restriction mapping and/or DNA sequencing.

3.2. E. coli *Host Strain Transformation*

3.2.1. Choosing the Appropriate Host Strain

As discussed in **Subheading 3.1.**, fusion proteins in all of the expression vectors described in this chapter are transcribed by the major leftward promoter of bacteriophage λ, called pL *(16)*. This promoter is controlled by the product of the λ repressor gene cI. The strains listed in **Table 1** constitute an array of host strains that have been developed for the expression of Trx fusion proteins at any growth temperature. All of these strains are based upon the *E. coli* K12 parent strain RB791 (*see* **ref. *26***; W3110 lacI^{q}lacPL8). Each strain contains a copy of the cI gene stably integrated into the chromosomal *amp*C locus under the transcriptional control of the *Salmonella typhimurium trp* promoter/operator inserted upstream of cI in the *amp*C locus. A variation of this type of control has been described by Mieschendahl *(27)*. The different strains are distinguished from each other by the efficiency of the Shine-Dalgarno sequence (ribosome-binding site) immediately upstream of the cI gene. As a result, levels of cI protein in these strains are controlled by both the amount of tryptophan in the growth media as well as the "strength" of the ribosome binding site.

Under conditions of low intracellular tryptophan levels, transcription of cI from the *trp* promoter is maximal, and the strains produce levels of cI protein in the order GI723 > GI724 > GI698. Growth in minimal media or in a tryptophan-free rich media (such as casamino acids in which the tryptophan has been destroyed by acid hydrolysis) results in repression of the pL promoter on the plasmid by the presence of cI protein in the host. Addition of tryptophan to the culture results in rapid cessation of cI expression, and slower induction of the pL promoter as cI is gradually depleted by cell doubling and degradation. The choice of strain depends upon the desired temperature of protein production; lower temperatures require strains that produce lower levels of cI protein (e.g., GI698) for maximal pL induction, while higher temperatures require strains that produce higher levels of cI protein (e.g., GI723 or GI724) to maintain pL in a repressed state when in the uninduced condition (*see* **Note 9**).

3.2.2. Preparation of "Electrocompetent" Cells

We choose to use electroporation because of the high transformation efficiencies that are attainable with this method. Stocks of "electrocompetent" GI723, GI724, and GI698 can be prepared beforehand and stored in small (e.g., 100 µL) aliquots at –80°C (*see* **Note 10**).

Use a fresh colony to inoculate 50 mL of SOB medium in a 500-mL flask. Grow cells overnight with vigorous aeration at 37°C.

1. Dilute 0.5 mL of cells from this overnight culture into 500 mL SOB in a 2.8-L flask. Grow at 37°C with vigorous aeration until A_{550} = 0.8.
2. Harvest cells by centrifugation at 2600*g* in GSA (or equivalent) rotor for 10 min at 4°C. Pour off supernatant carefully to minimize loss of cells and discard.
3. Place cells on ice and gently resuspend cell pellet in 500 mL of ice-cold sterile 10% glycerol solution by swirling. *Keep cells cold at all times.* Centrifuge at

2600g for 15 min at 4°C. *Carefully* pour off supernatant as soon as rotor stops. Cells do not pellet well from this step onward, so centrifugation times may need to be increased if supernatants are turbid. Also, aspiration may be a better method than pouring to remove supernatants.
4. Place cells back on ice and gently resuspend cell pellet again in 500 mL of ice-cold sterile 10% glycerol solution by swirling. Centrifuge at 2600g for 15 min at 4°C and carefully pour off or aspirate supernatant.
5. Resuspend the cell pellet in ice-cold, sterile 10% glycerol to a final volume of 2 mL. Often, enough 10% glycerol remains in the centrifuge bottle after pouring off or aspirating the supernatant that no additional need be added.
6. Aliquot into 1.5-mL microfuge tubes pre-chilled in dry ice, 100 µL per aliquot. Immediately store at –80°C.

3.2.3. Transformation of Host Strains

1. Thaw one aliquot of frozen electrocompetent GI724 by removing the tube from the –80°C freezer and immediately placing it on wet ice.
2. Unwrap a sterile microelectroporation chamber (e.g., Bio-Rad Gene-Pulser® cuvet, 0.1-cm electrode gap) and place it on ice so that the electrodes are chilled.
3. Place 1 µL of ligation mixture (or 1 µL of purified vector plasmid, 1 ng or more) into a sterile 1.5-mL microfuge tube and place on ice.
4. Add 40 µL of thawed cells to the tube containing the ligation or plasmid solution, mix by pipetting and keep on ice.
5. Pipet the cells + DNA into the chilled electroporation cuvet, taking care to dispense the mix evenly between the two electrodes. Tap the cuvet if necessary to evenly spread the cell mixture between the electrodes.
6. Place the electroporation cuvet into an electroporation pulser unit (e.g., BioRad E. coli Pulser®) set to 1.8 kV.
7. Pulse, and then *immediately* add 1 mL of SOC medium directly to the electroporation cuvet. Suspend cells in the cuvet and transfer to a sterile 18-mm culture tube (*see* **Note 11**).
8. Incubate for 1 h at 37°C with agitation (*see* **Note 12**).
9. Plate the culture onto IMC plates containing 100 µg/mL ampicillin. Grow at 30°C (*see* **Note 13**) overnight in a convection incubator.
10. Pick candidate colonies and use to inoculate 5 mL of CAA/glycerol/ampicillin 100 medium.
11. Perform small scale plasmid DNA isolation on each culture, and evaluate correct gene insertion by restriction mapping. Alternatively, a colony PCR approach may be used to screen candidate colonies.
12. Verify that the in-frame fusion is correct by DNA sequencing across the junction.

3.3. Induction of Thioredoxin Gene Fusion Expression

1. Streak IMC plates containing 100 µg/mL ampicillin with a scraping from a frozen stock culture of GI724 containing the thioredoxin expression plasmid. Grow for 20 h at 30°C in a convection incubator.

2. Pick a single fresh, well-isolated, colony from the plate and use it to inoculate 5 mL IMC medium containing 100 mg/mL ampicillin in an 18 × 150-mm culture tube. Grow to saturation by incubating overnight at 30°C on a roller drum.
3. Add 0.5 mL of the overnight culture to 50 mL of fresh IMC medium containing 100 µg/mL ampicillin in a 250-mL culture flask (1:100 dilution). Grow at 30°C with vigorous aeration until A_{550} = 0.4–0.6 OD/mL (~3.5 h).
4. Remove a 1-mL aliquot of the culture (uninduced cells). Measure the absorbance at 550 nm and harvest the cells by microcentrifugation for 1 min at maximum speed, room temperature. Carefully remove all of the supernatant with a pipet and store the cell pellet at –80°C.
5. Induce expression from the pL promoter on the plasmid by adding 0.5 mL of 10 mg/mL tryptophan (100 µg/mL final concentration) to the remaining culture. Transfer the culture to 37°C.
6. Incubate the culture for 4 h at 37°C. At hourly intervals during this incubation, remove 1 mL aliquots of the culture and harvest the cells as in **step 4** above.
7. After 4 h, harvest the remaining cells from the culture by centrifugation for 10 min at 3500*g* (e.g., 4000 rpm in a Beckman J6 rotor), 4°C. Store the cell pellet at –80°C.

3.4. Characterization of Protein Production

Most thioredoxin fusion proteins are expressed well, usually at levels that vary from 5 to 20% of the total cell protein. When analyzing the expression of the desired fusion protein, the following characteristics should be noted: the fusion protein should migrate on the gel at the expected molecular weight; it should be absent prior to induction and gradually accumulate during induction; and maximum accumulation of fusion protein should occur at approximately 3 h post-induction at 37°C. This protocol evaluates total expression. Characterization of soluble expression will be performed in the next section.

1. Resuspend the cell pellets from the induction intervals in 200 µL of SDS-PAGE sample buffer per 1 OD_{550} cells. Heat the resuspended cells for 5 min at 70°C to completely lyse the cells and denature the proteins. Load 20 µL (= 0.1 OD_{550} cells) per lane on an SDS-polyacrylamide gel.
2. Soak the gel in Coomassie brilliant blue for 1 h. Destain the gel with DestainV and check the expression characteristics as described above (*see* **Note 14**).

3.5. Post-Induction Lysis and Protein Fractionation of E. coli Cells

3.5.1. E. coli Cell Lysis Using a French Pressure Cell

A small 3.5-mL French pressure cell can be used with a French hydraulic press (SLM Instruments, Inc.) to lyse *E. coli* cells. The whole-cell lysate then can be fractionated into soluble and insoluble fractions by microcentrifugation. Alternative methods include sonication, cell wall digestion using lysozyme, or mechanical disruption using a microfluidizer (Microfluidics, Newton, MA).

1. Resuspend the cell pellet from the 4-h post-induction culture in lysis buffer to a concentration of 10 OD_{550} per mL.
2. Place up to 1.5 mL of the resuspended cell pellet solution in the French pressure cell, and passage the lysate through the cell at 20,000 psi. Usually one passage is sufficient for total lysis.
3. Slowly open the outlet valve until lysate begins to trickle from the outlet. Take care to maintain pressure. The lysate should flow into the collection vessel slowly and smoothly.

3.5.2. Fractionation of Cell Lysate

This fractionation procedure is a fairly reliable indicator of whether a protein has folded correctly. Thioredoxin fusions that are found in the soluble fraction almost always have adopted a correct conformation and proteins in the insoluble fraction have not. This has *not* been our experience with other fusion systems, where solubility can sometimes be achieved but the resulting protein is not properly folded. It should be noted, however, that on rare occasions a thioredoxin fusion protein may be found in the soluble fraction using this protocol without being truly soluble; instead, it may have formed suspended aggregates that were not sedimented during the relatively low speed centrifugation. To test more stringently for solubility, the lysate can be clarified by ultracentrifugation at 35,000 rpm for 30 min in a Beckman Type 50 Ti (or equivalent) rotor (~80,000*g*). Conversely, infrequently we have also encountered proteins that fractionate to the insoluble fraction but are properly folded. In these instances, the protein may be associating with cell membrane components and cosedimenting. In such cases the protein may be recoverable from these insoluble fractions by first washing the cells with low levels of sarkosyl prior to lysing.

1. Remove a 100-µL aliquot of the whole-cell lysate and mix with 100 µL of 2X SDS-PAGE loading buffer. Heat at 70°C for 5 min.
2. Remove a second 100-µL aliquot of the lysate and centrifuge for 10 min at maximum speed in a microfuge at 4°C.
3. Remove the supernatant (soluble fraction) and mix with 100 µL of 2X SDS-PAGE loading buffer. Heat at 70°C for 5 min.
4. Resuspend the pellet (insoluble fraction) in 200 µL of 1X SDS-PAGE loading buffer. Heat at 70°C for 5 min.
5. After heating, allow the samples to cool to room temperature and load 20 µL of each fraction (whole cell lysate, soluble fraction, and insoluble fraction) onto an SDS-polyacrylamide gel (*see* **Note 15**).

3.6. Purification of Thioredoxin Fusion Proteins

3.6.1. Release of Thioredoxin Fusion Proteins by Osmotic Shock

An alternative to total cell lysis for the liberation of *some* thioredoxin fusion proteins from the cytoplasm of *E. coli* is osmotic shock. While well known as a method for releasing proteins from the periplasmic space, osmotic shock release of cytoplasmic proteins is an unusual attribute displayed by only a few proteins. Thioredoxin happens to be one such protein *(28)*, and some fusions to thioredoxin retain this characteristic. For those that do, osmotic shock release provides an advantage because the fusion will be greatly purified away from other cellular proteins by the procedure.

1. Resuspend the cell pellet from the 4-h post-induction culture in ice-cold osmotic shock buffer (containing 20% sucrose) to a density of 50 OD_{550}/mL.
2. Incubate the cell suspension on ice for 20 min.
3. Pellet the cells by centrifugation at 3000*g* for 10 min at 4°C.
4. Gently resuspend the cells in osmotic release buffer (no sucrose).
5. Incubate the cell suspension on ice for 20 min.
6. Re-pellet the cells by centrifugation at 3000*g* for 30 min at 4°C. The supernatant fraction contains proteins released from the periplasm, a small subset of cytoplasmic proteins (for instance, Ef-Tu), and the thioredoxin fusion protein.

3.6.2. Heat Treatment

Thioredoxin is a remarkably heat-stable protein *(29)* and retains its conformation and solubility at high temperature while most other *E. coli* proteins denature and precipitate. This feature allows thioredoxin to be fractionated from most other cellular proteins that do not share this characteristic. We have found that some proteins adopt this capability when tethered to thioredoxin, and for those that do, the following simple purification method can be adopted.

1. Resuspend the cell pellet from the 4-h post-induction culture in cell lysis buffer to a density of 100 OD_{550}/mL. This high density is critical to the success of the procedure because it maximizes the precipitation of heat-denatured contaminants.
2. Lyse the cells in a French pressure cell as described in **Subheading 3.5.1.**
3. Place 2 mL of the crude lysate into a thin-walled 13 × 100 mm borosilicate glass tube. Place the tube in an 80°C water bath.
4. Remove 100-μL aliquots after 80°C incubation of 30 s, 1 min, 2 min, 5 min, and 10 min. Place each aliquot into a separate, labeled glass tube and immediately immerse the tubes in ice water.
5. Transfer the chilled aliquots to microcentrifuge tubes and centrifuge at maximum speed for 10 min (~14,000*g*).
6. Analyze the insoluble and soluble fractions from each heating time point by loading the equivalent of 0.1 OD_{550} per lane on an SDS-polyacrylamide gel (*see* **Note 16**).

3.6.3. Purification of Thioredoxin Fusions with Metal Affinities

Histidine-patch thioredoxin fusions and DsbA fusions can be purified by virtue of their engineered metal-binding properties. Immobilized metal affinity chromatography (IMAC) resin *(22)*, specifically iminodiacetic acid (IDA), preloaded with nickel has performed well in our hands for this purpose.

1. Lyse the cells in a French pressure cell as described in **Subheading 3.5.1.** Alternatively, osmotic shockates of hp-Trx and DsbA fusions can be purified further on IMAC, but the shockate contains EDTA which must be removed completely (e.g., by dialysis) prior to IMAC purification.
2. Adjust the lysate to 500 m*M* NaCl and 4 m*M* imidazole and incubate on ice for 30 min.
3. Clarify the lysate by ultracentrifugation at 80,000*g* for 30 min as described in **Subheading 3.5.2.** to pellet cellular debris.
4. Load the clarified supernatant onto a Ni^{2+}-IDA column equilibrated in IDA equilibration buffer A. Wash the column with buffer A until the A_{280} of the column elution drops to background levels.
5. Elute the bound proteins by applying a linear gradient of elution buffer B. Fusion proteins typically elute between 30 and 60 m*M* imidazole.

3.6.4. Purification of BIOTRX Fusions

Proteins fused to the BIOTRX and expressed under the appropriate conditions *(7)* can be purified by virtue of the biotin that is covalently attached to the amino-terminal biotinylation sequence on each protein chain. Immobilized avidin (or streptavidin) matrices bind these proteins with high affinity and provide a highly efficient and highly specific purification. The drawback to this method is that due to the extremely high affinity of avidin for biotin, it is very difficult to elute the BIOTRX fusion from the avidin beads (*see* **Note 17**).

1. Prepare a clarified lysate of the BIOTRX fusion-expressing cells as described in **Subheading 3.5.2.**, or shockate as described in **Subheading 3.6.1.**
2. Pre-equilibrate avidin-agarose beads (Sigma) with cell lysis buffer containing 200 m*M* NaCl and 0.1% Triton X-100. Mix the beads with the clarified lysate or shockate and allow to bind by incubating for 1 h at 4°C. Mix the slurry gently during the binding period.
3. Centrifuge the slurry to pellet the beads with the bound BIOTRX fusion protein.
4. Remove the unbound fraction, and wash the beads 3X with lysis buffer containing 200 m*M* NaCl and 0.1% Triton X-100.
5. Bound BIOTRX fusion protein can be released for analysis by heating the washed beads at 70°C in SDS-PAGE buffer and loading onto an SDS-polyacrylamide gel.

3.7. Site-Specific Cleavage of Fusion Proteins

All of the thioredoxin fusion expression vectors described in this chapter (with the exception of the non-fusion vector pALtrxA-781) encode the protein sequence "-DDDDK-" in a spacer region between the amino-terminal thioredoxin (or DsbA) and the carboxyl-terminal fusion partner. This sequence is recognized by the mammalian intestinal serine protease enterokinase (EK; also known as enteropeptidase) which cleaves on the carboxyl-terminal side of the lysine in the recognition sequence ***(30)***. EK is effective at cleaving fusion proteins under a wide range of pH, temperature, and non-ionic detergents, so that reaction conditions can be tailored to the properties of the particular fusion protein. Adding to EK's flexibility as a cleavage reagent is the fact that it is very permissive to the nature of the amino acid residue in the P1' position, tolerating any residue except proline while retaining its ability to cleave (E. R. LaVallie and L. A. Collins-Racie, unpublished data). In the past, the only drawback to the use of EK as a universal fusion protein cleaving reagent was the fact that EK purified from duodena was contaminated with similar but non-specific serine proteases such as trypsin and chymotrypsin which degrade the fusion protein. We cloned and expressed a cDNA for bovine enterokinase light chain ***(31)*** which exhibits superior fusion protein cleavage properties compared to the native enzyme, and this recombinant EK is commercially available from numerous molecular biology supply companies.

In the protocol below, the optimum ratio of EK to fusion protein is determined empirically by titrating the amount of enzyme in several pilot digestions. Different fusion proteins cleave with different efficiencies, with EK:fusion protein ratios ranging from 1:5000 to 1:100,000 (w/w) for a 16 h digestion at 37°C.

1. Prepare purified fusion protein (*see* **Note 18**) ≥ 1 mg/mL in 50 m*M* Tris-HCl, pH 8.0, 1 m*M* $CaCl_2$.
2. In separate tubes, prepare cleavage reactions of 20 µg of fusion protein with 0.1, 0.5, 1, 2, 5, and 10 U of EK. Adjust the total volume of each digest to 30 µL by adding the requisite amount of 50 m*M* Tris-HCl, pH 8.0, 1 m*M* $CaCl_2$.
3. Incubate the pilot digestions at 37°C for 16 h.
4. Add 30 µL of 2X SDS-PAGE buffer and heat to 70°C for 10 min.
5. Load 10 µL of each reaction onto an SDS-polyacrylamide gel alongside protein molecular weight markers, run the gel, stain, and destain. Assess the cleavage efficiency of each reaction by monitoring the conversion of the full-length fusion protein band to two faster migrating bands upon cleavage with increasing amounts of EK—a band of ~14 kDa corresponding to the Trx/spacer (or ~23 kDa for mature DsbA/spacer) and a second band corresponding to the M_r of the fusion protein partner. Choose the reaction conditions that result in 100% cleavage with the least amount of enzyme and scale up the reaction linearly (*see* **Note 19**).

4. Notes

1. Prerinsing the filter unit ensures that any substances that may leach off of the filter and reduce electroporation efficiency have been removed.
2. It is very important to use Difco Bacto-Casamino Acids (cat. no. 0230-17-3) with low sodium chloride. Other grades of casamino acids may contain higher levels of sodium chloride, which will be detrimental to fusion protein expression levels.
3. Addition of $CaCl_2$ will result in the formation of a visible precipitate as the calcium combines with the phosphate in the M9 salts to form calcium phosphate. This is normal.

4. Tryptophan is very difficult to dissolve. We find that the easiest way to prepare the 10 mg/mL solution is to heat the water to 50–60°C prior to adding the tryptophan powder, stir until dissolved, then allow it to cool before filter sterilization. Alternatively, the solution can be autoclaved. This solubilizes and sterilizes in one step.
5. Purchase only *recombinant* EK light chain enzyme. Even highly purified preparations of non-recombinant native EK contain contaminating proteolytic activities that will cause undesirable proteolysis of the fusion protein. Also, recombinant EK consists of only the catalytic subunit of the enzyme, which has superior cleaving characteristics on fusion proteins (*see* **ref. 5**; E. R. LaVallie, unpublished results).
6. Biotinylated fusion proteins produced by pBIOTRXFUS-BirA have extremely high affinity for avidin or streptavidin matrices. This affinity can be used as a purification method (*see* **Subheading 3.6.4.**), but in our experience it is very difficult to elute the bound fusion proteins from the matrix under non-denaturing conditions. This problem can be averted by using a matrix with less affinity for biotin; for instance, monomeric avidin (e.g., Soft-Link™, Promega; K_d for biotin is 10^{-7} *M* vs tetrameric avidin's K_d for biotin of 10^{-15} *M*) binds biotinylated BIOTRX fusion proteins strongly enough for purification purposes, but weakly enough that mild elution conditions can release the protein from the matrix *(7)*.
7. We have developed a version of the DsbA fusion vector, called pDsbAmatFUS, in which the DsbA signal peptide has been removed. Protein fusions to "mature" DsbA are localized to the cytoplasm and appear to be as good or better than thioredoxin in terms of producing soluble protein, at least for the limited number of genes tested (E. R. LaVallie, unpublished data).
8. When designing carboxyl-terminal thioredoxin fusions it is important to note that the cleavage specificity of enteropeptidase is --D-D-D-D-K\X--, and that enteropeptidase can cleave any K-X bond *except* when X is a proline.
9. The choice of *E. coli* host strain is dictated by the desired temperature of growth and fusion protein production because there is an apparent effect of temperature on pL in this system; it is unclear if this occurs at the level of cI repressor binding or further "upstream" with the tryptophan repressor/operator interaction which governs cI transcription. In any event, the net result is that at lower temperatures it is difficult to achieve full derepression of the pL promoter, so strains that make low levels of cI (such as GI698) are required for optimum expression levels at temperatures below 30°C . But strains which produce less cI repressor are "leaky" at temperatures above 30°C; that is, the pL promoter is partially derepressed even in the absence of tryptophan. This results in plasmid instability and loss of cell viability because pL is a very strong promoter which subjugates other cellular transcription. So at higher temperatures, strains that produce more cI (such as GI723 or GI724) are required. **Table 1** should be consulted to aid in the selection of the proper strain for the desired growth and induction temperature.
10. The strains are grown in a tryptophan-containing media (SOC) during the preparation of electrocompetent cells. This is acceptable because the strains do not harbor a pL-containing plasmid at this stage. However, once the cells are transformed with a pL-containing expression vector, it is imperative that they be grown on tryptophan-free media (e.g., casamino acids) to prevent loss of cellular viabil-

ity and/or selection for plasmid mutations. We find it easiest to create vector constructions in GI724 because transformants can be selected at 30°C and the correct construct will be in a host strain that can immediately be used to produce protein at 37°C and at 30°C. The verified constructs can then be quickly moved into the other expression strains by electroporation for expression at other temperatures.

11. Two important factors that contribute to the overall efficiency of electroporation are (1) the ionic strength of the DNA sample and (2) the length of time between pulsing and resuspending the cells in SOC. With regard to (1), no more than 1 µL of ligation reaction should be added to the electroporation cuvet to minimize the ionic strength of the solution; more than 1 µL will decrease the transformation efficiency and increase the risk of arcing. If it becomes necessary to transform more of the ligation to generate a sufficient number of transformants, then the ligation reaction should be extracted with phenol/chloroform and precipitated with 2 vol of ethanol after adjusting the solution to 3.5 *M* ammonium acetate and adding 10 µg of yeast tRNA as carrier. The precipitate is then resuspended in 2 µL of deionized H_2O and added to the electrocompetent cells for electroporation. Factor (2) refers to fact that cell viability (and resulting transformation efficiency) drops dramatically with a delay in adding the SOC medium to the cells after the pulse *(32)*.
12. Technically, this step contradicts the instruction in **Note 10**. However, we have found that this step, which allows time for β-lactamase expression before plating on ampicillin, does not result in pL induction. This is presumably because the plasmid copy number is very low at this stage, and endogenous levels of cI are probably adequate to keep the pL promoter repressed until plating.
13. GI724 transformed with a pL expression plasmid should be cultured at *no higher* than 30°C until induction is desired. Even lower pre-induction temperatures (which keep pL more tightly repressed) are acceptable and sometimes desirable if the fusion protein product is toxic to the cells.
14. This procedure should produce maximal levels of fusion protein, without regard for protein solubility. If the protein is totally or partially insoluble (as determined in **Subheading 3.5.2.**), then induction at lower temperature should be attempted (*see* **Note 15**).
15. The proportion of fusion protein that fractionates to the soluble fraction can be readily assessed by this procedure. If the protein is not soluble, then a number of additional experiments can be performed. The approach most likely to succeed in increasing the level of soluble expression is to decrease the temperature of expression to as low as 15°C *(4,33)*. This will require moving the expression construct into GI698 (*see* **Table 1**). If this is unsuccessful, extractions of the insoluble fraction with salt or prewashing the cells with a buffer containing a small amount of sarkosyl prior to cell lysis may help to convert the protein to the soluble fraction (E. DiBlasio-Smith and J. M. McCoy, unpublished data).
16. This protocol is designed for analytical scale preparation of heated lysates. Larger, preparative scale heat denaturation can be accomplished, but the larger volumes will require that even greater care must be devoted to ensuring good heat transfer and mixing during the heating and cooling steps.

17. As an alternative to eluting carboxyl-terminal BIOTRX fusion proteins from the avidin matrix, digestion with EK can be performed on the protein while it remains bound. The fusion protein:avidin agarose complex can be equilibrated in EK digestion buffer and incubated with EK. Upon cleavage the carboxyl-terminal protein should be released into the unbound fraction where it can be easily recovered.
18. EK is inactive in crude *E. coli* lysates. Fusion proteins must be at least partially purified prior to EK digestion. However, EK is sensitive to ionic strength, and its proteolytic activity decreases rapidly with increasing salt concentration. Fusion proteins purified on IMAC must be dialyzed against EK cleavage buffer to remove imidazole prior to EK digestion.
19. Occasionally, varying degrees of secondary cleavages are seen with some protein substrates. Typically these occur at subsites that resemble an EK site (a basic amino acid residue with some acidic residues on its amino-terminal side). Some approaches to minimize this unwanted proteolysis are:
    a. Lower the reaction temperature. EK will cleave partially denatured substrates with less specificity. Lower temperatures may stabilize the tertiary structure of the substrate and decrease secondary cleavages.
    b. Eliminate Ca^{2+} from the reaction buffer. We have seen that secondary cleavages can sometimes be alleviated by removing the $CaCl_2$ from the reaction, or even by adding Na_2EDTA (5 m*M* final concentration) to the digest.
    c. Decrease the amount of EK enzyme in the digest. Longer incubations with less enzyme sometimes help decrease secondary cleavages. EK is very stable, and retains >90% of its activity after 16 h at 37°C (unpublished observation).

References

1. Stormo, G. D., Schneider, T. D., and Gold, L. (1982) Characterization of translation initiation sites in *E. coli*. *Nucleic Acids Res.* **10,** 2971–2996.
2. Hirel, P. H., Schmitter, M. J., Dessen, P., Fayat, G., and Blanquet, S. (1989) Extent of N-terminal methionine excision from *Escherichia coli* proteins is governed by the side-chain length of the penultimate amino acid. *Proc. Natl. Acad. Sci. USA* **86,** 8247–8251.
3. Mitraki, A. and King, J. (1989) Protein folding intermediates and inclusion body formation. *Bio/Technology* **7,** 690–697.
4. LaVallie, E. R., DiBlasio, E. A., Kovacic, S., Grant, K. L., Schendel, P. F., and McCoy, J. M. (1993) A thioredoxin gene fusion expression system that circumvents inclusion body formation in the *E. coli* cytoplasm. *Bio/Technology* **11,** 187–193.
5. Collins-Racie, L. A., McColgan, J. M., Grant, K. L., DiBlasio-Smith, E. A., McCoy, J. M., and LaVallie, E. R. (1995) Production of recombinant bovine enterokinase catalytic subunit in *Escherichia coli* using the novel secretory fusion partner DsbA. *Bio/Technology* **13,** 982–987.
6. Lu, Z., DiBlasio-Smith, E. A., Grant, K. L., et al. (1996) "Histidine-patch" thioredoxins: Mutant forms of thioredoxin with metal chelating affinity that provide for convenient purifications of thioredoxin fusion proteins. *J. Biol. Chem.* **271,** 5059–5065.

7. Smith, P. A., Tripp, B. C., DiBlasio-Smith, E. A., Lu, Z., LaVallie, E. R., and McCoy, J. M. (1998) A plasmid expression system for quantitative *in vivo* biotinylation of thioredoxin fusion proteins in *Escherichia coli. Nucleic Acids Res.* **26,** 1414–1420.
8. Lunn, C. A., Kathju, S., Wallace, B. J., Kushner, S. R., and Pigiet, V. (1984) Amplification and purification of plasmid-encoded thioredoxin from *Escherichia coli* K12. *J. Biol. Chem.* **259,** 10,469–10,474.
9. Katti, S. K., LeMaster, D. M., and Eklund, H. (1990) Crystal structure of thioredoxin from *Escherichia coli* at 1.68 angstroms resolution. *J. Mol. Biol.* **212,** 167–184.
10. Bardwell, J. C. A., McGovern, K., and Beckwith, J. (1991) Identification of a protein required for disulfide bond formation in vivo. *Cell* **67,** 581–589.
11. Mazzarella, R. A., Srinivasan, M., Haugejorden, S. M., and Green, M. (1990) ERp72, an abundant luminal endoplasmic reticulum protein, contains three copies of the active site sequences of protein disulfide isomerase. *J. Biol. Chem.* **265,** 1094–1101.
12. LaVallie, E. R., Lu, Z., DiBlasio-Smith, E. A., Collins-Racie, L. A., and McCoy, J. M. (2000) Thioredoxin as a fusion partner for soluble recombinant protein production in *Escherichia coli. Methods Enzymol.* **326,** 322–340.
13. Silen, J. L., Frank, D., Fujishige, A., Bone, R., and Agard, D. A. (1989) Analysis of prepro-α-lytic protease expression in *Escherichia coli* reveals that the pro region is required for activity. *J. Bacteriol.* **171,** 1320–1325.
14. Shinde, U., Chatterjee, S., and Inouye, M. (1993) Folding pathway mediated by an intramolecular chaperone. *Proc. Natl. Acad. Sci. USA* **90,** 6924–6928.
15. Norrander, J., Kempe, T., and Messing, J. (1983) Construction of improved M13 vectors using oligonucleotide-directed mutagenesis. *Gene* **26,** 101–106.
16. Shimatake, H. and Rosenberg, M. (1981) Purified λ regulatory protein cII positively activates promoters for lysogenic development. *Nature* **292,** 128–132.
17. Colas, P., Cohen, B., Jessen, T., Grishina, I., McCoy, J., and Brent, R. (1996) Genetic selection of peptide aptamers that recognize and inhibit cyclin-dependent kinase 2. *Nature* **380,** 548–550.
18. Lu, Z., Murray, K. S., Van Cleave, V., LaVallie, E. R., Stahl, M. L., and McCoy, J. M. (1995) Expression of thioredoxin random peptide libraries on the *Escherichia coli* cell surface as functional fusions to flagellin: a system designed for exploring protein-protein interactions. *Bio/Technology* **13,** 366–372.
19. Tripp, B. C., Lu, Z., Bourque, K., Sookdeo, H., and McCoy, J. M. (2001) Investigation of the 'switch-epitope' concept with random peptide libraries displayed as thioredoxin loop fusions. *Protein Eng.* **14,** 367–377.
20. Smith, D. B. and Johnson, K. S. (1988) Single-step purification of polypeptides expressed in *Escherichia coli* as fusions with glutathione S-transferase. *Gene* **67,** 31–40.
21. di Guan, C., Li, P., Riggs, P. D., and Inouye, H. (1988) Vectors that facilitate the expression and purification of foreign peptides in *Escherichia coli* by fusion to maltose-binding protein. *Gene* **67,** 21–30.
22. Porath, J. (1992) Immobilized metal ion affinity chromatography. *Protein Expr. Purif.* **3,** 263–281.

23. Schatz, P. J. (1993) Use of peptide libraries to map the substrate specificity of a peptide-modifying enzyme: a 13 residue consensus peptide specifies biotinylation in *Escherichia coli. Bio/Technology* **11,** 1138–1143.
24. Hoffman, C. S. and Wright, A. (1985) Fusions of secreted proteins to alkaline phosphatase: an approach for studying protein secretion. *Proc. Natl. Acad. Sci. USA* **82,** 5107–5111.
25. Sambrook, J., Fritsch, E. F., and Maniatis, T. (1989) *Molecular cloning. A laboratory manual.* Cold Spring Harbor Laboratory , Cold Spring Harbor, NY.
26. Brent, R. and Ptashne, M. (1981) Mechanism of action of the lexA gene product. *Proc. Natl. Acad. Sci. USA* **78,** 4204–4208.
27. Mieschendahl, M., Petri, T., and Hanggi, U. (1986) A novel prophage independent trp regulated lambda pL expression system. *Bio/Technology* **4,** 802–808.
28. Lunn, C. A. and Pigiet, V. P. (1982) Localization of thioredoxin from *Escherichia coli* in an osmotically sensitive compartment. *J. Biol. Chem.* **257,** 11,424–11,430.
29. Holmgren, A. (1985) Thioredoxin. *Ann. Rev. Biochem.* **54,** 237–271.
30. Maroux, S., Baratti, J., and Desnuelle, P. (1971) Purification and specificity of porcine enterokinase. *J. Biol. Chem.* **246,** 5031–5039.
31. LaVallie, E. R., Rehemtulla, A., Racie, L. A., et al. (1993) Cloning and functional expression of a cDNA encoding the catalytic subunit of bovine enterokinase. *J. Biol. Chem.* **268,** 23,311–23,317.
32. Dower, W. J., Miller, J. F., and Ragsdale, C. W. (1988) High efficiency transformation of *E. coli* by high voltage electroporation. *Nucl. Acids Res.* **16,** 6127–6145.
33. Schein, C. H. and Noteborn, M. H. M. (1988) Formation of soluble recombinant proteins in *Escherichia coli* is favored by lower growth temperature. *Bio/Technology* **6,** 291–294.

9

Discovery of New Fusion Protein Systems Designed to Enhance Solubility in *E. coli*

Gregory D. Davis and Roger G. Harrison

1. Introduction

Fusion protein technology has been creatively applied to solve many problems encountered in the study of protein structure and function *(1,2)*. One prevalent application is the use of fusion proteins to improve protein expression in *Escherichia coli* and provide convenient methods for affinity-based protein purification. In many instances, when a foreign protein is overexpressed at high levels in *E. coli*, the majority of the protein is present as insoluble inclusion bodies. Previous research experience with fusion proteins containing glutathione *S*-transferase (GST) *(3)*, maltose binding protein (MBP) *(4)*, and thioredoxin *(5)* has shown that these proteins can improve the expression and solubility of many heterologous proteins. However, improvements in protein expression and solubility are not always guaranteed with these systems. The purpose of our recent research *(6)* has been to use a combination of protein solubility modeling, bioinformatics, and molecular biology techniques to systematically identify native *E. coli* proteins which have maximal potential for increasing recombinant fusion protein solubility.

In this chapter, we present the convenient internet-based use of a solubility model used to identify NusA and other proteins as solubilizing components of fusion proteins *(6)*. The solubility model predicts the solubility of a recombinant *E. coli* protein based on the number of turn-forming residues and on the net charge of the protein relative to the total number of residues in the protein *(7)*. The description and utility of the solubility model is intended to be as simple and straightforward as possible to help other life science researchers design more effective fusion protein systems for soluble protein expression.

From: *Methods in Molecular Biology, vol. 205, E. coli Gene Expression Protocols*
Edited by: P. E. Vaillancourt © Humana Press Inc., Totowa, NJ

In addition, we describe our cloning, expression, and cell fractionation methods for rapidly creating fusion proteins and evaluating their solubility. The three-fragment directional cloning protocol for creating binary gene fusions involves the use of PCR to add restriction sites and, if desired, DNA sequences coding for protease cleavage sites. After ligation of the DNA fragments, competent cells are transformed, and colonies are screened by one or more methods. Once clones are identified that produce a new protein of the size of the desired fusion protein (and also possibly that react with an antibody specific for the target protein), the solubility of the fusion protein can be evaluated by sodium dodecyl sulfate-polyacrylamide gel electrophoresis (SDS-PAGE) and/or Western blot procedures, which show the proteins in the same relative amounts for the soluble and insoluble fractions as they are in the cell.

2. Materials

2.1. Computer Requirements for Use of Solubility Model

The only software required for evaluating proteins using the solubility model is a web-based internet browser. The www.biotech.ou.edu web site has been constructed and implemented primarily with Netscape Communicator 4.7. However, there are no unique requirements, and any other internet browser should be compatible.

2.2. Construction of Fusion Protein Expression Vectors

1. The expression vector pKK223-3 (Amersham Phamacia Biotech, Piscataway, NJ), with the strong *tac* promoter, is used in this example. Other vectors with strong promoters that do not rely on raising the temperature can be used as well.
2. Plasmids or other DNA templates encoding the two proteins to be fused by subcloning.
3. Relevant restriction enzymes, T4 DNA ligase, and incubation buffers. Reagents used here were from New England BioLabs (Beverly, MA).
4. A high fidelity PCR system. Expand High Fidelity polymerase from Boehringer Mannheim (now Roche Molecular Biochemicals, Indianapolis, IN) was used in this example.
5. 1 *M* Ammonium acetate in ethanol.
6. 80% Ethanol.
7. Reagents or kits for purification of restricted DNA fragments by either agarose gel purification or gel filtration spin columns.
8. Ligation buffer: 50 m*M* Tris-HCl pH 7.5, 10 m*M* $MgCl_2$, 10 m*M* DTT, 1 m*M* ATP, and 25 µg/mL bovine serum albumin.
9. *n*-Butanol and isopropanol.
10. Electrocompetent *E. coli* JM105 cells.

2.3. Screening Recombinants

1. Luria broth (LB) with 100 µg/mL ampicillin.
2. Glycerol.
3. Isopropylthiogalactoside (IPTG): 1 *M* stock for inductions, stored at –20°C.
4. 50 m*M* Tris-HCl, pH 7.0, 10% SDS.
5. Nitrocellulose membrane.
6. Western blotting reagents.

2.4. Evaluating Protein Solubility by Cell-Lysate Fractionation

1. LB with 100 µg/mL ampicillin.
2. 13 100-mm glass test tubes for small scale cultures.
3. IPTG (1 *M*).
4. 50 m*M* NaCl, 1 m*M* EDTA, pH 8.0.

3. Methods

3.1. Protein Solubility Prediction and Fusion Protein Design

This section outlines how the Wilkinson-Harrison solubility model can be used to estimate the solubility of a particular protein or fusion protein when expressed in *E. coli*. The web site www.biotech.ou.edu has recently been constructed for researchers interested in directly using the solubility model with minimal "number crunching" effort. All that is required is to cut and paste a protein sequence into an input window, and the solubility probability of the protein will be calculated automatically. For researchers interested in the detailed theoretical and statistical basis for the model, please refer to previous literature *(6,7)*. A brief description follows.

The solubility model was developed in an effort to understand what causes proteins to form inclusion bodies when overexpressed in *E. coli*. The two factors found to be most directly associated with inclusion body formation were the net protein charge and the number of turn-forming residues in the primary amino acid sequence. Roughly stated, the model says that a highly charged protein (positive or negative) with very few turn residues (glycine, proline, serine, and asparagine) has a good chance of being soluble in *E. coli*. This description can be quantified as a "solubility or insolubility probability" by knowing only the amino acid composition of the protein. The two equations that constitute the model are given in the footnote to **Table 1**. Since the amino acid composition is the only input requirement, the model lends itself to be applied directly to the vast amount of DNA sequence data currently being generated from genome sequencing projects. Primary amino acid sequence data contained in the SwissProt (www.expasy.ch) and NCBI Entrez (www.ncbi.nlm.nih.gov) databases are convenient sources of sequence data for input into the www.biotech.ou.edu web site.

Table 1
***E. coli* Proteins of 100 Amino Acids in Length or Greater in the SWISS-PROT Protein Databank (www.expasy.ch) Predicted by the Wilkinson-Harrison Solubility Model to Have a Solubility Probability of 90% or Greater When Expressed in the *E. coli* Cytoplasm**[a]

Amino Acid length	SWISS-PROT sequence ID	Solubility probability (%)	Amino Acid length	SWISS-PROT sequence ID	Solubility probability (%)
613	RPSD_ECOLI	94	157	SMG_ECOLI	>98
497	FTSY_ECOLI	97	156	HYCI_ECOLI	94
495	AMY2_ECOLI	91	155	SECB_ECOLI	91
495	NUSA_ECOLI	95	155	YBEY_ECOLI	>98
294	YRFI_ECOLI	95	153	ELAA_ECOLI	93
263	MAZG_ECOLI	97	152	YFJX_ECOLI	93
263	S3AD_ECOLI	93	146	MIOC_ECOLI	91
261	SSEB_ECOLI	92	138	YJGD_ECOLI	>98
252	YCHA_ECOLI	93	137	HYFJ_ECOLI	94
243	YAGJ_ECOLI	92	136	RL16_ECOLI	93
238	YFBN_ECOLI	95	135	RS6_ECOLI	97
236	NARJ_ECOLI	93	133	YHHG_ECOLI	91
231	NARW_ECOLI	93	129	GCSH_ECOLI	>98
221	YECA_ECOLI	94	129	TRD5_ECOLI	>98
214	CHEZ_ECOLI	94	124	MSYB_ECOLI	>98
197	GRPE_ECOLI	93	123	RS12_ECOLI	91
196	SLYD_ECOLI	97	120	RL7_ECOLI	93
196	YJAG_ECOLI	98	120	YACL_ECOLI	>98
195	YIEJ_ECOLI	96	120	YBFG_ECOLI	97
194	YGFB_ECOLI	98	117	RL20_ECOLI	>98
191	YJDC_ECOLI	90	116	HYPA_ECOLI	92
184	YCDY_ECOLI	97	116	PTCA_ECOLI	95
177	AADB_ECOLI	95	115	YZPK_ECOLI	96
175	FLAV_ECOLI	>98	113	HYBF_ECOLI	96
173	FLAW_ECOLI	98	110	FER_ECOLI	98
173	YCED_ECOLI	94	110	YR7J_ECOLI	92
171	YFHE_ECOLI	95	108	GLPE_ECOLI	93
169	ASR_ECOLI	95	108	YGGL_ECOLI	91
169	YGGD_ECOLI	96	106	CYAY_ECOLI	>98
167	YHBS_ECOLI	93	105	YEHK_ECOLI	94
165	FTN_ECOLI	91	105	YR7G_ECOLI	>98
161	MENG_ECOLI	93	103	YQFB_ECOLI	96
160	YBEL_ECOLI	90	101	YCCD_ECOLI	94
158	BFR_ECOLI	95	100	RS14_ECOLI	91

3.1.1. Estimating the Solubility Probability of a Single Protein

For example, to calculate the solubility probability of the *E. coli* NusA protein, first go to the NCBI Entrez web site (www.ncbi.nlm.nih.gov). Follow the web page links to the Entrez section, and then to the Protein section. In the search field type in "P03003" which is the accession number for the NusA protein. Open the retrieved sequence, select the entire NusA amino acid sequence with the cursor, and copy it (this command is usually under "Edit" menu of most web browsers) (*see* **Note 1**). Now go to the File menu of your browser, open an additional browser window (*see* **Note 2**), and direct it to the www.biotech.ou.edu web site. Paste the copied NusA sequence into the input window and click on "Submit Query". The result will show that NusA has a 95% chance of solubility in *E. coli*. The output of the model is formatted so that the range of solubility extends from "greater than 98% chance of solubility" to "greater than 98% chance of insolubility". The solubility transition occurs between "50% chance of solubility" and "50% chance of insolubility". When the solubility probability is somewhere between 60% soluble and 60% insoluble, the predictions of the model are less reliable. However, predictions of 70–100% soluble or insoluble are more likely to be accurate.

The www.biotech.ou.edu web site is convenient for calculating solubilities of individual proteins of interest. However, for discovery purposes, the solubility

---

[a]Any of these proteins is likely have a high potential for creating soluble fusion proteins. NusA (495 aa), GrpE (197 aa), and BFR (158 aa) have been shown to significantly improve the solubility of human interleukin-3 as fusion proteins *(6)*; however, the remaining proteins have yet to be characterized as soluble fusion partners. The upper limit of quantification by the model is a 98% chance of solubility.

The revised Wilkinson-Harrison solubility model involves calculating a canonical variable (CV) or composite parameter for the protein for which the solubility is being predicted. The canonical variable in the two-parameter model is defined as:

$$CV = \lambda_1 \, [(N + G + P + S)/n] + \lambda_2 \, |[\{(R + K) - (D + E)\}/n] - 0.03|$$

where:

n = number of amino acids in protein
N,G,P,S = number of Asn, Gly, Pro, or Ser residues, respectively.
R,K,D,E = number of Arg, Lys, Asp, or Gln residues, respectively.
λ_1, λ_2 = coefficients (15.43 and –29.56, respectively)

The probability of the protein solubility is based on the parameter CV – CV', where CV' is the discriminant equal to 1.71. If CV – CV' is positive, the protein is predicted to be insoluble, while if CV – CV' is negative, the protein is predicted to be soluble. The probability of solubility or insolubility can be predicted from the following equation:

$$\text{Probability of solubility or insolubility} = 0.4934 + 0.276|CV - CV'| - 0.0392(CV - CV')^2$$

model was incorporated into a Unix-based bioinformatics program which scanned through all the available protein sequences in the *E. coli* genome, looking for the most soluble proteins. **Table 1** shows the solubility probability calculated for several of the most soluble *E. coli* proteins. This list has been used to select and evaluate the NusA, GrpE, and BFR proteins, all of which have shown favorable solubilizing characteristics as fusions with human interleukin-3 ***(6)***. It should be emphasized that this list has not been extensively evaluated by cloning and expression experiments, and might contain other proteins which possess excellent expression and solubility characteristics when used as fusion partners.

The NusA protein has been further evaluated beyond its ability to solubilize human interleukin-3 and shown to solubilize human interferon-γ, bovine growth hormone, and tyrosinase ***(6,8)***. Because of these consistently favorable solubility characteristics, the NusA protein has been incorporated into the pET vector system (Novagen Inc., Madison, WI) as the pET-43 and pET-44 series of vectors. These vectors provide a multiple cloning site downstream of the NusA gene which is driven by the strong T7 promoter. The vectors also provide various affinity purification tags and protease cleavage sites to facilitate protein purification.

3.1.2. Estimating the Solubility Probability of Fusion Proteins

The previous section shows how to estimate the solubility of a particular protein of interest. If the protein is predicted to be insoluble (or if experimental data has already demonstrated insolubility), the model can be used to estimate the solubility benefits provided by potential fusion partners. Once the protein of interest has been copied and pasted into the input window of the www.biotech.ou.edu web site, simply copy and paste the sequence of another protein directly below it (NusA or one of the sequences from **Table 1**, for instance). The solubility of the fusion protein will then be calculated. If the solubility probability is increased to 70% soluble or above, then it is probably worth evaluating the fusion protein experimentally by cloning and overexpression studies, as outlined in the next section.

3.2. Construction of Vectors for Fusion Protein Expression

This section describes a rapid three-fragment cloning method which is widely applicable to the combination of any carrier gene (coding for a solubilizing protein), target gene (coding for the protein-of-interest), and an expression vector backbone. This protocol is designed to provide flexibility for creating various fusion protein combinations with NusA, GrpE, BFR, or other proteins listed in **Table 1**.

1. The general cloning scheme for this protocol is shown in **Fig. 1**, using the NusA/hIL-3 fusion protein as an example. Choose the restriction sites to use for ligating

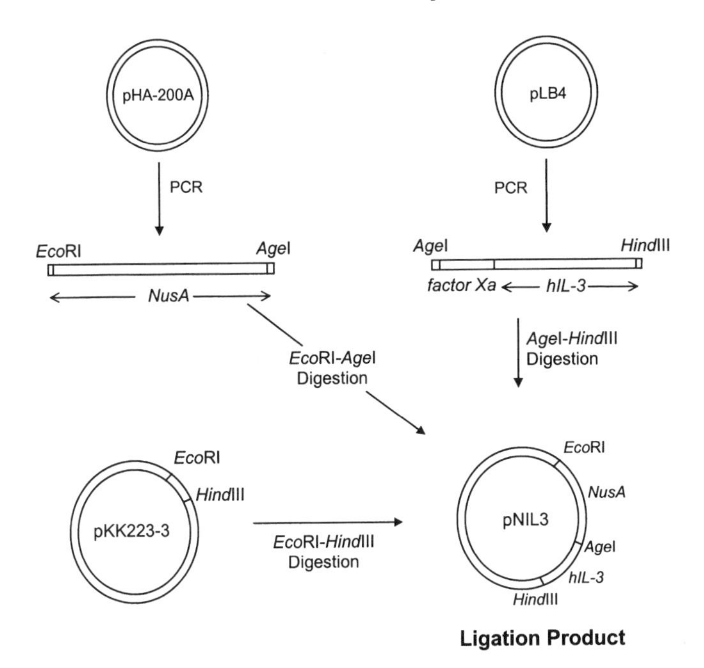

Fig. 1.Three-fragment subcloning scheme for creating binary gene fusions. First, the genes are amplified by PCR from their parental plasmids using primers that insert *Eco*RI, *Age*I, and *Hin*dIII sites. The DNA fragments are then digested with the appropriate restriction enzymes and ligated together in a single reaction with plasmid pKK223-3.

DNA fragments together. Convenient sites in the pKK223-3 multiple cloning site are *Eco*RI and *Hin*dIII. Next, choose the restriction site for joining the ends of the carrier and target genes. The *Age*I restriction enzyme recognizes a very unique sequence not commonly found in most genes. Check to be sure that any chosen restriction sites are not present internally in the gene sequences.

2. Generate linear DNA fragments of the carrier gene and the target gene by PCR (typically a 100 μL reaction per gene) using primers which incorporate restriction sites with sticky 5' and 3' ends. Be sure that the stop codon of the carrier gene (NusA in this example) is removed by primer design to permit translation of the target gene. Also, the 5'-end of each primer should contain about eight nonspecific bases to ensure the efficiency of restriction enzyme cleavage near the ends of the PCR products. For example, the primers used to create the NusA/hIL-3 fusion gene were:

   NusA forward primer (*EcoR*I - Start):
   5'-CGTTAGCC<u>GAATTC</u>**ATG**AACAAAGAAATTTTGGC

   NusA reverse primer (*Age*I):
   5'-CGCGCATT<u>ACCGGT</u>CGCTTCGTCACCGAACCAGC

hIL-3 forward primer (*Age*I - Factor Xa):
5'-CGCGCATTACCGGT**ATCGAAGGTCGA**GCTCCCATGACCCAGACAACG

hIL-3 reverse primer (*Hin*dIII - Stop):
5'-CGATTCGCAAGCTT**TCA**AAAGATCGCGAGGCTCAAAG

3. Purify the PCR products by ethanol/ammonium acetate precipitation. Add 3 vol of 1 *M* ammonium acetate (in 100% ethanol) to the finished PCR reaction. Incubate in the –80°C freezer for 10 min, and then pellet the DNA by spinning the tube for 15 min in a microfuge at maximum speed. Remove the supernatant with a micropipet. Wash the DNA in 200 µL of 80% ethanol. Centrifuge at maximum speed for 2 min. Remove as much of the supernatant as possible with a micropipet. Invert the tube and air dry for 10 min. Be sure that no ethanol droplets remain before proceeding. Dry for a longer period or under vacuum if needed. Resuspend the DNA in the desired volume of dH_2O or TRIS-EDTA (TE) buffer, pH 8.0.
4. Cut 500 ng of the pKK223-3 vector simultaneously with 20 U of *Eco*RI and 20 U of *Hin*dIII in a 20 µL reaction using New England Biolabs (NEB) buffer #2 for 2 h at 37°C. Cut 500 ng of the carrier gene (NusA) simultaneously with 20 U of *Eco*RI and 2 U of *Age*I in a 20 µL reaction using NEB buffer #1 for 2 h at 37°C. Cut 500 ng of the target gene (hIL-3 in this case) simultaneously with 2 U of *Age*I and 20 U of *Hin*dIII in a 20 µL reaction using NEB buffer #2 for 2 h at 37°C.
5. Purify the digested DNA PCR products and linearized pKK223-3 by agarose gel purification or by gel filtration using spin columns (*see* **Note 3**).
6. Combine roughly 100 ng of linearized pKK223-3, 100 ng of carrier gene, and 100 ng of target gene together in a 20 µL volume with ligation buffer. *Prior to adding the ligase*, heat the combined DNA fragments at 65°C for 2 min (*see* **Note 4**). Cool the mixture on ice for 2 min and add 1 U of T4 DNA ligase. Incubate the ligation at 15°C for 16 h (*see* **Note 5**).
7. The ligation reaction must be de-salted prior to electroporation. For de-salting, dilute the 20 µL ligation reaction to 50 µL with dH_20, add 500 µL of *n*-butanol, and vortex for 5 s. Centrifuge at maximum speed (~10,000*g*) for 10 min. Pipet off the supernatant and dry the pellet (dry under vacuum or invert the tube under an air flow for 10 min). Add 50 µL of dH_20 and vortex for 30 s to redissolve the pellet (note: it is likely that the pellet is invisible at this point). Add 50 µL of isopropanol and vortex to mix. Incubate for 15 min at room temperature and then centrifuge for 15 min at maximum speed. Dry the pellet under vacuum or air flow.
8. Resuspend the dry pellet in 40 µL of electrocompetent cells (*see* **Note 6**), transform by electroporation, and plate the transformation using media with the appropriate antibiotic. Incubate the plates overnight at 37°C.

3.3. Screening of Recombinants for Protein Expression

If blue-white colony screening is not available for your particular expression vector, there are several other options for screening colonies:

1. Mini-prep the plasmid DNA and cut with restriction enzymes that indicate the presence of the cloned gene.

2. Perform colony PCR using primers which specifically indicate the presence of the cloned gene (*see* **Note 7**).
3. Dot-blot a small amount of lysed culture that was induced and detect with an antibody specific for the target protein.
4. Grow cultures and perform SDS-PAGE or a Western blot after induction.
5. Observe the growth (e.g., OD_{600}) differences between induced and uninduced cultures *(9)*.

Methods 1 and 4 are fairly common protocols and method 2, colony PCR, has been described elsewhere (10). Methods 2 and 3 are the most rapid and accurate at indicating the presence of the correct gene or expressed protein. A brief procedure for method 3, immunoblotting, is outlined below:

1. Pick 30 colonies into 1 mL of media each (LB with 100 µg/mL ampicillin) and grow for 2 h in a 1.5-mL Eppendorf tube at 37°C with shaking at 250 rpm (*see* **Note 8**). As a negative control, also grow a 1 mL culture of *E. coli* containing the original cloning vector *without* the target gene. This will confirm that the antibody is not detecting non-specific *E. coli* proteins.
2. At 2 h, for each culture, mix 50 µL of cells with 50 µL of glycerol and store the mixture at –80°C for future use. In the remaining 950 µL of culture, induce the cells by adding IPTG to 1 m*M*. Continue shaking at 37°C for 1 h.
3. Centrifuge the cells in a microcentrifuge for 1 min at maximum speed. Resuspend the cells in 100 µL of 10% SDS with 50 m*M* Tris-HCl at pH 7.0, and heat at 100°C for 2 min. Allow the mixture to cool to room temperature (about 5 min). The liquid should be clarified at this point.
4. Cut out a piece of nitrocellulose membrane and mark the edges to make a grid. Place 2 µL of lysate from each colony on the grid and allow the membrane to air dry for 30 min.
5. To detect the presence of the target protein, follow the Western blot procedure developed for your particular target protein and antibody. Run an SDS-PAGE and Western blot of positive colonies to confirm the size of the expressed protein.

3.4. Evaluating Protein Solubility by Cell Lysate Fractionation

This method is designed to assess the solubility of an expressed fusion protein. The protocol is optimized for 4 mL expression cultures which are prepared as follows (*see* **Fig. 2** for a flow chart summary):

1. Colonies from a plate should be picked into 1 mL of media and incubated for 2 h at 37°C with shaking at 250 rpm (*see* **Note 8**).
2. For a typical solubility analysis, inoculate three 13 × 100-mm test tubes containing 4 mL media each with 300 µL of the 2 h culture.
3. At an OD_{600} of 0.4, induce two of the cultures with 1 m*M* IPTG (*see* **Note 9**). Do not induce the third tube. At 3 h post-induction, pellet the cells at 1000*g* for 10 min, discard the supernatant, and freeze the pellets at –20°C (*see* **Note 10**).

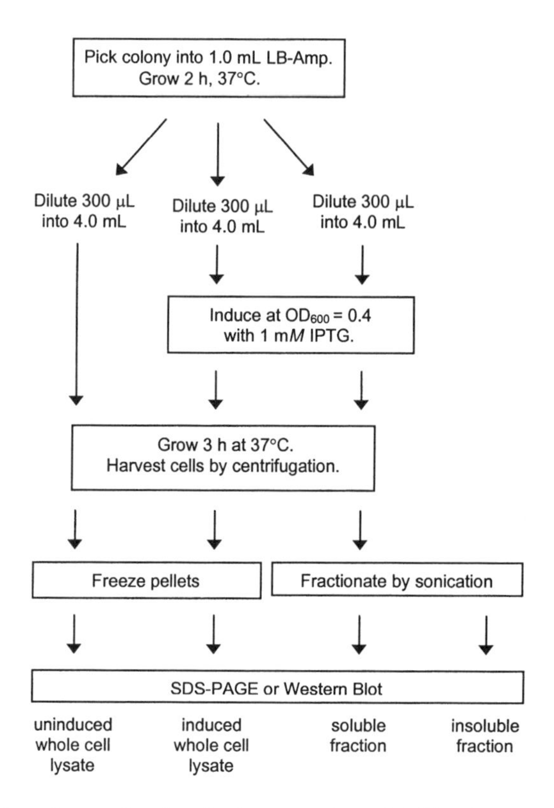

Fig. 2. Flow chart for evaluating fusion protein solubility by cell fractionation.

One of the cell pellets from a 4 mL induced culture is then treated according to the protocol that follows. By using the volumes in this protocol, the relative amounts of the proteins for the cell lysate soluble and insoluble fractions are the same as they are in the cell, as indicated by SDS-PAGE or Western blotting.

1. Resuspend a frozen cell pellet from an induced culture in 10 mL of sonication buffer (50 m*M* NaCl and 1 m*M* EDTA, pH 8.0) in a small 20-mL beaker, measure the OD_{600}, and place it on an ice water bath.
2. Place the sonication horn into the solution as far down as possible without touching the bottom of the beaker.
3. Sonicate for 30 s at 90 W and then allow the solution to cool for 30 s on ice. Repeat 3 more times for a total sonication time of 2 min (2 min sonication + 2 min total cooling time = 4 total min for the procedure). Measure the OD_{600} of the lysate: it should be less than the original OD_{600} (*see* **Note 11**).
4. Centrifuge the lysate at 12,000*g* for 30 min at 4°C.
5. At this point, take care not to disturb the inclusion body pellet. Carefully take the top 5 mL of the supernatant by drawing it into the pipet just under the liquid

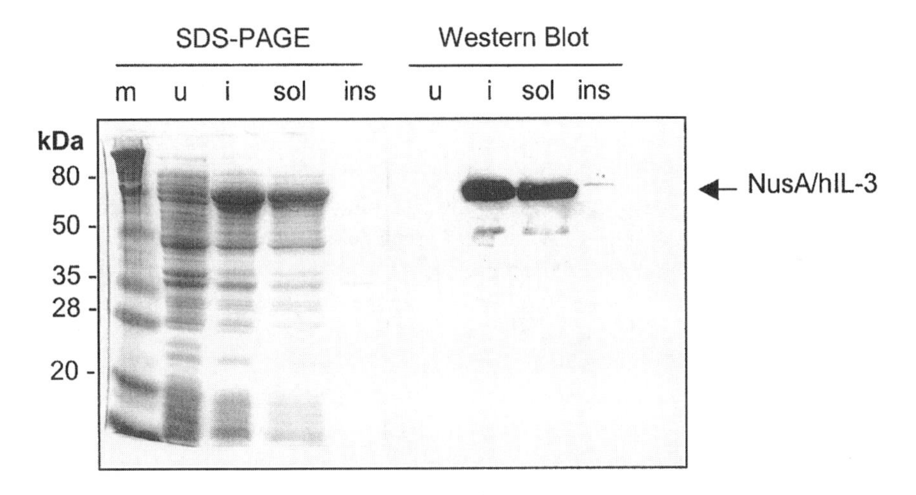

Fig. 3. SDS-PAGE and Western Blot of cell fractions containing the NusA/hIL-3 fusion protein. Equal portions of cell lysate, soluble fraction, and insoluble fraction were loaded in each lane. *(m)* markers, *(u)* uninduced whole cell lysate, *(i)* induced whole cell lysate, *(sol)* soluble fraction, *(ins)* insoluble fraction. Fusion proteins were expressed from plasmid pKK223-3 under control of the *tac* promoer in *E. coli* JM105 at 37°C. Cells were induced with 1 m*M* IPTG and grown for 3 h post-induction. The Western blot was probed with mouse anti-hIL-3 monoclonal antibody and visualized using chemiluminescence. The percentage of soluble fusion protein was 97% based on the Western blots (density of soluble band divided by the density of the soluble plus insoluble bands).

surface as far away from the insoluble pellet as possible and place in a labeled freeze drying flask. Immediately freeze the supernatant in the –80°C refrigerator. Discard the remaining supernatant and freeze the inclusion body pellet at –20°C.
6. Place the frozen soluble lysate supernatant samples on a freeze dryer overnight for sublimation (*see* **Note 12**).
7. Store the freeze dried solid at –20°C until SDS-PAGE analysis.
8. For SDS-PAGE analysis resuspend the freeze dried supernatant in 400 µL of SDS-PAGE gel loading buffer and the inclusion body pellet, the uninduced pellet, and the induced pellet in 800 µL of SDS-PAGE gel loading buffer. Run an SDS-PAGE of the uninduced whole cell lysate, the induced whole cell lysate, the soluble fraction, and insoluble fraction. Western blotting can be performed to get a clear picture of the distribution of soluble and insoluble fusion protein.

Solubility results for the NusA/hIL-3 fusion protein analyzed by this method are shown in **Fig. 3**. The corresponding Western blot of the SDS-PAGE shows that the uninduced culture is very well repressed and that 97% of the fusion protein is soluble, as determined by densitometry analysis.

4. Notes

1. Most Entrez or Swiss-Prot protein sequence files contain numbers to annotate the sequence. These numbers will be copied along with the sequence, but they will be ignored by the calculation and need not be removed. However, if any other non-protein characters which correspond to the single letter amino acid code are present, they will be regarded as residues and must be removed prior to calculation.
2. For the Netscape browser, go to "File", then "New", then "Navigator Window". For the Microsoft Internet Explorer browser, go to "File", then "New", then "Window". This command can be repeated several times to open several independent windows of the www.ncbi.nlm.nih.gov site, so that a different window is open for each different protein you wish to evaluate.
3. The GeneClean kit protocol from BIO101 (Vista, CA) was used in this particular experiment. In general, the yields from agarose gel purification are quite low and can considerably decrease the cloning efficiency (e.g., the total number of colonies obtained after transformation). One alternative is to use larger amounts of DNA in the digestions. Another, more rapid option is to use gel filtration spin-columns (e.g., Chromaspin columns from Clontech, Palo Alto, CA) which can be used to simultaneously remove both the salts and the small molecular weight DNA end fragments from the restriction enzyme digestion of PCR products.
4. Following restriction digestion and purification, a large fraction of the DNA fragments may be self-annealed. Heating the DNA fragments prior to adding the ligase "shuffles" the sticky ends of all the fragments and increases the ligation-recombination efficiency.
5. Required ligation times may vary. Ligation times of 10 min at room temperature have been routinely successful for two-fragment ligations (e.g., one gene, one vector). If three-fragment ligations are to be performed on a routine basis, it may be possible to decrease the time from 16 h at 16°C to 2 h at room temperature to speed up the cloning procedure.
6. In this experiment, *E. coli* JM105 was used as an expression host, since it is compatible with the pKK223-3 vector. The primary reason for using strain JM105 is that it contains a higher level of the *lacI^q^* repressor gene which is generally effective at silencing the *tac* promoter. Previous experience has shown that using strains that do not contain the *lacI^q^* gene, such as DH5α, with pKK223-3 are likely to cause random frame-shift mutations in the open reading frame of the cloned gene which ultimately prevent expression.
7. Colony PCR is especially efficient for screening for gene fusions. The 5'-primer of the carrier gene can be used with the 3'-primer of the target gene so that only recombinants containing the entire gene fusion will be detected.
8. A convenient method for growing 1 mL cultures is to put the cultures in 1.5-mL Eppendorf tubes, tightly close the lids, and drop all the tubes into the bottom of a 500-mL Erlenmyer flask which can fit in an orbital shaker. The tubes will bounce around the bottom of the flask for good mixing. This works well for growing up culture stocks from colonies to inoculate larger volumes, but should not be used to grow cultures longer than 2 h since there is no aeration in the Eppendorf tube.

9. At the time of induction, an additional 100 µg/mL of ampicillin is usually added to ensure an effective selection since ampicillin is constantly being degraded by the culture. This practice has been passed down by protocol, and may not be entirely necessary, especially if you are sure your plasmid is very stable.
10. It was noticed that when a significant overexpression of protein occurred (e.g., at least 5% of the total cell protein), the induced and uninduced cell pellets could be somewhat differentiated by color just after centrifugation and removal of the supernatant. Induced cell pellets typically were near-white and uninduced cell pellets were a light brownish-yellow color. In addition, induced cell pellets were typically smaller in size. For clones which did not overexpress recombinant protein, these differences were not observed, ruling out any general effects of IPTG on the cell. This was observed with *E. coli* strain JM105.
11. The first time the sonication is performed, the OD_{600} should be monitored after each pulse. When the OD_{600} fails to decrease any further, lysis is near complete. In our experience, this typically required 4 cycles of pulsing.
12. Freeze-drying was performed to simply re-concentrate the soluble lysate so the protein can be visualized by SDS-PAGE. If a freeze dryer is not available, a spin-vac is a useful alternative. If neither piece of equipment is available, alter the protocol at **step 1** by resuspending the pellet in a smaller volume of sonication buffer and using a 5X concentrated stock of SDS-PAGE gel loading buffer.

Acknowledgments

We would like to gratefully acknowledge Vinod Asundi for kindly providing the starting CGI computer code on which our subsequent web-programming efforts were based.

References

1. LaVallie, E. R. and McCoy, J. M. (1995) Gene fusion expression systems in *Escherichia coli. Curr. Opin. Biotechnol.* **6,** 501–506.
2. Uhlen, M., Forsberg, G., Moks, T., Hartmanis, M., and Nilsson, B. (1992) Fusion proteins in biotechnology. *Curr. Opin. Biotechnol.* **3,** 363–369.
3. Smith, D. B. and Johnson, K. S. (1988) Single-step purification of polypeptides expressed in *Escherichia coli* as fusions with glutathione *S*-transferase. *Gene* **67,** 31–40.
4. di Guan, C., Li, P., Riggs, P. D., and Inouye, H. (1988) Vectors that facilitate the expression and purification of foreign peptides in *Escherichia coli* by fusion to maltose-binding protein. *Gene* **67,** 21–30.
5. LaVallie, E. R., DiBlasio, E. A., Kovacic, S., Grant, K. L., Schendel, P. F., and McCoy, J. M. (1993) A thioredoxin gene fusion expression system that circumvents inclusion body formation in the *E. coli* cytoplasm. *Bio/Technology* **11,** 187–193.
6. Davis, G. D., Elisee, C., Newham, D. M., and Harrison, R. G. (1999) New fusion protein systems designed to give soluble expression in *Escherichia coli. Biotech. Bioeng.* **65,** 382–388.
7. Wilkinson, D. L. and Harrison, R. G. (1991) Predicting the solubility of recombinant proteins in *Escherichia coli. Bio/Technology* **9,** 443–448.

8. Harrison, R. G. (2000) Expression of soluble heterologous proteins via fusion with NusA protein. *inNovations* (newsletter of Novagen, Inc., Madison, WI), June, 4–7.
9. Davis, G. D. and Harrison, R. G. (1998) Rapid screening of fusion protein recombinants by measuring effects of protein overexpression on cell growth. *BioTechniques* **24,** 360–362.
10. Hofmann, M. A. and Brian, D. A. (1991) Sequencing PCR DNA amplified directly from a bacterial colony. *Biotechniques* **11,** 30–31.

10

Assessment of Protein Folding/Solubility in Live Cells

Rhesa D. Stidham, W. Christian Wigley, John F. Hunt, and Philip J. Thomas

1. Introduction

The folding of a protein to its final native state involves a series of complex steps of intra- and intermolecular interactions between the nascent polypeptide chain, its solvent environment, and the quality control machinery of the cell *(1)*. These steps are often halting, as the unfolded protein visits a series of intermediates states en route to its final native structure. In the cell, undesirable reactions such as aggregation and proteolytic degradation compete for these folding intermediates, shuffling them off the productive folding pathway (*see* **Fig. 1**). Thus, in order to properly fold, many proteins require the oversight of molecular chaperones that bias intermediates towards productive folding rather than off-pathway self-associations *(2)*. Changes in this delicate balance between on- and off-pathway reactions have ramifications both for human health and the study of proteins of structural interest or of commercial utility.

Protein misfolding is the underlying pathology of a number of human diseases *(3,4)*. In certain instances, the disease phenotype resulting from misfolding stems from the lack of sufficient functional, folded protein *(5–7)*; in others, the process of off-pathway aggregation or the aggregate itself is thought to be toxic *(8–15)*. Moreover, protein misfolding and poor chemical solubility of the native state often lead to the formation of inclusion bodies in bacterial expression systems *(16)*. This reduction in the yield of functional, soluble protein presents a barrier to the structural characterization of such proteins and to the large scale production of those with clinical relevance.

Fortunately, some improvement in the folding yield of a heterologously expressed protein can be mediated by alterations in the primary amino acid

From: *Methods in Molecular Biology, vol. 205, E. coli Gene Expression Protocols*
Edited by: P. E. Vaillancourt © Humana Press Inc., Totowa, NJ

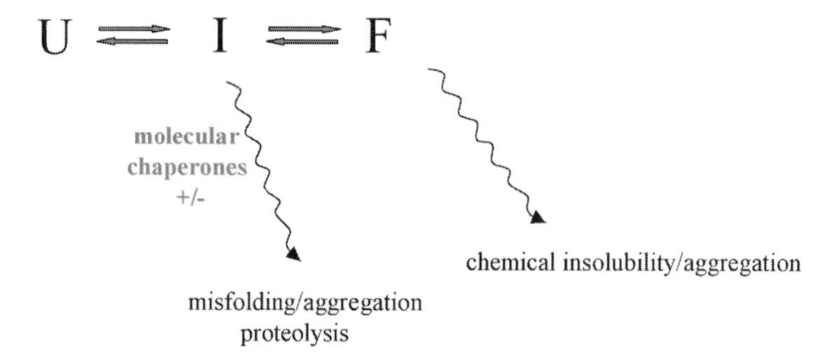

Fig. 1. Schematic of the in vivo protein folding process. Protein folding is depicted as a pathway along which the nascent, unfolded protein (U) moves through a series of requisite intermediate states (I) in order to reach the final, folded native state (F). This process is driven by the difference in thermodynamic stability between the unfolded and folded protein; the final yield of folded protein, however, will be determined by a variety of other factors as well as the native state stability. Chemical insolubility of the native state and misfolded species generated along the folding pathway lead to aggregation and reduction in the yield of soluble, folded protein. The quality control machinery of the cell also affects the yield of folded protein. The proteolytic machinery reduces this yield through degradation of intermediates; molecular chaperones may act to facilitate either this degradation or productive folding by preventing off-pathway aggression.

sequence of the target protein ***(17,18)***, co-expression with molecular chaperones ***(19)***, or changes in the growth conditions/genetic background for expression ***(20–22)***. More recently, small molecules have been shown to bind to, stabilize, and increase the folding yield of several clinically relevant proteins both in vitro and in cell culture models ***(23,24)***, suggesting that similar molecules may someday be used for increasing the folding yield of these disease-related proteins in patients. There also remains the intriguing possibility that a small molecule could be isolated that inhibits the toxic formation of aggregates associated with several important neurodegenerative diseases.

Traditional biochemical and immunocytochemical methods for evaluating the increase in folding yield under different expression conditions are time-consuming and impractical for large-scale screening of folding conditions or point mutations in the target protein. Such difficulties have led to the development of several general assays aimed at allowing the rapid assessment of a target protein's solubility/folding ***(25–27)***. Here we review an assay that we developed based upon structural complementation of a genetic marker protein, β-galactosidase (β-gal), and contrast it with other solubility/folding assays.

1.1. Structural Complementation to Monitor Solubility/Folding

The β-gal assay utilizes structural complementation as an indicator of the solubility/folding of a target protein *(27)*. The assay will report on changes in the steady state level of soluble protein, which is affected by both the chemical solubility properties of the native state and misfolding events leading to aggregation (*see* **Fig. 1**). In that the system can be used to monitor these types of misfolding events, it is also a "folding" assay. Henceforth, the term solubility assay will be used, with its utility as a folding assay left implied. Structural complementation involves the separation of a protein into two fragments, which are capable of associating to form a functional protein when expressed in trans *(28)*. The degree of complementation is governed by the association constant that defines the interaction; thus, the amount of complementation will be sensitive to the concentration of the expressed fragments. A target protein with poor solubility reduces the concentration of soluble fusion, with a concomitant reduction in the amount of complementation and measurable functional protein (*see* **Fig. 2A**).

Any protein with measurable activity and for which there are known complementing fragments could be amenable to adaptation as a solubility assay; in its current form the assay utilizes the classic complementation system of β-gal *(29)*. Each monomer of β-gal can be divided into the small α fragment (50–90 residues from the N-terminus of the protein) and the larger ω fragment (typically the ΔM15 mutant containing a deletion of residues 11–41 from the 135 kDa monomer) *(30)*. Complementation of the two fragments allows the formation of a functional homotetramer capable of hydrolyzing lactose and other substrates (*see* **Fig. 2A**). Recently, we demonstrated that the amount of complementation of ω by a fusion of the α fragment to the C-terminus of a target protein, as predicted, reflects the solubility properties of the target *(27)*.

1.2. In vivo Assay

The assessment of β-gal activity on indicator plates provides a rapid, qualitative means of screening a large library of point mutants of a target protein and changes in expression conditions which might affect target protein solubility. Complementation is detected by plating bacteria expressing both the target-α fusion and the ω fragment on plates containing the chromogenic substrate 5-bromo-4-chloro-3-indolyl-β-D-galactoside (X-gal) that forms a blue precipitate upon hydrolysis by β-gal. The correlation between solubility of the target-α fusion and the intensity of blue color on X-gal plates has been demonstrated for a number of target proteins *(27)*. In particular, the expression of a fusion of the maltose binding protein (MBP) to the α fragment in DH5α bacteria gives intense blue colony color on indicator plates. This correlates with data

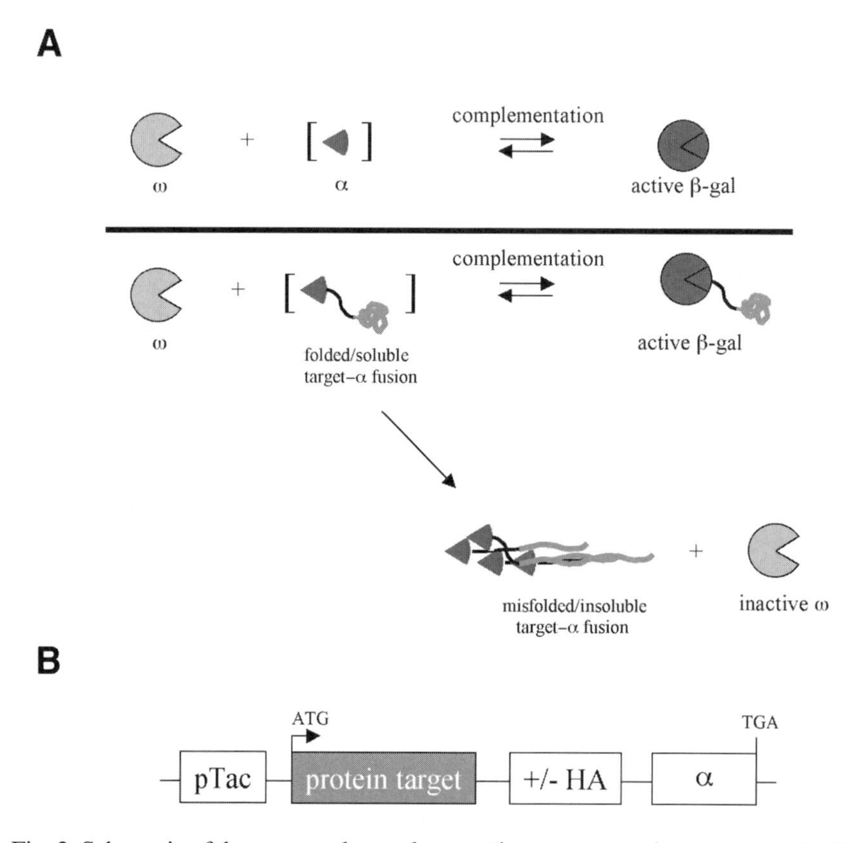

Fig. 2. Schematic of the structural complementation system used to measure solubility/folding. **(A)** *Top:* Structural complementation of β-galactosidase. β-gal is divided into two fragments, α and ω. Neither fragment has activity alone; upon association, however, the two fragments form an active enzyme capable of hydrolyzing the substrate X-gal, producing a blue precipitate. The amount of active enzyme depends upon the degree of association. This association is governed by the affinity constant (K_d) for the interaction and is, thus, dependent on the concentration of the complementing fragments. *Bottom:* Adaptation of β-gal complementation to monitor folding/solubility. Events, such as aggregation, which reduce the concentration of one of the fragments lead to a decrease in the amount of complementation and detected β-gal activity. Aggregation of an insoluble target protein, when in the context of a fusion, will remove the α fragment from solution as well. The subsequent decrease in the amount of enzymatic activity can then be used as an indicator of target protein solubility. **(B)** Diagram of the target-α fusion construct used to demonstrate the utility of the folding/solubility assay *(27)*. The expression plasmid was modified from the pMal.c2x plasmid from New England Biolabs. Expression of the fusion is driven from a pTac promoter. An HA tag was inserted in the polylinker region between the target and the α fragment in order to detect fusions with low expression levels (*see* **Subheading 2.1.**). Figure 2B was modified from **ref.** ***27***.

that shows a significant fraction of the fusion protein is soluble (*see* **Fig. 3A, C**). In contrast, the G32D/I33P double mutant of MBP, which has drastically reduced periplasmic folding yield in vivo *(31)*, gives colonies which are much less intensely blue on indicator plates and little soluble fusion protein after fractionation (*see* **Fig. 3A, C**) *(27)*.

The inherent enzymatic amplification of the signal allows small changes in solubility to be detected; this amplification has its greatest utility in monitoring the increase in solubility of a target-α fusion that initially is almost completely insoluble, the case of greatest practical value. Expression of the complementing fragments is induced during overnight growth of the bacteria on indicator plates, and the signal generated from the soluble target-α fusion is integrated over this time course. As a result, the visual distinction of small changes in signal generated from target proteins which are already partially soluble is difficult due to saturation of the signal. This is the case with the MBP single mutants (G32D, I33P), which have intermediate folding yield between wild type MBP and the double mutant, but are very similar to colonies expressing the wild type MBP-α fusion on indicator plates *(27)*.

The method is readily adaptable to a 96-well plate format for screening expression conditions in suspension culture. This format may overcome some of the range limitations encountered using solid media indicator plates, since the signal need not be measured during the whole growth phase of the bacteria (*see* **Subheading 3.1.2.**).

1.3. Cell Lysate Assay

The cell lysate assay allows a rigorous, quantitative comparison of β-gal activity and, thus, solubility. This method involves the induction of expression of the complementing fragments in liquid culture, lysis by several freeze/thaw cycles after expression, and assessment of β-gal activity by monitoring hydrolysis of the substrate analog *o*-nitrophenyl-β-D-galactopyranoside (ONPG) spectrophotometrically. In this form, the assay of β-gal activity is less convenient for high-throughput screening. However, relatively small changes in solubility over a wide range of initial target solubility levels are readily detected. For example, the difference in enzymatic activity between the wild type MBP-α fusion and the single mutant MBP-α fusions is measurable and significant, in contrast to the plate assay. In fact, there is a linear correlation between the fractional biochemically determined solubility of each mutant and the amount of enzymatic activity that is detected in cell lysates (*see* **Fig. 3B**).

1.4. Comparison with Other Folding Assays

The structural complementation system offers several advantages over other reported folding assays, which rely on the misfolding of the target protein to

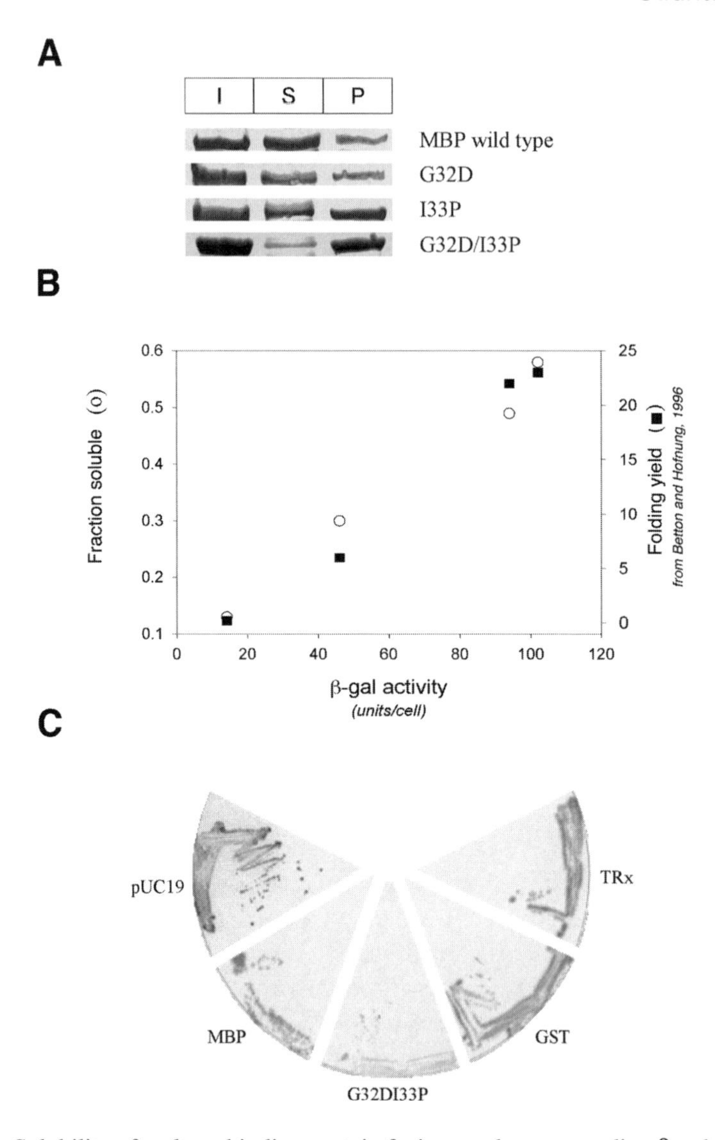

Fig. 3. Solubility of maltose binding protein fusions and corresponding β-galactosidase activity. **(A)** Fractionation of MBP fusions two hours after expression. Samples were run on SDS-PAGE and stained with Coomassie. The total induced lane (I) provides an internal standard to account for loss of protein during fractionation. The pellet lane (P) contains insoluble fusion; the supernatant lane (S) contains the soluble fusion. The wild type MBP clearly fractionates primarily with the supernatant, while the G32D/I33P double mutant fractionates primarily with the insoluble fraction. The single mutants of MBP, G32D, and I33P, show intermediate levels of solubility. **(B)** The fraction of soluble fusion protein determined by densitometry of fractionation results demonstrated in (A) is plotted against the β-gal activity determined by the cell lysate

induce the misfolding of a C-terminally fused indicator protein such as the green fluorescent protein (GFP) and chloramphenicol acetyltransferase (CAT) ***(25,26)***. While these assays would be expected to be most sensitive to misfolding events which occur early in the folding process, perhaps even cotranslationally, it is less clear that they will reliably report on slower aggregation events such as those which occur in many neurodegenerative diseases. In fact, GFP fused to the C-terminus of expanded polyglutamine tracts as well as to a disease-associated mutant of the huntingtin exon 1 containing an increased number of glutamines, gives rise to fluorescent inclusions ***(32–35)***. Thus slow, post-translational aggregation does not always prevent the folding of the green fluorescent reporter. In contrast, the association of the α and ω fragments of β-gal reflects a dynamic equilibrium which is dependent upon the concentration of the soluble fusion at any given time. This was demonstrated using the Aβ peptide which is associated with Alzheimer's disease ***(11)***. In the case of a fusion of the Aβ peptide to the α fragment, at early time points after expression there is soluble fusion protein and enzymatic activity. As the fusion forms aggregates over several hours, there is a concomitant decrease in the measured β-gal activity, indicating that the aggregation process can compete for the α-fragment previously associated with ω ***(27)***.

Both the CAT fusion system and the α complementation system allow for phenotypic selection of improved solubility. Mutations in the target protein or alterations in the cellular environment that increase fusion solubility yield greater complementation of β-gal activity and would, thus, allow the expressing bacteria to utilize lactose as the sole carbon source in minimal media.

While the utility of the α complementation system has been demonstrated using several target proteins fused to the α fragment ***(27)***, it is clear that the

---

Fig. 3. *(continued)* assay (*see* **Subheading 1.3.**). Clearly, there is a linear correlation between the amount of soluble fusion and the amount of measured β-gal activity. MBP fusions in these assays are lacking the periplasmic targeting sequence. However, the reported in vivo periplasmic folding yield of the wild type MBP and point mutants also shows a linear correlation with measured β-gal activity, indicating that the relative solubility levels of the MBP fusions reflect the normal in vivo levels. **(C)** Glycerol stocks of DH5α harboring the fusion expression plasmid with the indicated target protein streaked to single colonies on indicator plates. Those colonies which are most intensely blue are the darkest in this representation. The α fragment alone is produced from pUC19 and serves as a positive control for color development. The MBP fusion to the α fragment complements to give intense blue color, although it does not reach the intensity seen in the pUC19 control. The level of color development is drastically reduced for the G32D/I33P double mutant of MBP. Two other soluble target fusions are also represented, GST (glutathione S-transferase) and TRx (thioredoxin). Figures reproduced from *(27)*. Used with permission.

assay must be tailored for the particular target protein of interest. Expression levels will affect the range of detectable enzymatic activity, and thus the assay will find its greatest utility in comparison of the solubility of target proteins of similar expression levels or in comparison of sequence variants/expression conditions for a single protein. It is also important to consider that the accessibility of the α fragment will depend upon its context in the folded target protein; in one instance, the α fragment is known to be occluded from complementation due to oligomerization of a fused partner ***(36)***. Finally, it is also possible that the α fragment may retain some ability to complement even in the form of an aggregate. These caveats aside, in many cases the structural complementation assay for protein solubility offers a potent means of investigating the folding of a variety of target proteins in multiple cellular contexts.

2. Materials

2.1. Expression System

1. Expression plasmid (*see* **Note 1**).
2. Competent DH5α *E. coli* (*see* **Note 2**).

2.2. In vivo Assay

2.2.1. Detection on Indicator Plates

1. Luria-Bertani (LB) agar plates containing 100 μg/mL ampicillin (or concentration of appropriate antibiotic if using a different expression plasmid than described) (*see* **Subheading 2.1.**).
2. LB agar indicator plates containing 80 μg/mL X-gal (5-bromo-4-chloro-3-indolyl-β-D-galactoside), 0.1 m*M* isopropylthio-β-D-galactoside (IPTG), 100 μg/mL ampicillin.
3. Replica-plating tool.

2.2.2. Detection in 96-Well Plates

1. Bacterial growth media (LB, or the like).
2. Overnight culture of DH5α/fusion expression plasmid.
3. Ampicillin: 50 mg/mL stock in sterile dH_2O.
4. IPTG: 1 *M* stock in sterile dH_2O.
5. 5-Bromo-4-chloro-3-indolyl-β-D-galactoside (X-gal): 20 mg/mL stock in *N,N*-dimethylformamide.
6. Flat-bottom 96-well plate.
7. 37°C Shaker.

2.3. Cell Lysate Assay

2.3.1. In vitro Detection of β-Galactosidase Activity

1. Bacterial growth media (LB, or the like).
2. Overnight culture of DH5α/fusion expression plasmid.

3. Z buffer: 10 m*M* KCl, 2.0 m*M* $MgSO_4$, 60 m*M* Na_2HPO_4, 40 m*M* Na_2PO_4, pH 7.0.
4. o-Nitrophenyl-β-D-galactopyranoside (ONPG) solution (*see* **Note 3**): 4 mg/mL in Z buffer.
5. Z buffer with 0.27% β-mercaptoethanol (*see* **Note 4**).
6. 1 *M* Na_2CO_3.
7. Liquid nitrogen.
8. 37°C Water bath.
9. Spectrophotometer.

2.3.2. Fractionation

1. Lysis buffer: 100 m*M* NaCl, 1 m*M* ethylenediaminetetraacetic acid (EDTA), 50 m*M* Tris-HCl, pH 7.6.
2. 2X sample buffer: 2.4 mL glycerol, 4.5 mL 2 *M* Tris-HCl, pH 8.45, 0.8 g sodium dodecylsulfate (SDS), 1.5 mL 0.1% Serva Blue G, 0.5 mL 0.1% phenol red, 0.5 mL β-mercaptoethanol, 0.6 mL dH_2O.
3. Coomassie stain: 0.25 g of Coomassie brilliant blue R250 in 90 mL of methanol:H_2O (1:1 v/v) and 10 mL of glacial acetic acid
4. Destain: 90 mL of methanol: H_2O (1:1 v/v) and 10 mL of glacial acetic acid
5. Polyacrylamide gel (4–20% gradient gel)
6. Sonicator (settings outlined in **Subheading 3.3.2.** are for use with a Branson 450 model sonicator using a microtip probe)

3. Methods

3.1. *In vivo* Assay

3.1.1. Detection on Indicator Plates

1. To screen a pool of mutated target plasmids (changes in amino acid sequence) or for colonies of DH5α (changes in genetic background) that give increased folding yield, transform the plasmid or plasmid pool into DH5α and plate on LB-agar plates containing ampicillin. Grow overnight at 37°C. The colony count should be low enough that the colonies are well separated on the plate; this can be determined by serial dilution. For screening different conditions of expression (pharmacological agents, osmolytes, etc.) for a single target-α construct, skip to **step 4**.
2. Contingent upon the toxicity of the expressed product, make a replica of the original transformation plate on an LB-agar indicator plate (*see* **Note 5**). Mark a reference point on each plate so that colonies of interest on the indicator plates can be identified on the original LB-ampicillin plates for subsequent isolation and characterization. Grow both the original and the replica plate at 37°C overnight.
3. Screen colonies on the indicator plates for increase/decrease in intensity of blue color (*see* **Note 6**). Identify the location of the colonies of interest on the original LB-ampicillin plate (may need to restreak on LB-ampicillin to ensure single colony isolates). Set up an overnight culture of promising colonies and make glycerol stocks.
4. Streak isolates from glycerol stocks onto LB indicator plates to confirm phenotype (*see* **Note 7**). To assay different expression conditions for a single target-α

fusion, streak from glycerol stock onto LB indicator plates containing the appropriate concentration of additive. Incubate plates overnight at 37°C and verify/detect changes in the intensity of blue color compared to a wild type/no additive control (*see* **Note 6**).

5. Confirm that the change in colony color correlates with a change in the fraction of soluble protein (*see* **Subheading 3.2.2.**)

3.1.2. Detection in 96-Well Plates

1. Inoculate 5 mL of LB containing 100 µg/mL ampicillin with overnight culture of DH5α/expression plasmid (1:1000 dilution). Grow culture at 37°C to mid-log phase (OD_{600} ~ 0.5).
2. Add 125 µL of LB containing 0.6 m*M* IPTG and 100 µg/mL ampicillin to each well of the 96-well plate designated for a sample. Add the appropriate concentration of osmolyte/pharmaceutical agent to be screened to each well.
3. Add 125 µL of the mid-log culture from **step 1** above to each well. Incubate at 37°C with shaking for 1 h.
4. Add X-gal to 80 µg/mL in each well. Grow overnight in a 37°C shaker. Compare the intensity of blue colony color between samples and no additive control (*see* **Note 8**).

3.2. Cell Lysate Assay

3.2.1. In vitro Detection of b-Galactosidase Activity

1. Inoculate 10 mL of LB containing 100 µg/mL ampicillin with overnight culture of DH5α/expression plasmid (1:1000 dilution).
2. Grow culture in a 37°C shaker to mid-log (OD_{600} ~ 0.5). Add IPTG to 0.3 m*M* final concentration. Add concentration of appropriate additive to be tested for increasing/decreasing folding yield.
3. Incubate in a 37°C shaker for another two hours after induction (*see* **Note 9**).
4. Record the OD_{600} absorbance for each culture. Pellet the bacteria from 1.5 mL of the culture. Discard supernatant.
5. Resuspend the bacterial pellet in 1 mL of Z buffer. Pellet the bacteria. Discard supernatant.
6. Resuspend the bacteria in 300 µL of Z buffer. Snap freeze the samples in liquid nitrogen (*see* **Note 10**). Thaw samples in 37°C water bath. Repeat freeze/thaw cycle two times.
7. Remove 100 µL of each sample to a fresh microfuge tube. Add 100 µL of Z buffer alone to one microfuge tube as a blank.
8. Add 700 µL of Z buffer containing β-mercaptoethanol to a reaction. Start a timer. Immediately add 160 µL of ONPG solution. Invert tube to mix. Place tube at 37°C.
9. Incubate each tube for 10 min (*see* **Note 11**). Hydrolysis of ONPG by β-gal gives a visible yellow sample color.
10. Add 400 µL of 1 *M* Na_2CO_3 to quench the reaction. Vortex to mix.
11. Pellet the cell debris for 10 min at 16,000*g* in a microcentrifuge.

12. Transfer supernatants to cuvets (*see* **Note 12**).
13. Use the blank to zero the spectrophotometer at 420 nm. Measure the OD_{420} of each of the samples.
14. Calculate the β-gal units per cell for each sample:

$$\beta\text{-gal units*/cell} = (1000 \times OD_{420})/(t \times V \times OD_{600})$$

where:

t = time interval of the hydrolysis (in min) (10 min, *see* **step 9**)

V = 0.1 mL × concentration factor

(In this case 1.5 mL of culture was concentrated to 0.3 mL in **step 6**, so the concentration factor is 1.5/0.3 = 5)

OD_{600} = A_{600} of culture measured in **step 4**.

*A unit is defined as the amount which hydrolyzes 1 μmol of ONPG to o-nitrophenol and D-galactose/min. Procedure modified from Clontech Laboratories, Inc. Protocol #PT3024-1.

15. Confirm that the change in detected activity correlates with a change in the fraction of soluble protein (*see* **Subheading 3.2.2.**).

3.2.2. Fractionation

1. Pellet 3 mL of each culture from **step 4** of **Subheading 3.2.1.** for subsequent sonication. Pellet 1 mL of each culture for running a total induced protein control. Discard supernatants. Set the 1 mL pellets aside.
2. Resuspend the bacteria to be sonicated in 1 mL of lysis buffer. Pellet the bacteria. Discard supernatant.
3. Resuspend bacteria from **step 2** in 600 μL of lysis buffer. Sonicate samples three times for 30 s using a power output of 4 and a duty cycle of 50%. Incubate samples on ice for several minutes between sonicating in order to reduce heating of the sample.
4. Pellet the insoluble material for 10 min at 18,000*g* at 4°C in a microcentrifuge.
5. Carefully remove the supernatant to a clean microfuge tube (*see* **Note 13**).
6. Resuspend the pellet in 300 μL of 2X sample buffer. Add 300 μL of dH_2O. This is the pellet sample (P) and contains the insoluble target-α fusion.
7. Add 100 μL of 2X sample buffer to 100 μL of the supernatant from **step 5**. This is the supernatant sample (S) and contains the soluble target-α fusion.
8. Resuspend the pellet from 1 mL of culture (*see* **step 1**) in 100 μL of sample buffer. Add 100 μL of dH_2O. This is the induced sample (I) and contains total protein.
9. Heat samples at 95°C for 5 min. Load samples on a polyacrylamide gel. The amount of protein from an equivalent number of cells should be loaded on the gel. For example, the induced sample represents 1 mL of culture resuspended in a 200 μL volume, and thus has been concentrated 5 fold. The supernatant sample has been concentrated 2.5 fold (3 mL concentrated to 0.6 mL, then diluted 2 fold

in sample buffer) and the pellet sample 5 fold. Twice the volume of the supernatant sample must be loaded compared to the induced (I) or pellet (P) samples in order to have protein from an equivalent amount of bacteria in each lane (*see* **Note 14**)
10. Run gel and incubate in Coomassie stain for several hours. In order to remove excess dye, incubate in destain for several hours, changing the destaining solution several times (*see* **Note 15**).

4. Notes

1. The α complementation system has been developed using an expression plasmid with the target-α fusion under control of the pTac promoter (plasmid modified from pMal.c2x from New England Biolabs [Beverly, MA]). MBP was replaced with other targets using the *Nde*I restriction site just upstream of the start codon and one of the sites in the polylinker region between the end of the MBP coding region and the α fragment. The choice of restriction site in the polylinker region will determine the spacing between the target and the α fragment. A linker of eight amino acids (cloning into the *Nde*I/*Sal*I sites) and a linker of thirty-five amino acids (cloning into the *Nde*I/*Sac*I sites) have been used with success for different target proteins. The α fragment used in this vector is comprised of residues 6–59 of β-galactosidase. An HA tag was inserted into the *Sal*I site of the pMal.c2x vector and, for target proteins with low expression, used to probe the amount of the fusion in the soluble fraction by Western blot analysis (*see* **Fig. 2B**).

   In theory, the choice of promoter, linker length, and even the specific residues of the α fragment fused to the target may be varied. In some cases, it may be desirable to have higher expression levels from a stronger promoter. Increasing the length of the linker region may also aid in the accessibility of the α fragment for complementation in those cases where it is occluded in the context of a soluble target (*see* **Subheading 1.4.**). As well, there are several α fragments which have been utilized for α complementation in standard blue/white screening; typically, the α fragment is 50–90 residues in length. The smallest one was chosen in the development of the solubility assay in order to minimize the effect that the α fragment has on the folding of the target protein.
2. The DH5α strain of *E. coli* contains a chromosomal copy of the ω fragment (ΔM15 deletion mutant of β-galactosidase). Another strain could potentially be used if both the target-α fusion and the ω fragment were expressed from a plasmid.
3. ONPG solution should be made fresh before use. ONPG dissolves slowly at room temperature.
4. β-mercaptoethanol should be added to Z buffer just prior to use.
5. The expression of the target-α fusion can mediate some degree of toxicity to the bacteria, and thus there is selective pressure for the bacteria to reduce expression of the gene (point mutations/rearrangements in the plasmid). Replica plating allows colonies to be screened on indicator plates and the colonies of interest isolated from the original LB-ampicillin plates where they have not been subjected to IPTG induction.

6. For more highly expressing fusions, color development may be sufficient for comparison after 18 h of growth; for fusions with lower levels of expression, color development may need to be evaluated after 48 h of growth.
7. Colony color can appear more intensely blue in regions of high cell density; comparison of color development under different conditions of expression should be carried out at the single colony level.
8. For more highly expressing fusions, color development may be sufficient for comparison 2–3 h after adding X-gal; for fusions with lower levels of expression, color development may need to be evaluated after overnight induction. Care must also be taken that small changes in the intensity of blue color are not attributable to differences in colony number between wells.
9. For constructs with lower levels of expression, it may be necessary to allow expression for longer than 2 h to get a measure of the β-gal activity well above background. Alternatively, more bacteria can be resuspended in the same final volume (*see* **Subheading 3.2.1.**, **step 6**) so that the measured signal is greater.
10. Freeze the sample with the microfuge tube turned upright so that the bacteria are at the bottom of the tube. Upon thawing, pressure may force the cap of the microfuge tube open, resulting in loss of sample that has been frozen around the cap of the tube.
11. For constructs with lower levels of expression, the samples may need to be incubated for longer than 10 min in order to get a signal significantly above background.
12. Pellet withdrawn with the supernatant can significantly alter the OD_{420} detected for the sample.
13. Pellet withdrawn with the supernatant will give an inflated estimate of the amount of the fusion that is soluble. In order to avoid this, remove the supernatant immediately after pelleting the insoluble material. Since only a portion of the supernatant is required for SDS-PAGE, the supernatant well above the pellet can be removed and saved. The supernatant close to the pellet can be discarded.
14. The protein in the supernatant and the pellet lanes should account for the protein in the induced lane, providing an internal control for loss of protein sample during fractionation and sample preparation.
15. Instead of Coomassie staining, the gel can be transferred to nitrocellulose and probed by Western blot analysis for fusions that express at low levels.

References

1. Wickner, S., Maurizi, M. R., and Gottesman, S. (1999) Posttranslational quality control: folding, refolding, and degrading proteins. *Science* **286,** 1888–1893.
2. Frydman, J. (2001) Folding of newly translated proteins in vivo: the role of molecular chaperones. *Annu. Rev. Biochem.* **70,** 603–647.
3. Thomas, P. J., Qu, B. H., and Pedersen, P. L. (1995) Defective protein folding as a basis of human disease. *Trends Biochem. Sci.* **20,** 456–459.
4. Dobson, C. M. (1999) Protein misfolding, evolution and disease. *Trends Biochem. Sci.* **24,** 329–332.

5. Thomas, P. J., Ko, Y. H., and Pedersen, P. L. (1992) Altered protein folding may be the molecular basis of most cases of cystic fibrosis. *FEBS Lett.* **312,** 7–9.
6. Brown, C. R., Hong-Brown, L. Q., and Welch, W. J. (1997) Correcting temperature-sensitive protein folding defects. *J. Clin. Invest.* **99,** 1432–1444.
7. Rao, V. R., Cohen, G. B., and Oprian, D. D. (1994) Rhodopsin mutation G90D and a molecular mechanism for congenital night blindness. *Nature* **367,** 639–642.
8. Harper, J. D. and Lansbury, P. T., Jr. (1997) Models of amyloid seeding in Alzheimer's disease and scrapie: mechanistic truths and physiological consequences of the time-dependent solubility of amyloid proteins. *Annu. Rev. Biochem.* **66,** 385–407.
9. Bruijn, L. I., Houseweart, M. K., Kato, S., Anderson, K. L., et al. (1998) Aggregation and motor neuron toxicity of an ALS-linked SOD1 mutant independent from wild-type SOD1. *Science* **281,** 1851–1854.
10. Prusiner, S. B. (1998) Prions. *Proc. Natl. Acad. Sci. USA* **95,** 13,363–13,383.
11. Hind, C. R., Tennent, G. A., Evans, D. J., and Pepys, M. B. (1983) Demonstration of amyloid A (AA) protein and amyloid P component (AP) in deposits of systemic amyloidosis associated with renal adenocarcinoma. *J. Pathol.* **139,** 159–166.
12. Colon, W. and Kelly, J. W. (1992) Partial denaturation of transthyretin is sufficient for amyloid fibril formation in vitro. *Biochemistry* **31,** 8654–8660.
13. The Huntington's Disease Collaborative Research Group. (1993) A novel gene containing a trinucleotide repeat that is expanded and unstable on Huntington's disease chromosomes. *Cell* **72,** 971–983.
14. Davies, S. W., Turmaine, M., Cozens, B. A., et al. (1997) Formation of neuronal intranuclear inclusions underlies the neurological dysfunction in mice transgenic for the HD mutation. *Cell* **90,** 537–548.
15. Martin, J. B. and Gusella, J. F. (1986) Huntington's disease. Pathogenesis and management. *N. Engl. J. Med.* **315,** 1267–1276.
16. King, J., Haase-Pettingell, C., Robinson, A. S., Speed, M., and Mitraki, A. (1996) Thermolabile folding intermediates: inclusion body precursors and chaperonin substrates. *FASEB J.* **10,** 57–66.
17. Huang, B., Eberstadt, M., Olejniczak, E. T., Meadows, R. P., and Fesik, S. W. (1996) NMR structure and mutagenesis of the Fas (APO-1/CD95) death domain. *Nature* **384,** 638–641.
18. Sugihara, J. and Baldwin, T. O. (1988) Effects of 3' end deletions from the Vibrio harveyi luxB gene on luciferase subunit folding and enzyme assembly: generation of temperature-sensitive polypeptide folding mutants. *Biochemistry* **27,** 2872–2880.
19. Wynn, R. M., Davie, J. R., Cox, R. P., and Chuang, D. T. (1992) Chaperonins groEL and groES promote assembly of heterotetramers (alpha 2 beta 2) of mammalian mitochondrial branched-chain alpha-keto acid decarboxylase in *Escherichia coli. J. Biol. Chem.* **267,** 12,400–12,403.
20. Bourot, S., Sire, O., Trautwetter, A., et al. (2000) Glycine betaine-assisted protein folding in a lysA mutant of Escherichia coli. *J. Biol. Chem.* **275,** 1050–1056.
21. Brown, C. R., Hong-Brown, L. Q., Biwersi, J., Verkman, A. S., and Welch, W. J. (1996) Chemical chaperones correct the mutant phenotype of the delta F508 cys-

tic fibrosis transmembrane conductance regulator protein. *Cell Stress.Chaperones.* **1,** 117–125.
22. Blackwell, J. R. and Horgan, R. (1991) A novel strategy for production of a highly expressed recombinant protein in an active form. *FEBS Lett.* **295,** 10–12.
23. Foster, B. A., Coffey, H. A., Morin, M. J., and Rastinejad, F. (1999) Pharmacological rescue of mutant p53 conformation and function. *Science* **286,** 2507–2510.
24. Morello, J. P., Salahpour, A., Laperriere, A., et al. (2000) Pharmacological chaperones rescue cell-surface expression and function of misfolded V2 vasopressin receptor mutants. *J. Clin. Invest* **105,** 887–895.
25. Waldo, G. S., Standish, B. M., Berendzen, J., and Terwilliger, T. C. (1999) Rapid protein-folding assay using green fluorescent protein. *Nat. Biotechnol.* **17,** 691–695.
26. Maxwell, K. L., Mittermaier, A. K., Forman-Kay, J. D., and Davidson, A. R. (1999) A simple in vivo assay for increased protein solubility. *Protein Sci.* **8,** 1908–1911.
27. Wigley, W. C., Stidham, R. D., Smith, N. M., Hunt, J. F., and Thomas, P. J. (2001) Protein solubility and folding monitored in vivo by structural complementation of a genetic marker protein. *Nat. Biotechnol.* **19,** 131–136.
28. Zabin, I. and Villarejo, M. R. (1975) Protein complementation. *Annu. Rev. Biochem.* **44,** 295–313.
29. Ullmann, A., Jacob, F., and Monod, J. (1967) Characterization by in vitro complementation of a peptide corresponding to an operator-proximal segment of the beta-galactosidase structural gene of *Escherichia coli. J. Mol. Biol.* **24,** 339–343.
30. Jacobson, R. H., Zhang, X. J., DuBose, R. F., and Matthews, B. W. (1994) Three-dimensional structure of beta-galactosidase from *E. coli. Nature* **369,** 761–766.
31. Betton, J. and Hofnung, M. (1996) Folding of a mutant maltose-binding protein of *Escherichia coli* which forms inclusion bodies. *J. Biol. Chem.* **271,** 8046–8052.
32. Krobitsch, S. and Lindquist, S. (2000) Aggregation of huntingtin in yeast varies with the length of the polyglutamine expansion and the expression of chaperone proteins. *Proc. Natl. Acad. Sci. USA* **97,** 1589–1594.
33. Satyal, S. H., Schmidt, E., Kitagawa, K., et al. (2000) Polyglutamine aggregates alter protein folding homeostasis in *Caenorhabditis elegans. Proc. Natl. Acad. Sci. USA* **97,** 5750–5755.
34. Kazantsev, A., Preisinger, E., Dranovsky, A., Goldgaber, D., and Housman, D. (1999) Insoluble detergent-resistant aggregates form between pathological and nonpathological lengths of polyglutamine in mammalian cells. *Proc. Natl. Acad. Sci. USA* **96,** 11,404–11,409.
35. Senut, M. C., Suhr, S. T., Kaspar, B., and Gage, F. H. (2000) Intraneuronal aggregate formation and cell death after viral expression of expanded polyglutamine tracts in the adult rat brain. *J. Neurosci.* **20,** 219–229.
36. Luzzago, A. and Cesareni, G. (1989) Isolation of point mutations that affect the folding of the H chain of human ferritin in *E. coli. EMBO J.* **8,** 569–576.

11

Improving Heterologous Protein Folding via Molecular Chaperone and Foldase Co-Expression

François Baneyx and Joanne L. Palumbo

1. Introduction

1.1. In vivo Protein Folding: Molecular Chaperones vs Foldases

Protein folding in the viscous and crowded environment of the cell is very different from in vitro processes in which a single protein is allowed to refold at low concentration in an optimized buffer. Although Anfinsen's observation that all the information necessary for a protein to reach a proper conformation is contained in the amino acid sequence *(1)* remains unchallenged, it has recently become obvious that the efficient in vivo folding of subsets of cellular proteins, as well as that of most recombinant proteins, requires the assistance of folding modulators that can be broadly classified as molecular chaperones and foldases.

Molecular chaperones are ubiquitous and highly conserved proteins that help other polypeptides reach a proper conformation without becoming part of the final structure. They are however not true folding catalysts since they do not accelerate folding rates. Rather, they prevent "off-pathway" aggregation reactions by transiently binding hydrophobic domains in partially folded polypeptides, thereby shielding them from each other and from the solvent. Molecular chaperones also facilitate protein translocation, participate in proteolytic degradation, and help proteins that have been damaged by heat shock or other types of stress regain an active conformation *(2)*. Most—but not all—cytoplasmic molecular chaperones are heat shock proteins (Hsps) whose high level transcription depends on the alternative sigma factor σ^{32}. By contrast, foldases are true catalysts that accelerate rate-limiting—and typically late steps—along the folding pathway. These enzymes include cytoplasmic and periplasmic peptidyl-prolyl

From: *Methods in Molecular Biology, vol. 205, E. coli Gene Expression Protocols*
Edited by: P. E. Vaillancourt © Humana Press Inc., Totowa, NJ

cis/trans isomerases (PPIases) that catalyze the *trans* to *cis* isomerization of X-Pro peptide bonds, and periplasmic thiol/disulfide oxidoreductases that promote the formation and isomerization of disulfide bonds.

1.2. Cytoplasmic Folding Modulators

1.2.1. *The* DnaK-DnaJ-GrpE *and* GroEL-GroES *Systems*

The most extensively characterized molecular chaperones in the *E. coli* cytoplasm are the ATP-dependent DnaK-DnaJ-GrpE and GroEL-GroES systems *(3–5)*. These proteins are required for growth at high or all temperatures *(2,6)*, suggesting that they play a key role in the folding of at least one—and most likely several—essential host protein.

DnaK is a 69-kDa monomeric protein composed of a N-terminus ATPase domain and a C-terminus substrate binding region. It recognizes heptameric stretches of amino acids consisting of a 4–5 residues-long hydrophobic core flanked by basic residues *(7)*. The fact that such motifs arise every 36 residues on the average protein *(8)* explains the promiscuity of DnaK (*see* **Note 1**). DnaJ, which is independently able to associate with nonnative proteins via hydrophobic interactions *(9)*, activates DnaK for tight binding and directs it to high affinity sites on substrate polypeptides. The nucleotide exchange factor GrpE completes the triad by mediating complex resolution. Current models for chaperone-assisted protein folding hold that substrate proteins ejected from DnaK either fold into a proper conformation, are recaptured by DnaK-DnaJ for additional cycles of binding and release, or are transferred in a partially folded form to the "downstream" GroEL-GroES chaperonins for subsequent folding (*see* **Fig. 1**).

GroEL, which belongs to the Hsp60 family of chaperonins, is organized as an ≈800-kDa hollow cylinder formed by two homoheptameric rings stacked back to back. The central chambers defined by each ring are physically separated by centrally projecting C-terminal extensions. GroEL interacts with both substrate proteins and the cochaperonin GroES (a 70-kDa dome-shaped homoheptamer) through a ring of hydrophobic residues located in its apical domains *(10)*. In vivo, the ring of GroEL that is not associated with GroES preferentially binds nonnative proteins consisting of two or more domains with α/β-folds enriched in hydrophobic and basic residues *(11,12)*. ATP binding and hydrolysis in the polypeptide bound ring leads to a conformational change that vastly expands the size of the cavity (it becomes large enough to accommodate a partially folded protein 50–60 kDa in size) and promotes GroES docking. The substrate is concomitantly released and folds in a capped and hydrophilic environment. Quantized ATP hydrolysis in the opposite ring of GroEL leads to the ejection of GroES and substrate protein and "resets" the chaperonin for additional cycles of folding.

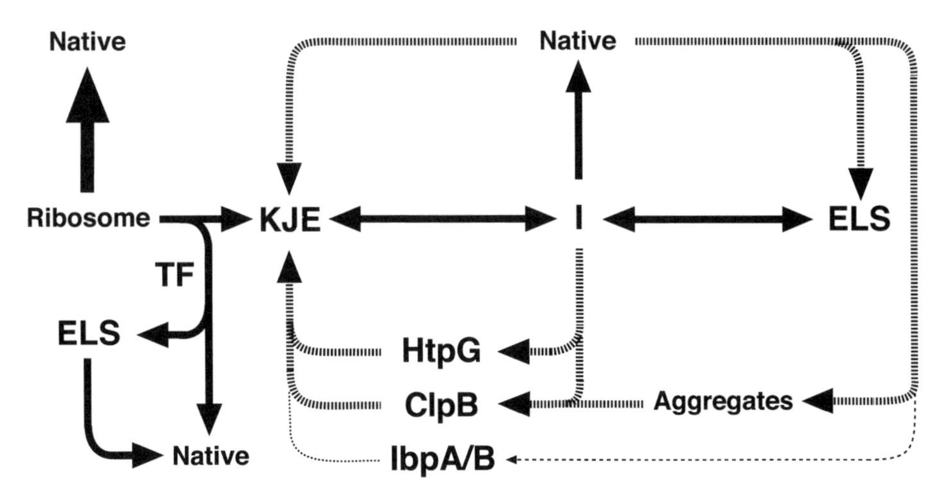

Fig. 1. A model for chaperone-assisted protein folding in the cytoplasm of *E. coli*. As they emerge from ribosomes, proteins requiring the assistance of chaperones to reach a proper conformation interact with either trigger factor (TF) or the DnaK-DnaJ-GrpE (KJE) system, depending on their size. TF-dependent proteins may be released in a native form or transferred to the GroEL-GroES chaperonins (ELS) for additional folding. Under non-stressful conditions (solid lines), KJE-dependent proteins enter the chaperone cycle. Folding intermediates (I) are released into a native conformation following cycles of binding and release by KJE and/or ELS. Upon heat shock or other stressful conditions (e.g., recombinant protein expression), alternate pathways (dashed lines) become important. Misfolded proteins may re-enter the KJE-ELS pathway directly or associate with additional chaperones. ClpB breaks down small aggregates of newly synthesized proteins or large aggregates of thermally-inactivated proteins and return them to KJE. HtpG acts as a "holder" chaperone, maintaining subsets of newly synthesized proteins in a conformation that is accessible for subsequent KJE-dependent folding. Although IbpA/B are capable of accomplishing the same task in vitro, their primary in vivo role appears to be in inner membrane protein folding (unpublished data).

1.2.2. ClpB, HtpG and IbpA/B as DnaK-DnaJ-GrpE Partners

In addition to the DnaK-DnaJ-GrpE and GroEL-GroES, a number of other Hsps including the ClpB ATPase, the Hsp90 homolog HtpG, and the small heat-shock proteins IbpA and IbpB have long been suspected of functioning as molecular chaperones ***(2,13,14)***. All of these proteins are dispensable up to 44–45°C. However, their absence exerts a deleterious effect on bacterial growth at very high temperatures ***(15–18)***, and *clpB* mutants die much more quickly than wild type cells following prolonged exposure at 50°C ***(15–18)***. Deletions in *clpB*, *htpG,* or in the *ibpAB* operon have been found to exacerbate the growth defects of certain *dnaK* mutants at 42°C *(18)*. Thus, ClpB, HtpG and IbpA/B may play a significant role in folding events under stressful conditions and/or when the Hsp70 team becomes unable to efficiently handle this task (*see* **Fig. 1**). Overloading of the DnaK-DnaJ-GrpE system is expected to occur upon high level recombinant protein overexpression, when slow folding proteins are unable to undergo multiple cycles of interaction with DnaK, or when proteins requiring GroEL-GroES for proper folding are not transferred to the chaperonins in a timely and efficient manner.

ClpB belongs to the Hsp100 family of heat shock proteins, a set of ATPases whose *E. coli* homologs include ClpA, ClpX, and ClpY. Whereas the latter proteins appear to promote the net unfolding of proteins targeted for degradation and transfer these substrates to an associated proteolytic component (ClpP or ClpQ), ClpB is the only family member whose primary function appears to be in protein folding rather than in proteolysis. The *clpB* gene contains an internal translation initiation site that leads to the synthesis of two gene products: a 93– and a 79-kDa polypeptide known as ClpB95 and ClpB80, respectively *(17,19)*. ClpB95 is the major translation product, and represents 70–90% of the total cellular ClpB. In vitro, ClpB95 and ClpB80 associate to form mixed hexamers provided that the protein concentration is high and that ATP is present in the buffer *(20)*. While the precise function of the truncated form of ClpB remains unclear, it has been proposed to act as a regulator of ClpB95 function based on the fact that an increase in the fraction of ClpB85 in ClpB95-ClpB80 oligomers leads to a decrease in ATPase activity *(21)*.

ClpB collaborates with the DnaK-DnaJ-GrpE system to reactivate large insoluble aggregates in vitro *(22,23)* and thermally aggregated proteins in vivo *(24)*. The process is thought to involve: (i) ClpB binding to structured hydrophobic domains exposed to the solvent by non-native proteins via the N-terminus of the Hsp *(25)*; (ii) shearing of the aggregates into smaller species by the ATP-dependent remodeling activity of ClpB; and (iii) transfer of the "disaggregated" proteins to the DnaK-DnaJ-GrpE team for subsequent refolding (either in a GroEL-GroES dependent or independent fashion). In addition to functioning as a disaggregase, ClpB facilitates *de novo* protein folding in the cytoplasm of *E. coli* *(26)*, presumably by using its remodeling activity to disentangle newly synthesized chains (and/or early aggregates consisting of a few associated proteins) that remain partially folded, but have packed hydrophobic DnaK recognition sequences away from the solvent after failing to reach a proper conformation.

Eukaryotic Hsp90 associates with a number of chaperones and foldases to control the activation, and prevent the misfolding of steroid receptors and kinases *(27)*. HtpG, the *E. coli* Hsp90 homolog, is a 71-kDa protein that forms homodimers at physiological temperatures *(28,29)*. It exhibits chaperone activity in vitro and appears to bind substrate proteins via its N-terminus ***(28)***. In vivo, HtpG has been shown to participate in *de novo* folding events (*see* **Note 2**), apparently by enhancing the ability of the DnaK-DnaJ-GrpE system to interact with partially folded proteins ***(26)***. At present, however, little is known about its cellular function.

IbpA and IbpB are two highly homologous 16-kDa polypeptides belonging to the small heat shock protein (sHsp) family and believed to originate from a gene duplication event. They were first identified as contaminants associated with heterologous protein inclusion bodies ***(30)***. The *ibpAB* operon is transcribed from a σ^{32}-dependent promoter that undergoes the highest level of heat shock induction upon culture transfer to 50°C ***(31)***. Purified IbpB exhibits chaperone function and forms large amorphous aggregates that dissociate into ≈600-kDa oligomers following exposure to high temperatures ***(32)***. The bacterial sHsps have been proposed to act as "holder" chaperones that bind partially folded polypeptides on their surfaces until stress has abated and "folding" chaperones become available. Indeed, IbpB-bound proteins are efficiently reactivated by DnaK-DnaJ-GrpE (but not GroEL-GroES) in vitro ***(33)***. Yet, there is no obvious increase in host protein misfolding when *ibpAB* null mutants are exposed to high temperatures ***(18,24)***, and the absence or overproduction of IbpA/B does not impact the folding/aggregation behavior of a number of model substrates in vivo ***(26)***. At present, the precise function of the bacterial sHsps remains unclear.

1.2.3. Trigger Factor: A PPIase and Molecular Chaperone

The *trans* conformation of X-Pro bonds is energetically favored in nascent protein chains. However, about 5% of all prolyl peptide bonds are found in a *cis* conformation in native proteins. The *trans* to *cis* isomerization of X-Pro bonds is a late, rate-limiting step in the folding of many polypeptides and is catalyzed in vivo by peptidyl prolyl *cis/trans* isomerases (PPIases). To date three PPIases—trigger factor (TF), SlyD and SlpA—have been identified in the cytoplasm of *E. coli* ***(34,35)***. Among these, the most extensively characterized is TF (reviewed in **ref. *36***), a 48-kDa protein which associates at 1:1 stoichiometry with about 30–40% of the cell ribosomes and is specific for their 50S subunit. TF is a modular protein consisting of three independent folding units. The N-terminus domain is necessary and sufficient for ribosome binding. The central domain shares weak homology with the FK506-binding proteins (FKBP) family and contains the PPIase active site. The C-terminal domain is more poorly characterized and may play a role in mediating the association of ribosome-bound TF with nascent polypeptides. Although the isolated central fragment displays high PPIase activity against short peptides, it does not efficiently catalyze the isomerization of peptidyl-prolyl bonds on large proteins

(37). This is presumably due to the fact that either (or both) the N- or C-terminal domain of TF contain polypeptide binding sites that play an important role by presenting portions of a large protein chain to the PPIase domain. Thus, TF exhibits both PPIases and chaperone activity and these functions are associated with different physical locations.

Mutations in the trigger factor gene (*tig*) and the *dnaKJ* operon are synthetically lethal ***(38,39)***. suggesting that there is an overlap between DnaK and TF in *de novo* protein folding (*see* **Fig. 1**). TF appears to preferentially interacts with short (<30-kDa) polypeptides emerging from the ribosomes, while larger proteins are more likely to be engaged by the DnaK-DnaJ-GrpE system ***(39)***. Since the fraction of newly synthesized proteins that co-immunoprecipitates with DnaK increases in Δ*tig* cells, the DnaK-DnaJ-GrpE team can however compensate for TF absence. It should finally be noted that TF has been found in association with GroEL in vivo ***(40)***, suggesting that it may also function in delivering partially folded substrates to the chaperonins (*see* **Fig. 1**).

1.2.4. Secretory Chaperones: SecB and SRP

To be efficiently engaged by the Sec translocation machinery ***(41)***, secreted proteins must be maintained in an extended conformation. The vast majority of exported proteins (those synthesized with a cleavable signal sequence) rely on either generic molecular chaperone systems, such as DnaK-DnaJ and GroEL-GroES, or specialized chaperones to maintain export-competence. The most extensively studied secretory chaperone is SecB, a non-Hsp present at low levels in the cell cytoplasm and implicated in the export of about 10 periplasmic and outer membrane host proteins ***(42)***. SecB is organized as a tetramer of 16-kDa identical subunits and binds positively charged and flexible sites in its substrates at multiple locations. During the process, SecB undergoes a conformational change that exposes structured hydrophobic domains to the solvent and allows the formation of a tight complex with the mature region of precursor proteins, maintaining the signal sequence available for interactions with the export machinery. SecB-bound proteins are transferred to SecA which directs them to the SecYE pore and energizes translocation using ATP hydrolysis ***(41)***.

By contrast, most integral membrane proteins make use of the bacterial signal recognition particle (SRP), a hybrid complex consisting of a 48-kDa GTPase termed Ffh and a 4.5S RNA about 100 nt in length. SRP interacts with the non-cleavable and highly hydrophobic transmembrane segments (TMS) of nascent inner membrane proteins and promotes their cotranslational targeting to the SecYE pore in a process that involves FtsY, the SRP receptor ***(43)***. How does the cell discriminate between SecB-dependent and SRP-dependent proteins? Based on crosslinking experiments, it has been proposed that TF preferentially interacts with SecB-dependent proteins, thereby precluding SRP binding ***(44)***. However, a recent study performed under more physiologically relevant conditions ***(45)*** indicates that SRP has the greatest affinity for highly hydrophobic targeting signals (e.g., the TMS of inner membrane proteins), since SecB-dependent proteins can be rerouted to the SRP pathway when the hydrophobicity of their signal peptide is increased or when their leader peptide is replaced by a TMS. Thus, the composition of the export targeting signal appears to play a dominant role in determining whether SecB- or SRP-dependent export will take place.

1.3. Periplasmic Folding Modulators

1.3.1. Chaperones and PPIases: Skp, FkpA, SurA, PpiA and PpiD

While many cytoplasmic chaperones rely on ATP-driven conformational changes to energize protein folding or refolding, the periplasm of *E. coli* does not contain a pool of ATP, and the existence of periplasmic molecular chaperones was originally doubted. However, it is now clear that, in addition to specialized chaperones (e.g., the PapD and FimC proteins which are involved in the assembly of the pilus and fimbriae, respectively), the periplasmic space contains folding helpers that have evolved to assist outer membrane biogenesis ***(46)***.

Among these is Skp (OmpH), a highly basic 17-kDa periplasmic protein that was originally believed to be a DNA binding protein (*see* **Note 3**). *skp* is the first gene of a dicistronic operon transcribed by RNA polymerase (E) complexed with the extracytoplasmic alternative sigma factor, σ^E ***(47)***. The principal physiological role of Skp appears to be in the transport/assembly of outer membrane proteins ***(48)*** and/or lipopolysaccharides ***(49)***. However, this protein also functions as a general molecular chaperone as evidenced by the fact that it facilitates the phage display and folding of single chain antibody fragments targeted to the periplasmic space ***(49–51)***.

To date four PPIases, FkpA, SurA, PpiA (RotA), and PpiD have been identified in the *E. coli* periplasm. Both FkpA and SurA belong to the σ^E regulon ***(47)*** while *ppiD* is transcribed from a σ^{32}-dependent promoter, a surprising observation in view of the fact that $E\sigma^{32}$ is normally responsible for the transcription of cytoplasmic Hsps ***(52)***. PpiA is dispensable and does not appear to play a crucial role in folding events, since disruption of its structural gene does not affect the folding of periplasmic and outer membrane proteins, and PpiA overproduction has little influence on the recovery yields of recombinant proteins targeted to the periplasm ***(49,53–55)***. The principal function of PpiD and SurA is to catalyze the isomerization of peptidyl-prolyl bonds in outer membrane proteins ***(52,56)***. However, SurA overexpression also improves the folding of unstable or aggregation-prone proteins in the periplasm ***(46)***. Among periplasmic PPIases, FkpA has the broadest folding helper role ***(46,49)***. This is most likely due to the fact that like TF, this homodimer of 26-kDa chains combines PPIase and chaperone activities ***(57,58)***.

1.3.2. Cysteine-Thiol Oxidoreductases

Stable disulfide bridges do not form in the cytoplasm of wild type *E. coli* cells, since they are rapidly reduced by the catalytic action of thioredoxins and glutaredoxins. By contrast, the periplasm is much more oxidizing and contains several enzymes (the Dsb proteins) responsible for the formation and reshuffling of disulfide bonds ***(59,60)***. The 21-kDa protein DsbA is the primary catalyst of disulfide bond formation in the periplasm. It efficiently donates its unstable active site disulfide to reduced substrates that are bound in a deep groove running along the accessible cysteine. Once reduced, DsbA is reoxidized by the inner membrane protein DsbB which exposes four cysteine residues to the periplasm. Isomerization of incorrect disulfide bonds is carried out mainly by DsbC, a 24-kDa homodimer that also contains a very reactive and unstable disulfide bond. To carry out nucleophilic attack of an erroneous disulfide and catalyze its rearrangement, DsbC must be maintained in a reduced form. This function is performed by the inner membrane protein DsbD which uses the reducing power of thioredoxin to keep DsbC reduced.

2. Materials

2.1. Plasmids for Chaperone and Foldase Co-Expression

A selection of ColE1 compatible plasmids suitable for the co-expression of various molecular chaperones and foldases using different induction strategies is compiled in **Table 1**.

2.2. Growth Media and Inducers

1. LB broth: Mix 10 g of Difco tryptone peptone, 5 g Difco yeast extract, and 10 g of NaCl in 950 mL of ddH_2O. Shake to dissolve all solids, adjust the pH to 7.4 with 5 *N* NaOH, and the volume to 1 L with ddH_2O; autoclave. If desired, add 5 mL of 20% (wt/vol) glucose from a filter sterilized stock (*see* **Note 6**).
2. Superbroth: Mix 32 g of Difco tryptone peptone, 20 g of Difco yeast extract, and 5 g of NaCl in 950 mL of ddH_2O. Shake to dissolve all solids, adjust the pH to 7.4 with 5 *N* NaOH, and the volume to 1 L with ddH_2O; autoclave. If desired, add 5 mL of 20% (wt/vol) glucose from a filter sterilized stock (*see* **Note 6**).
3. Terrific broth: Mix 12 g of Difco tryptone peptone, 24 g of Difco yeast extract, and 4 mL of glycerol in 900 mL of ddH_2O until all solids have dissolved; autoclave. After cooling to 60°C, add 100 mL of a sterile solution containing 170 m*M* KH_2PO_4, 720 m*M* K_2HPO_4 (made by mixing 2.31 g of KH_2PO_4 and 12.54 g of K_2HPO_4 into 90 mL of ddH_2O, adjusting the volume to 100 mL and autoclaving).
4. Isopropyl-β-D-thiogalactopyranoside (IPTG). For a 1 *M* stock, dissolve 2.38 g of IPTG into 8 mL of ddH_2O in a graduated tube and adjust the volume to 10 mL.

Filter sterilize through a 0.22 μm filter, dispense 250 μL aliquots into sterile microcentrifuge tubes and store at –20°C (*see* **Note 7**).

5. Arabinose stock: For a 20% (wt/vol) stock, dissolve 0.2 g of L(+) arabinose into 800 μL of ddH_2O in a graduated Eppendorf tube and adjust the volume to 1 mL. Filter sterilize. Make fresh as needed.
6. Anhydrotetracycline is available from Clonetech (Palo Alto, CA) as a 2 mg/mL stock solution in ethanol that should be stored at –20°C (*see* **Note 8**).

2.3. Cell Fractionation

2.3.1. Soluble and Insoluble Fractions

1. Potassium phosphate monobasic: For a 50 m*M* stock, dissolve 6.8 g of KH_2PO_4 in 950 mL of ddH_2O, adjust the pH to 6.5 and the volume to 1 L. Store at 4–8°C.
2. Upper Tris buffer: Dissolve 15 g of Tris base in 200 mL of ddH_2O. Adjust the pH to 6.8 and the volume to 250 mL with ddH_2O. Store at 4°C
3. 1X Sodium dodecyl sulfate (SDS) loading buffer: Dissolve 182 mg of dithiothreitol (DTT) into 5.8 mL ddH_2O. Add 1.8 mL of upper Tris buffer, 1 mL of 100% glycerol, 3.0 mL of 20% (wt/vol) SDS and 0.4 mL 0.05% (wt/vol) Bromphenol blue. Store at room temperature.
4. Methanol.
5. Chloroform.
6. SLM-Aminco French pressure cell and press or equivalent.

2.3.2. Periplasmic Fractions

1. Buffer A: (0.03 *M* Tris-HCl, pH 7.3): Dissolve 0.189 g of Tris-HCl in 30 mL of sterile ddH_2O. Adjust the pH to 7.3 and the volume to 40 mL with sterile ddH_2O. Filter sterilize and store at room temperature.
2. Buffer B (osmotic shock solution): Mix 4 g sucrose, 30 μL 0.5 *M* ethylene diaminetetraacetic acid (EDTA) pH 8.0, and 8 mL of Buffer A in a graduated tube. Adjust the volume to 10 mL with Buffer A, filter sterilize and store at room temperature.
3. Sterile, ice cold ddH_2O (*see* **Note 9**).
4. Rubber policeman.

2.3.3. Membrane Fractions

1. Potassium phosphate monobasic (*see* **Subheading 2.3.1.**).
2. Metrizamide (Sigma).
3. Ultracentrifuge with Beckman SW-55 rotor or equivalent.
4. Bausch and Lomb refractometer or equivalent.

3. Methods

3.1. Co-Expressing Folding Modulators

Although it is now clear that molecular chaperones associate with partially folded substrate proteins via hydrophobic and charge interactions, it remains impossible to predict whether the co-expression of a particular molecular chaperone or foldase will improve the folding of a recombinant polypeptide since

Table 1
Selected ColE1-Compatible Plasmids for Molecular Chaperone and Foldase Co-Expression

Plasmid[a]	Chaperone or Foldase genes or Operons[b]	Promoter[c]	Resistance[d]	Source or reference
Sigma32 Co-Expression				
pσ^{32}	*rpoH*	*lac*	Cm	***(61)***
DnaK-DnaJ and DnaK-DnaJ-GrpE Co-Expression				
pDnaK/J	*dnaK-dnaJ*	*lac* and native	Cm	A. A. Gatenby
pOFXlac-KJ1, 2, or 3	*dnaK-dnaJ*	*lac*	1 = Sp-Km 2 = Sp-Cm 3 = Sp	***(62)***
pOFXtac-KJ1, 2, or 3	*dnaK-dnaJ*	*tac*	1 = Sp-Km 2 = Sp-Cm 3 = Sp	***(62)***
pOFXbad-KJ1, 2, or 3	*dnaK-dnaJ*	*araB*	1 = Km 2 = Tc 3 = Ap	***(62)***
pOFXlac-KJE1, 2, or 3	*dnaK-dnaJ-grpE*	*lac*	1 = Sp-Km 2 = Sp-Cm 3 = Sp	***(62)***
pOFXtac-KJE1,2,3	*dnaK-dnaJ-grpE*	*tac*	1 = Sp-Km 2 = Sp-Cm 3 = Sp	***(62)***
pOFXbad-KJE1, 2, or 3	*dnaK-dnaJ-grpE*	*araB*	1 = Km 2 = Tc 3 = Ap	***(62)***

pKJE5 or 8	*dnaK-dnaJ-grpE*	5 = *trp* 8 = *araB*	Km	***(63)***
pKJE7	*dnaK-dnaJ-grpE*	*araB*	Cm	***(63)***
GroEL-GroES Co-Expression				
pGroESL	*groES-groEL* operon	*lac* and native	Cm	***(64)***
pOFXlac-SL1, 2, or 3	*groES-groEL* operon	*lac*	1 = Sp-Km 2 = Sp-Cm 3 = Sp	***(62)***
pOFXtac-SL1, 2, or 3	*groES-groEL* operon	*tac*	1 = Sp-Km 2 = Sp-Cm 3 = Sp	***(62)***
pOFXbad-SL1, 2, or 3	*groES-groEL* operon	*araB*	1 = Km 2 = Tc 3 = Ap	***(62)***
pGro6, or 12	*groES-groEL* operon	6 = *trp* 12 = *araB*	Km	***(63)***
pGro7, or 11	*groES-groEL* operon	7 = *araB* 11 = *Pzt-1* (*tetA*)	Cm	***(63)***
DnaK-DnaJ-(GrpE) and GroEL-GroES Co-Expression				
pG-KJE3	*groES-groEL-dnaK-dnaJ-grpE*	*araB*	Cm	***(63)***
pG-KJE7	*groES-groEL-dnaK-dnaJ-grpE*	*araB*	Km	***(63)***
pG-KJE8 (*see* **Note 4**)	*dnaK-dnaJ-grpE* *groES-groEL*	*araB* *Pzt-1* (*tetA*)	Km	***(65)***
Trigger Factor Co-Expression				
pTf16	*tig*	*araB*	Cm	***(65)***

(continued)

Table 1 *(continued)*

Plasmid[a]	Chaperone or Foldase genes or Operons[b]	Promoter[c]	Resistance[d]	Source or reference
Trigger Factor and GroEL-GroES Co-Expression				
pG-Tf2	*groES-groEL-tig*	*Pzt-1 (tetA)*	Cm	***(65)***
pG-Tf3	*groES-groEL*	*Pzt-1 (tetA)*	Cm	***(65)***
	tig	*araB*		
ClpB Co-Expression				
pClpB	*clpB*	native	Cm	***(18)***
HtpG Co-Expression				
pHtpG	*htpG*	native	Cm	***(18)***
IbpA/B Co-Expression				
pIbp	*ibpA-ibpB*	*lac* and native	Cm	***(18)***
SecB Co-Expression				
pAB	*secB*	*araB*	Cm	***(66)***
SRP Co-Expression				
pHDB7	*ffh*	native	Cm	***(45)***
pHQ3	*ffh*	native	Cm	***(45)***
	ffs	native		
PHQ4	*ffh*	native	Cm	***(45)***
	ffs	native		
	ftsY	native		
SecB and Skp Co-Expression				
pHELP1	*secB-skp*	*araB*	Cm	***(51)***

FkpA Co-Expression				
pSR3170	*fkpA*	native	Ap	***(46)***
SurA Co-Expression				
pDM1554	*surA*	native	Ap	***(46)***
DsbA Co-Expression				
pBADdsbA	*dsbA*	*araB*	Cm	***(67)***
DsbC Co-Expression				
pBADdsbC	*dsbC*	*araB*	Cm	***(67)***
DsbA-DsbB Co-Expression				
pDbAB1	*dsbA-dsbB*	*araB*	Cm	***(68)***
DsbA-DsbC Co-Expression				
pDbAC1	*dsbA-dsbC*	*araB*	Cm	***(68)***
DsbC-DsbD Co-Expression				
pDbCD1 (*see* **Note 5**)	*dsbC-dsbD*	*araB*	Cm	***(68)***
DsbA-DsbB-DsbC-DsbD Co-Expression				
pDbABCD1 (*see* **Note 5**)	*dsbA-dsbB-dsbC-dsbD*	*araB*	Cm	***(68)***

[a]All plasmids listed in Table 1 contain a p15A origin of replication.

[b]Only *dnaK-dnaJ*, *groES-groEL* and *ibpA-ibpB* are authentic operons. All other polycistrons (indicated by hyphenated genes) are artifical with gene order as indicated. Operons and genes listed on different lines within an entry are located on distal regions of the plasmid and are typically under transcriptional control of different promoters.

[c]Promoters listed on different lines within an entry control the transcription of the corresponding genes or operons in the "Chaperone and Foldase Gene and Operons" column. Numbers correspond to the different plasmid names in column 1.

[d]Abbreviations are: Ap, ampicillin resistance; Cm, chloramphenicol resistance; Km, kanamycin resistance; Sp, spectinomycin resistance; Tc, tetracycline resistance. Numbers correspond to the different plasmid names in column 1.

the binding of folding helpers to their substrates appear to largely depend on the folding pathway rather than on the primary structure. Thus, the best approach to examine the influence of folding modulators co-expression is to select a standard genetic background (e.g., MC4100, W3110 or BL21), transform the cells with ColE1-compatible plasmids encoding various chaperones and foldases (*see* **Table 1**) thereby generating a set of test strains, and further introduce the plasmid encoding the gene of interest in these cells using a ColE1-based construct. Additional comments about the usefulness and limitations of specific folding modulators are provided in the following sections.

3.1.1. DnaK-DnaJ and GroEL-GroES Co-Expression

A large number of studies have shown that an increase in the intracellular concentration of the major cytoplasmic folding machines, DnaK-DnaJ-GrpE (*see* **Notes 10** and **11**) or GroEL-GroES can greatly improve the solubility of aggregation-prone proteins expressed in the cytoplasm of *E. coli* (*see* **Note 12** and **ref.** ***69*** for a review). Certain proteins need high levels of DnaK-DnaJ to reach a proper conformation, but will not respond to an increase in GroEL-GroES concentration (e.g., **ref.** ***70***). Others appear to transit rapidly (if at all) through the DnaK-DnaJ-GrpE system, but exhibit a strong requirement for high levels of GroEL-GroES to reach a soluble form (e.g., **ref.** ***26***). In some rare cases (and usually for unstable polypeptides), overproduction of either DnaK-DnaJ-GrpE or GroEL-GroES has a similar beneficial effect ***(63)***. Nevertheless, the combined expression of both major chaperone systems does not usually lead to a synergistic improvement in the folding of aggregation-prone proteins, although there are some exceptions ***(65)***. Since the separate co-overproduction of DnaK-DnaJ or GroEL-GroES reduces the metabolic burden on the cell, it is recommended to first examine the effect of each major chaperone system independently before co-expression of both operons is attempted (*see* **Note 13**). It should finally be noted that although GroES-capped GroEL can only encapsulate proteins smaller than 60-kDa, GroEL-GroES overproduction may still improve the folding of larger proteins (e.g., **ref.** ***71***; *see* **Note 14**).

3.1.2. Trigger Factor Co-Expression

As discussed in **Subheading 1.2.3.**, TF and DnaK-DnaJ-GrpE appear to carry out redundant but somewhat distinct functions in the cell. Although it has been argued that TF interacts preferentially with small polypeptides (Mr < 30-kDa; **ref.** ***39***), co-expression of this foldase can also increase the solubility of large proteins (e.g., the 150-kDa human oxygen-regulated protein; **ref.** ***65***). For the few documented cases in which TF co-expression has been shown to exert a beneficial effect (*see* **Note 15**), a large improvement in solubility was also observed upon DnaK-DnaJ-GrpE overproduction ***(65)***. Thus, it is unlikely that TF overproduction will alleviate the misfolding of proteins that do not respond to an increase in the intracellular concentration of DnaK-DnaJ-GrpE and/or GroEL-GroES (unpublished data). On the other hand, combining TF and GroEL-GroES co-expression led to significant increase in the solubility of human lysozyme and human oxygen-regulated protein relative to overproduction of TF alone. This is presumably due to the fact that TF-bound proteins are more efficiently delivered to the downstream GroEL-GroES system under these conditions ***(65)***.

3.1.3. ClpB and HtpG Co-Expression

As discussed in **Subheading 1.2.2.**, ClpB—and to a lesser extent HtpG—become important for the optimal folding of proteins under conditions of stress (e.g., at high temperatures and when recombinant proteins are massively over-produced; **ref. *26***). However, we have found that overproducing these Hsps fails to improve the folding of several aggregation-prone polypeptides, and that ClpB does not disaggregate preformed recombinant protein inclusion bodies (*see* **ref. *26***; unpublished data). Although it remains possible that the coordinated expression of DnaK-DnaJ and ClpB/HtpG may have a beneficial effect on the misfolding of certain proteins, independent co-expression of these chaperones may not be useful in the majority of cases.

3.1.4. Overexpressing all Cytoplasmic Hsps

Since a large number of cytoplasmic molecular chaperones are transcribed by $E\sigma^{32}$, it is possible to globally increase their synthesis by supplying the cells with a plasmid bearing the *rpoH* gene (which encodes σ^{32}; **Table 1**). This approach is typically slightly less effective than the direct overproduction of a single necessary chaperone ***(61,72)*** and will lead to the induction of heat shock proteases (e.g., Lon, ClpAP, ClpXP and ClpYQ) which may be a problem if the protein of interest is unstable. On the other hand, σ^{32} overproduction provides a rapid means of assessing whether overproduction of DnaK-DnaJ-GrpE or GroEL-GroES will meet with success. It is also possible to induce the synthesis of both cytoplasmic and periplasmic Hsps by supplementing the cultures with 3% (v/v) ethanol prior to induction of heterologous protein production ***(72)***. Furthermore, combining ethanol supplementation and direct chaperone co-expression may synergistically improve the folding of certain—but not all—aggregation-prone proteins (*see* **Note 16** and **ref.** ***72***).

3.1.5. SecB and SRP Co-Expression

A number of groups have reported that SecB co-expression facilitates the secretion of proteins targeted to the *E. coli* periplasm (reviewed in ***(69)***). This effect is however highly dependent on the identity of the exported polypeptide and that of the signal sequence ***(73)***. Thus, as with other chaperones, the usefulness of SecB co-expression should be investigated on a case-by-case basis (*see* **Note 17**). To date, there is no published information on how an increase in SRP concentration affects the expression/insertion/stability of heterologous membrane proteins expressed in *E. coli*. This approach may prove fruitful and will likely require that Ffh, the 4.5S RNA (encoded by the *ffs* gene), and possibly the FtsY receptor, be simultaneously overproduced.

3.1.6. Skp and FkpA Co-Expression

Both Skp and FkpA can independently improve the folding of periplasmic proteins and the phage display of antibody fragments ***(49–51)***. Depending on the identity of the substrate (and most likely on its folding pathway), either Skp or FkpA co-expression may prove more effective ***(49)***. Thus, the influence of each of these folding helpers should be investigated when dealing with a recombinant protein that misfolds in the periplasm. Although effects are not additive, marginal improvement in folding may occur when both Skp and FkpA are co-expressed relative to Skp or FkpA overproduction alone ***(49)***.

3.1.7. Co-Expression of the Dsb Proteins

Although co-expression of DsbA alone can improve the solubility of secreted proteins containing a single (or possibly a simple set) of disulfide bridges ***(74)***, an increase in the intracellular concentration of the protein disulfide isomerase DsbC appears to be much more effective in promoting the efficient folding (and/or limiting the aggregation) of more complex disulfide-bonded proteins produced in the periplasm of *E. coli* ***(68,75,76)***. For proteins that form aberrant disulfide bridges between non-consecutive cysteine residues, DsbD, the cognate reducer of DsbC, should be simultaneously co-expressed to maintain high protein disulfide isomerase activity ***(68)***. Finally, joint overproduction of the DsbA-DsbB-DsbC-DsbD set can facilitate both the transport and folding of proteins targeted to the periplasm of *E. coli* ***(75)***.

3.2. Assessing the Influence of Folding Modulators Co-Expression

The most sensitive technique to determine if co-expression of a specific folding modulator exerts a beneficial effect on the folding of a target protein involves biological activity assays. However, these may not be available or may be difficult to perform. A simple alternative is to determine whether an increase in the intracellular concentration of molecular chaperones or foldases affects the partitioning of the recombinant protein of interest between soluble and insoluble cellular fractions.

3.2.1. Growth and Induction Conditions

1. Grow an overnight inoculum of cells harboring plasmids encoding the heterologous protein and the desired chaperone(s) (*see* **Table 1**) in 5 mL of LB medium

(*see* **Subheading 2.2.**) supplemented with the appropriate antibiotics at 37°C (*see* **Note 18**).

2. On the next day inoculate a 125-mL shake flask containing 25 mL of either LB, Superbroth, or Terrific broth (**Subheading 2.2.**; *see* **Note 19**) supplemented with the appropriate antibiotics. If the chaperone/foldase gene(s) are under transcriptional control of inducible promoters, the medium may be further supplemented (*see* **Subheading 2.2.**) with 0.2% arabinose, in the case of the *araB* (P_{BAD}) promoter (*see* **Note 6**); 10 ng/mL anhydrotetracycline (*see* **Note 8**) or tetracycline in the case of the *tetA* (*Pzt-1*) promoter; 250 µ*M* IPTG in the case of the *lac* promoter to accumulate chaperones and/or foldases (*see* **Note 20**).
3. Grow the cells to an optical density at 600 nm (A_{600}) of approx 0.5 (LB cultures) or 1.0 (Superbroth and Terrific broth). Record the exact A_{600} value and collect a pre-induction sample for solubility or cellular localization analysis (*see* **Subheadings 3.2.2.–3.2.4.**); induce heterologous protein expression (*see* **Note 21**).
4. Harvest samples 1, 3 and 24 h post-induction, record the exact A_{600} value (*see* **Note 22**) and determine solubility and cellular localization as described in the following paragraphs.

3.2.2. Preparing Whole Cells, Soluble and Insoluble Fractions

1. For each time point (including pre-induction) collect a 1 mL culture sample in an Eppendorf tube for whole cell analysis and a 3-mL sample in a Corex tube chilled on ice for fractionation of soluble and insoluble proteins.
2. Prepare whole cell fractions as follows. Centrifuge the 1 mL samples at 8000*g* for 2 min in a microfuge. Discard the supernatant and remove residual moisture using a Kimwipe. Resuspend the cell pellet into (165 × A_{600}) µL of 1X SDS loading buffer. Freeze at –20°C until the day of use.
3. Spin the 3-mL samples at 6500*g* for 10 min in a refrigerated centrifuge. Discard the supernatant and remove residual moisture with a Kimwipe. Resuspend the cell pellet in 3 mL of potassium phosphate monobasic (*see* **Subheading 2.3.1.**).
4. Disrupt the cells by French pressing at 10,000 psi and 4°C. Using a pipetman fitted with a 5-mL tip, estimate sample volume to account for losses during French pressing.
5. Centrifuge at 8000*g* for 10 min in a refrigerated centrifuge.
6. Save 1.5 mL of the supernatant (soluble fraction) in an Eppendorf tube and place on ice; discard the rest of the supernatant and remove excess moisture with a Kimwipe.
7. Calculate the volume (in µL) of 1X SDS buffer to be added to the pellet (this is the insoluble fraction which contains cytoplasmic or periplasmic inclusion bodies) using the following formula: (165 × A_{600} × Sample volume after French pressing [mL])/3.
8. Resuspend the insoluble material in the appropriate volume of 1X SDS buffer. Transfer to Eppendorf tube and freeze at –20°C until the day of use.
9. Transfer (90.91/A_{600}) µL of soluble fraction from **step 6** into a fresh, graduated Eppendorf tube and precipitate the proteins as follows (*see* **Note 23**).
10. Adjust the volume to 500 µL with methanol and vortex briefly.

11. Add 100 µL of chloroform, vortex briefly, and centrifuge at 16,000*g* for 30 s.
12. Add 350 µL of ddH_2O and vortex for 20 s. The sample should become opaque. If this is not the case add an additional 100–200 µL of ddH_2O.
13. Centrifuge for 2 min at 16,000*g* in a microfuge. A white protein pellet should be visible at the interface.
14. Remove the majority of the top layer using a Pasteur pipet and discard (*see* **Note 24**).
15. Adjust the volume to 750 µL with methanol and mix by inverting 4–5 times.
16. Centrifuge for 2 min at 16,000*g* in a microfuge.
17. Remove the majority of the methanol using a Pasteur pipet being careful not to disturb the precipitated protein pellet and lyophilize for 5–10 min.
18. Resuspend into 15 µL of 1X SDS loading buffer and freeze at –20°C until the day of use.
19. To visualize protein on minigels, use 15 µL of whole cell sample from **step 2**, 5 µL of insoluble fraction from **step 8** (add 10 µL of 1X SDS buffer), and 15 µL of soluble fraction from **step 18** (*see* **Note 25**). Heat all samples for 5 min at 95°C before loading onto SDS-polyacrylamide minigels. For immunoblots, use 3–4 times less material.

3.2.3. Preparing Periplasmic Fractions

If desired, periplasmic fractions can be efficiently separated from cytoplasmic fractions by osmotic shock *(77)*. Volumes may be scaled up if this procedure is used for purifying soluble recombinant proteins secreted in the periplasm of *E. coli*.

1. Immediately after collecting a 3 mL culture sample, centrifuge at 2500*g* for 5 min at room temperature. Discard the supernatant and remove excess moisture with a Kimwipe.
2. Resuspend the cell pellet in 1.2 mL of Buffer A (*see* **Subheading 2.3.2.**) using a rubber policeman.
3. Centrifuge at 5000*g* for 5 min at room temperature. Discard the supernatant and remove excess moisture with a Kimwipe.
4. Resuspend the cell pellet in 0.6 mL of Buffer A using a rubber policeman. Slowly add 0.6 mL of Buffer B (*see* **Subheading 2.3.2.**) while continuously agitating with the rubber policeman. Allow the cells to equilibrate for 15 min at room temperature.
5. Centrifuge at 2500*g* for 5 min at room temperature. Discard the supernatant and remove excess moisture with a Kimwipe.
6. Rapidly add 1.2 mL of ice cold ddH_2O (*see* **Note 9**), resuspend the cell pellet with the rubber policeman and incubate the sample on ice for 10 min.
7. Centrifuge at 3600*g* for 5 min at 4°C. Recover and save the supernatant (periplasmic fraction).
8. Resuspend the pellet in 3 mL of 50 m*M* potassium phosphate monobasic, pH 6.5, and prepare soluble and insoluble fractions as described in **Subheading 3.2.2.**

9. To load an equal amount of periplasmic proteins, transfer ($36.36/A_{600}$) μL of periplasmic fraction from **step 7** into a clean Eppendorf tube and perform methanol/chloroform precipitation as described in **Subheading 3.2.2., steps 10–18** (*see* **Note 25**).
10. Load 15 μL on minigels for Coomassie blue detection and 3–4 times less material for immunoblots.

3.2.4. Distinguishing Insoluble and Membrane-Associated Proteins

Insoluble fractions obtained by following the protocol of **Subheading 3.2.2.** consist of aggregated and membrane-associated proteins. If needed, sedimentation on metrizamide gradients *(78)* can be used to determine if a polypeptide accumulates as *bona fide* inclusion bodies or if it is associated with cell membranes.

1. Select a convenient time point (typically 2–3 h following induction of heterologous protein production) to carry out the analysis.
2. Harvest the whole culture, record the exact A_{600} value and sediment the cells by centrifugation at 6500*g* for 10 min in a refrigerated centrifuge. Discard the supernatant and remove residual moisture with a Kimwipe.
3. Resuspend the cell pellet in potassium phosphate monobasic (*see* **Subheading 2.3.1.**) so that the final A_{600} is 6.0.
4. Disrupt the sample by French pressing at 10,000 psi and bring 0.5 mL of lysate to a density of 1.29 g/mL by addition of 0.38 g of solid metrizamide.
5. Transfer 0.2 mL to the bottom of a soft ultracentrifuge tube and gently layer 1.8 mL of a metrizamide solution of density 1.27 g/mL (0.68 g metrizamide/mL) over the lysate.
6. Centrifuge at 35,000 rpm and 4°C for 16 h in a Beckman SW-55 rotor or equivalent.
7. Pierce the bottom of the ultracentrifuge tube with a needle and collect successive fractions (9 are usually sufficient) into clean Eppendorf tubes.
8. Measure the refractive index (c) of each sample and calculate the density (ρ) in g/mL using the formula *(79)*: $\rho = (3.350 \times c) - 3.462$.
9. Carry out methanol/chloroform extraction (*see* **Subheading 3.2.2., steps 10–17**).
10. Resuspend the protein pellets into 45 μL of 1X SDS loading buffer, and use 15 μL aliquots for SDS-PAGE analysis. Inclusion body proteins will be predominantly found at the bottom of the gradient (high density fractions) while membrane-associated polypeptides are present at the top of the gradient.

4. Notes

1. Although numerous DnaK binding sites may be exposed to the solvent by a partially folded protein, only one or two DnaK monomers associate with a substrate polypeptide.
2. HtpG appears to be less important in *de novo* folding events than ClpB *(26)*. This may be related to the fact that ClpB actively dissociates and unfolds proteins on a dead end branch of their folding pathway, while HtpG serves as a "holder"

chaperone that maintains a subset of newly synthesized proteins in a conformation that is accessible to DnaK ***(26)***.

3. Skp was also misidentified as an outer membrane-associated protein but it was later shown that the protein is periplasmic ***(80)***.
4. Plasmid pG-KJE8 is a derivative of pG-KJE6 ***(63)*** containing a *rrnB* terminator downstream of the *grpE* gene in the artificial *dnaK-dnaJ-grpE* operon; this allows better separate control of GroEL-GroES and DnaK-DnaJ-GrpE co-expression.
5. Plasmids pDbCD1 and pDbABCD1 lack the 76 N-terminal amino acids of authentic DsbD. This truncated form of DsbD remains functional and can complement the defective phenotype of *dsbD* null mutants ***(68)***.
6. Glucose represses the *araB* promoter and should not be included when chaperone or foldase are to be co-expressed from this promoter.
7. IPTG aliquots should be held on ice following thawing. Left-over inducer should be discarded after 4–5 freeze-thaw cycles.
8. Although the *tetA* promoter is inducible with tetracycline, anhydrotetracycline is a preferred inducer since it does not exhibit antibiotic activity and hence does not interfere with cell growth.
9. When scaling up the osmotic shock procedure, include 1 m*M* $MgCl_2$ to stabilize membranes.
10. Although a number of DnaK-DnaJ co-expression plasmids also include *grpE* either cloned on a different location of the plasmid or as part of an artificial operon, the nucleotide exchange factor GrpE appears to function in a catalytic fashion and the single chromosomal copy of the gene is usually sufficient to deal with higher intracellular concentrations of DnaK-DnaJ.
11. For helping the folding of heterologous proteins, DnaK should never be expressed in the absence of DnaJ since this leads to filamentation and cell death ***(81)***.
12. For unclear reasons, co-expression of molecular chaperones and foldases may reduce the overall yields of recombinant protein and one should carefully weigh the improvement in solubility against the reduction in steady-state accumulation levels.
13. DnaK-DnaJ overexpression reduces the production of chromosomal GroEL, an effect that becomes more pronounced as the growth temperature increases ***(61)***. Thus, simultaneous co-expression of DnaK-DnaJ and GroEL-GroES may be required for those polypeptides requiring interactions with both folding machines.
14. This presumably occurs in a GroES-independent fashion via passive binding of folding intermediates to the GroEL toroid (the so-called "buffering" effect; *see* **refs.** ***82,83***).
15. TF co-expression has been reported to lead to cell filamentation ***(84)***. This may be a problem in fermentors.
16. This approach is not suitable in fermentor set-ups since ethanol will be rapidly scrubbed from the medium in well aerated tanks.
17. In some cases, DnaK-DnaJ or GroEL-GroES co-expression can facilitate the translocation of secreted proteins across the inner membrane and this approach should not be discounted.

18. For plasmids encoding ampicillin or kanamycin resistance, use 50 μg/mL of carbenicillin or neomycin since these antibiotics are more stable than ampicillin or kanamycin. Working concentrations for other antibiotics are: chloramphenicol, 34 μg/mL (in ethanol); tetracycline, 25 μg/mL; and spectinomycin, 50 μg/mL.
19. The use of very rich media such as Terrific broth or Superbroth results in higher biomass and recombinant protein accumulation. It also allows the induction of recombinant protein synthesis to be carried out at a later stage, thereby increasing recovery yields. On the other hand, the demand of host proteins for chaperones and foldases may be higher in fast growing cells, and this may have a negative impact on the folding of the target polypeptide. It is therefore recommended to perform initial experiments in LB.
20. These conditions should result in full induction of the promoters at the onset of the culture. However, if the accumulation of chaperones or foldases has a deleterious effect on growth, an alternative is to induce their synthesis 30–60 min prior to the induction of heterologous protein expression.
21. Although initial experiments may be performed at 37°C, low temperature expression often has an additive or synergistic effect on solubility. Thus, it may be useful to shift the culture to 30 or 25°C prior to induction of heterologous protein expression.
22. To remain in the dynamic range of the spectrophotometer dilute samples taken after mid-exponential phase 10-fold using 10 g/L NaCl before measuring the absorbance at 600 nm and multiply the reading by ten for calculations.
23. The methanol/chloroform extraction protocol *(85)* is more efficient than traditional TCA precipitation. It is recommended to carry out the extraction on duplicate samples so that enough material is available to run two gels or one gel and three or four immunoblots.
24. To avoid protein loss, leave a small layer of fluid on top of the interface. This does not affect extraction efficiency.
25. These volumes correspond to approximately 20 μg of total proteins.

Acknowledgment

This work was supported by grants from the BES division of the US National Science Foundation and the MBC division of the US American Cancer Society.

References

1. Anfinsen, C. B. (1973) Principles that govern the folding of protein chains. *Science* **181,** 223–230.
2. Gross, C. A. (1996) Function and regulation of the heat shock proteins *in Escherichia coli* and *Salmonella* Cellular and Molecular Biology (Neidhardt, F. C., Curtiss III, R., Ingraham, et al., eds.), ASM Press, Washington, D. C., pp. 1382–1399.
3. Bukau, B. and Horwich, A. L. (1998) The Hsp70 and Hsp60 chaperone machines. *Cell* **92,** 351–366.

4. Gottesman, M. E. and Hendrickson, W. A. (2000) Protein folding and unfolding by *Escherichia coli* chaperones and chaperonins. *Curr. Opin. Microbiol.* **3,** 197–202.
5. Richardson, A., Landry, S. J., and Georgopoulos, C. (1998) The ins and outs of a molecular chaperone machine. *Trends Biochem. Sci.* **23,** 138–143.
6. Fink, A. L. (1999) Chaperone-mediated protein folding. *Physiol. Rev.* **79,** 425–449.
7. Zhu, X., Zhao, X., Burkholder, W. F., et al. (1996) Structural analysis of substrate binding by the molecular chaperone DnaK. *Nature* **272,** 1606–1614.
8. Rüdiger, S., Germeroth, L., Schneider-Mergener, J., and Bukau, B. (1997) Substrate specificity of the DnaK chaperone determined by screening cellulose-bound peptide libraries. *EMBO J.* **16,** 1501–1507.
9. Rüdiger, S., Schneider-Mergener, J., and Bukau, B. (2001) Its substrate specificity characterizes the DnaJ co-chaperone as a scanning factor for the DnaK chaperone. *EMBO J.* **20,** 1042–1050.
10. Fenton, A., Kashi, Y., Furtak, K., and Horwich, A. L. (1994) Residues in chaperonin GroEL required for polypeptide binding and release. *Nature* **371,** 614–619.
11. Coyle, J. E., Jaeger, J., Gross, M., Robinson, C. V., and Radford, S. E. (1997) Structural and mechanistic consequences of polypeptide binding by GroEL. *Fold. Des.* **2,** 93–104.
12. Houry, W. A., Frishman, D., Eckerskorn, C., Lottspeich, F., and Hartl, F. U. (1999) Identification of in vivo substrates of the chaperonin GroEL. *Nature* **402,** 147–154.
13. Jakob, U. and Buchner, J. (1994) Assisting spontaneity: the role of hsp90 and small hsps as molecular chaperones. *Trends Biochem. Sci.* **19,** 205–211.
14. Squires, C. and Squires, C. L. (1992) The Clp proteins: proteolysis regulators or molecular chaperones? *J. Bacteriol.* **174,** 1081–1085.
15. Bardwell, J. C. and Craig, E. A. (1988) Ancient heat shock gene is dispensable. *J. Bacteriol.* **170,** 2977–2983.
16. Katayama, Y., Gottesman, S., Pumphrey, J., Ridikoff, S., Clark, W. P., and Maurizi, M. R. (1988) The two-component, ATP-dependent Clp protease of *Escherichia coli*: Purification, cloning, and mutational analysis of the ATP-binding component. *J. Biol. Chem.* **263,** 15,226–15,236.
17. Squires, C. L., Pedersen, S., Ross, B. M., and Squires, C. (1991) ClpB is the *Escherichia coli* heat shock protein F84.1. *J. Bacteriol.* **173,** 4254–4262.
18. Thomas, J. G. and Baneyx, F. (1998) Roles of the *Escherichia coli* small heat shock proteins IbpA and IbpB in thermal stress management: comparison with ClpA, ClpB, and HtpG in vivo. *J. Bacteriol.* **180,** 5165–5172.
19. Woo, K. M., Kim, K. I., Goldberg, A. L., Ha, D. B., and Chung, C. H. (1992) The heat shock protein ClpB in *Escherichia coli* is a protein-activated ATPase. *J. Biol. Chem.* **267,** 20,429–20,434.
20. Zolkiewski, M., Kessel, M., Ginsburg, A., and Maurizi, M. R. (1999) Nucleotide-dependent oligomerization of ClpB from *Escherichia coli*. *Protein Sci.* **8,** 1899–1903.

21. Park, S. K., Kim, K. I., Woo, K. M., et al. (1993) Site-directed mutagenesis of the dual translational initiation sites of the *clpB* gene of *E. coli* and characterization of its gene products. *J. Biol. Chem.* **268,** 20,170–20,174.
22. Goloubinoff, P., Mogk, A., Ben Zvi, A. P., Tomoyasu, T., and Bukau, B. (1999) Sequential mechanism of solubilization and refolding of stable protein aggregates by a bichaperone network. *Proc. Natl. Acad. Sci. USA* **96,** 13,732–13,737.
23. Zolkiewski, M. (1999) ClpB cooperates with DnaK, DnaJ and GrpE in suppressing protein aggregation. *J. Biol. Chem.* **274,** 28,083–28,086.
24. Mogk, A., Tomoyasu, T., Goloubinoff, P., et al. (1999) Identification of thermolabile *Escherichia coli* proteins: prevention of aggregation by DnaK and ClpB. *EMBO J.* **18,** 6934–6949.
25. Barnett, M. E., Zolkiewska, A., and Zolkiewski, M. (2000) Structure and activity of ClpB from *Escherichia coli*. Role of the amino- and -carboxyl-terminal domains. *J. Biol. Chem.* **275,** 37565–37571.
26. Thomas, J. G. and Baneyx, F. (2000) ClpB and HtpG facilitate *de novo* protein folding in stressed *Escherichia coli* cells. *Mol. Microbiol.* **36,** 1360–1370.
27. Buchner, J. (1999) Hsp90 & Co. - a holding for folding. *Trends Biochem. Sci.* **24,** 136–141.
28. Nemoto, T. K., Ono, T., and Tanaka, K. (2001) Substrate-binding character istics of proteins in the 90 kDa heat shock protein family. *Biochem. J.* **354,** 663–670.
29. Spence, J. and Georgopoulos, C. (1989) Purification and properties of the *Escherichia coli* heat shock protein, HtpG. *J. Biol. Chem.* **264,** 4398–4403.
30. Allen, S. P., Polazzi, J. O., Gierse, J. K., and Easton, A. M. (1992) Two novel heat shock genes encoding proteins produced in response to heterologous protein expression in *Escherichia coli*. *J. Bacteriol.* **174,** 6938–6947.
31. Chuang, S.-E., Burland, V., Plunkett III, G., Daniels, D. L., and Blattner, F. R. (1993) Sequence analysis of four new heat-shock genes constituting the *hslTS/ibpAB* and *hslVU* operons in *Escherichia coli*. *Gene* **134,** 1–6.
32. Shearstone, J. R. and Baneyx, F. (1999) Biochemical characterization of the small heat shock protein IbpB from *Escherichia coli*. *J. Biol. Chem.* **274,** 9937–9945.
33. Veinger, L., Diamant, S., Buchner, J., and Goloubinoff, P. (1998) The small heat-shock protein IbpB from *Escherichia coli* stabilizes stress-denatured proteins for subsequent refolding by a multichaperone network. *J. Biol. Chem.* **273,** 11,032–11,037.
34. Hottenrott, S., Schumann, T., Plückthun, A., Fischer, G., and Rahfeld, J.-U. (1997) The *Escherichia coli* SlyD is a metal ion-regulated peptidyl-prolyl *cis/trans* isomerase. *J. Biol. Chem.* **272,** 15,697–15,701.
35. Stoller, G., Rücknagel, K. P., Nierhaus, K. H., Schmid, F. X., Fischer, G., and Rahfeld, J.-U. (1995) A ribosome-associated peptidyl-prolyl *cis/trans* isomerase identified as the trigger factor. *EMBO J.* **14,** 4939–4948.
36. Hesterkamp, T. and Bukau, B. (1996) The *Escherichia coli* trigger factor. *FEBS Lett.* **389,** 32–34.

37. Zarnt, T., Tradler, T., Stoller, G., Scholz, C., Schmid, F. X., and Fischer, G. (1997) Modular structure of the trigger factor required for high activity protein folding. *J. Mol. Biol.* **271,** 827–837.
38. Deuerling, E., Schulze-Specking, A., Tomoyasu, T., Mogk, A., and Bukau, B. (1999) Trigger factor and DnaK cooperate in folding of newly synthesized proteins. *Nature* **400,** 693–696.
39. Teter, S. A., Houry, W. A., Ang, D., et al. (1999) Polypeptide flux through bacterial Hsp70: DnaK cooperates with trigger factor in chaperoning nascent chains. *Cell* **97,** 755–765.
40. Krandor, A., Sherman, M., Moerschell, R., and Goldberg, A. L. (1997) Trigger factor associates with GroEL in vivo and promotes its binding to certain polypeptides. *J. Biol. Chem.* **272,** 1730–1734.
41. Economou, A. (1998) Bacterial preprotein translocase: mechanism and conformational dynamics of a processive enzyme. *Mol. Microbiol.* **27,** 511–518.
42. Kim, J. and Kendall, D. A. (2000) Sec-dependent protein export and the involvement of the molecular chaperone SecB. *Cell Stress Chaperones* **5,** 267–275.
43. Muller, M., Koch, H. G., Beck, K., and Schafer, U. (2000) Protein traffic in bacteria: multiple routes from the ribosome to and across the membrane. *Prog. Nucleic Acid Res. Mol. Biol.* **66,** 107–157.
44. Beck, K., Wu, L. F., Brunner, J., and Muller, M. (2000) Discrimination between SRP- and SecA/SecB-dependent substrates involves selective recognition of nascent chains by SRP and trigger factor. *EMBO J.* **19,** 134–143.
45. Lee, H. C. and Bernstein, H. D. (2001) The targeting pathway of *Escherichia coli* presecretory and integral membrane proteins is specified by the hydrophobicity of the targeting signal. *Proc. Natl. Acad. Sci. USA* **98,** 3471–3476.
46. Missiakas, D., Betton, J.-M., and Raina, S. (1996) New components of protein folding in extracytoplasmic compartments of *Escherichia coli* SurA, FkpA and Skp/OmpH. *Mol. Microbiol.* **21,** 871–884.
47. Dartigalongue, C., Missiakas, D., and Raina, S. (2001) Characterization of the *Escherichia coli* σ^E Regulon. *J. Biol. Chem.* **23,** 23.
48. Chen, R. and Henning, U. (1996) A periplasmic protein (Skp) of *Escherichia coli* selectively binds a class of outer membrane proteins. *Mol. Microbiol.* **19,** 1287–1294.
49. Bothmann, H. and Pluckthun, A. (2000) The periplasmic *Escherichia coli* peptidylprolyl *cis,trans*-isomerase FkpA. I. Increased functional expression of antibody fragments with and without cis-prolines. *J. Biol. Chem.* **275,** 17,100–17,105.
50. Bothmann, H. and Plückthun, A. (1998) Selection for a periplasmic factor improving phage display and functional periplasmic expression. *Nat. Biotechnol.* **16,** 376–380.
51. Hayhurst, A. and Harris, W. J. (1999) *Escherichia coli* Skp chaperone co-expression improves solubility and phage display of single-chain antibody fragments. *Protein Expr. Purif.* **15,** 336–343.
52. Dartigalongue, C. and Raina, S. (1998) A new heat-shock gene, *ppiD*, encodes a peptidyl-prolyl isomerase required for the folding of outer membrane proteins in *Escherichia coli. EMBO J.* **17,** 3968–3980.

53. Battistoni, A., Mazzetti, A. P., Petruzzelli, R., et al. (1995) Cytoplasmic and periplasmic production of human placental glutathione transferase in *Escherichia coli. Protein Expression Purif.* **6,** 579–587.
54. Kleerebezem, M., Heutink, M., and Tommassen, J. (1995) Characterization of an *Escherichia coli rotA* mutant, affected in periplasmic peptidyl-prolyl *cis/trans* isomerase. *Mol. Microbiol.* **18,** 313–320.
55. Knappik, A., Krebber, C., and Plückthun, A. (1993) The effect of folding catalysts on the in vivo folding of different antibody fragments expressed in *Escherichia coli. Bio/Technology* **11,** 77–83.
56. Lazar, S. W. and Kolter, R. (1996) SurA assists the folding of *Escherichia coli* outer membrane proteins. *J. Bacteriol.* **178,** 1770–1773.
57. Arie, J. P., Sassoon, N., and Betton, J. M. (2001) Chaperone function of FkpA, a heat shock prolyl isomerase, in the periplasm of *Escherichia coli. Mol. Microbiol.* **39,** 199–210.
58. Ramm, K. and Plückthun, A. (2000) The periplasmic *Escherichia coli* peptidyl-prolyl *cis,trans* isomerase FkpA. II. Isomerase-independent chaperone activity in vitro. *J. Biol. Chem.* **275,** 17,106–17,113.
59. Debarbieux, L. and Beckwith, J. (1999) Electron avenue: pathways of disulfide bond formation and isomerization. *Cell* **99,** 117–119.
60. Missiakas, D. and Raina, S. (1997) Protein folding in the bacterial periplasm. *J. Bacteriol.* **179,** 2465–2471.
61. Thomas, J. G. and Baneyx, F. (1996) Protein misfolding and inclusion body formation in recombinant *Escherichia coli* cells overproducing heat-shock proteins. *J. Biol. Chem.* **271,** 11,141–11,147.
62. Castanié, M.-P., Bergès, H., Oreglia, J., Prère, M.-F., and Fayet, O. (1997) A set of pBR322-compatible plasmids allowing the testing of chaperone-assisted folding of proteins overexpressed in *Escherichia coli. Anal. Biochem.* **254,** 150–152.
63. Nishihara, K., Kanemori, M., Kitagawa, M., Yanaga, H., and Yura, T. (1998) Chaperone co-expression plasmids: differential and synergistic roles of DnaK-DnaJ-GrpE and GroEL-GroES in assisting folding of an allergen of Japanese cedar pollen, Cryj2, in *Escherichia coli. Appl. Environ. Microbiol.* **64,** 1694–1699.
64. Goloubinoff, P., Gatenby, A. A., and Lorimer, G. H. (1989) GroE heat-shock proteins promote assembly of foreign prokaryotic ribulose bisphosphate carboxylase oligomers in *Escherichia coli. Nature* **337,** 44–47.
65. Nishihara, K., Kanemori, M., Yanagi, H., and Yura, T. (2000) Overexpression of trigger factor prevents aggregation of recombinant proteins in *Escherichia coli. Appl. Environ. Microbiol.* **66,** 884–889.
66. Pérez-Pérez, J., Martínez-Caja, C., Barbero, J. L., and Gutiérrez, J. (1995) DnaK/DnaJ supplementation improves the periplasmic production of human granulocyte-colony stimulating factor in *Escherichia coli. Biochem Biophys. Res. Commun.* **210,** 524–529.
67. Bessette, P. H., Aslund, F., Beckwith, J., and Georgiou, G. (1999) Efficient folding of proteins with multiple disulfide bonds in the *Escherichia coli* cytoplasm. *Proc. Natl. Acad. Sci. USA* **96,** 13,703–13,708.
68. Kurokawa, Y., Yanagi, H., and Yura, T. (2001) Overproduction of bacterial protein disulfide isomerase (DsbC) and its modulator (DsbD) markedly enhances

periplasmic production of human nerve growth factor in *Escherichia coli. J. Biol. Chem.* **276,** 14,393–14,399.
69. Thomas, J. G., Ayling, A., and Baneyx, F. (1997) Molecular chaperones, folding catalysts and the recovery of biologically active recombinant proteins from *E. coli*: to fold or to refold. *Appl. Biochem. Biotechnol.* **66,** 197–238.
70. Thomas, J. G. and Baneyx, F. (1996) Protein folding in the cytoplasm of *Escherichia coli*: requirements for the DnaK-DnaJ-GrpE and GroEL-GroES molecular chaperone machines. *Mol. Microbiol.* **21,** 1185–1196.
71. Roman, L. J., Sheta, E. A., Martasek, P., Gross, S. S., Liu, Q., and Masters, B. S. S. (1995) High-level expression of functional rat neuronal nitric oxide synthase in *Escherichia coli. Proc. Natl. Acad. Sci. USA* **92,** 8428–8432.
72. Thomas, J. G. and Baneyx, F. (1997) Divergent effects of chaperone overexpression and ethanol supplementation on inclusion body formation in recombinant *Escherichia coli*. *Prot. Expr. Purif.* **11,** 289–296.
73. Bergès, H., Joseph-Liauzun, E., and Fayet, O. (1996) Combined effects of the signal sequence and the major chaperone proteins on the export of human cytokines in *Escherichia coli. Appl. Environ. Microbiol.* **62,** 55–60.
74. Jeong, K. J. and Lee, S. Y. (2000) Secretory production of human leptin in *Escherichia coli. Biotechnol. Bioeng.* **67,** 398–407.
75. Kurokawa, Y., Yanagi, H., and Yura, T. (2000) Overexpression of protein disulfide isomerase DsbC stabilizes multiple- disulfide-bonded recombinant protein produced and transported to the periplasm in *Escherichia coli. Appl. Environ. Microbiol.* **66,** 3960–3965.
76. Qiu, J., Swartz, J. R., and Georgiou, G. (1998) Expression of active human tissue-type plasminogen activator in *Escherichia coli. Appl. Environ. Microbiol.* **64,** 4891–4896.
77. Nossal, N. and Heppel, L. (1966) The release of enzymes by osmotic shock from *Escherichia coli* in exponential phase. *J. Biol. Chem.* **241,** 3055–3062.
78. Betton, J. M., Boscus, D., Missiakas, D., Raina, S., and Hofnung, M. (1996) Probing the structural role of an alpha beta loop of maltose-binding protein by mutagenesis: heat-shock induction by loop variants of the maltose-binding protein that form periplasmic inclusion bodies. *J. Mol. Biol.* **262,** 140–150.
79. Birnie, G. D., Rickwood, D., and Hell, A. (1973) Buoyant densities and hydration of nucleic acids, proteins and nucleoprotein complexes in metrizamide. *Biochim. Biophys. Acta* **331,** 283–294.
80. Thome, B. M. and Muller, M. (1991) Skp is a periplasmic *Escherichia coli* protein requiring SecA and SecY for export. *Mol. Microbiol.* **5,** 2815–2821.
81. Blum, P., Ory, J., Bauernfeind, J., and Krska, J. (1992) Physiological consequences of DnaK and DnaJ overproduction in *Escherichia coli. J. Bacteriol.* **174,** 7436–7444.
82. Ayling, A. and Baneyx, F. (1996) Influence of the GroE molecular chaperone machine on the in vitro folding of *Escherichia coli* β-galactosidase. *Protein Sci.* **5,** 478–487.

83. Buchner, J., Schmidt, M., Fuchs, M., Jaenicke, R., Schmid, F. X., and Kiefhaber, T. (1991) GroE facilitates the refolding of citrate synthase by suppressing aggregation. *Biochemistry* **30,** 1586–1591.
84. Guthrie, B. and Wickner, W. (1990) Trigger factor deletion or overproduction causes defective cell division but does not block protein export. *J. Bacteriol.* **172,** 5555–5562.
85. Wessel, D. and Flügge, U. I. (1984) A method for the quantitative recovery of protein in dilute solution in the presence of detergents and lipids. *Anal. Biochem.* **138,** 141–143.

12

High-Throughput Purification of PolyHis-Tagged Recombinant Fusion Proteins

Thomas Lanio, Albert Jeltsch, and Alfred Pingoud

1. Introduction

Methods for the efficient overexpression and purification of recombinant proteins are of paramount importance for biotechnology. In particular, for the era of functional genomics that we have entered after sequencing complete genomes, this has become a routine matter. High-throughput protein purification will, therefore, become a key technology to unravel the function of gene products (**Fig. 1**). To facilitate the procedure of protein purification, several tags to generate fusion proteins are available (e.g., polyHis, GST, MBP, CBP, and the like) for parallel purification using matrices coupled with affinity anchors, like Ni^{2+}-nitrilotriacetic acid (Ni^{2+}-NTA) which is a powerful chelating ligand for the purification of His_6-tagged proteins under native conditions. Ni-NTA affinity matrices allow to purify the protein of interest contained in a crude protein mixture at a concentration of 1% in one step to more than 95% homogeneity *(1)*.

Mutational analysis of structure/function relationships in proteins is often carried out by site directed mutagenesis leading normally to a small number of variants which can be purified by conventional methods. Random mutagenesis methods, used to overcome the limitations of rational design, result in much larger libraries of $10^9 - 10^{10}$ variants *(2,3)*. Usually in vivo or in vitro selection systems are employed to screen the libraries for variants with the desired properties *(4,5)*. However, properties that can be measured only with purified enzyme preparations require a fast, efficient and reliable high throughput system for protein expression and purification *(6)*. The high throughput protein purification scheme presented here makes it possible to purify and analyze some 10^4 different protein variants in a reasonable period of time, provided

From: *Methods in Molecular Biology, vol. 205, E. coli Gene Expression Protocols*
Edited by: P. E. Vaillancourt © Humana Press Inc., Totowa, NJ

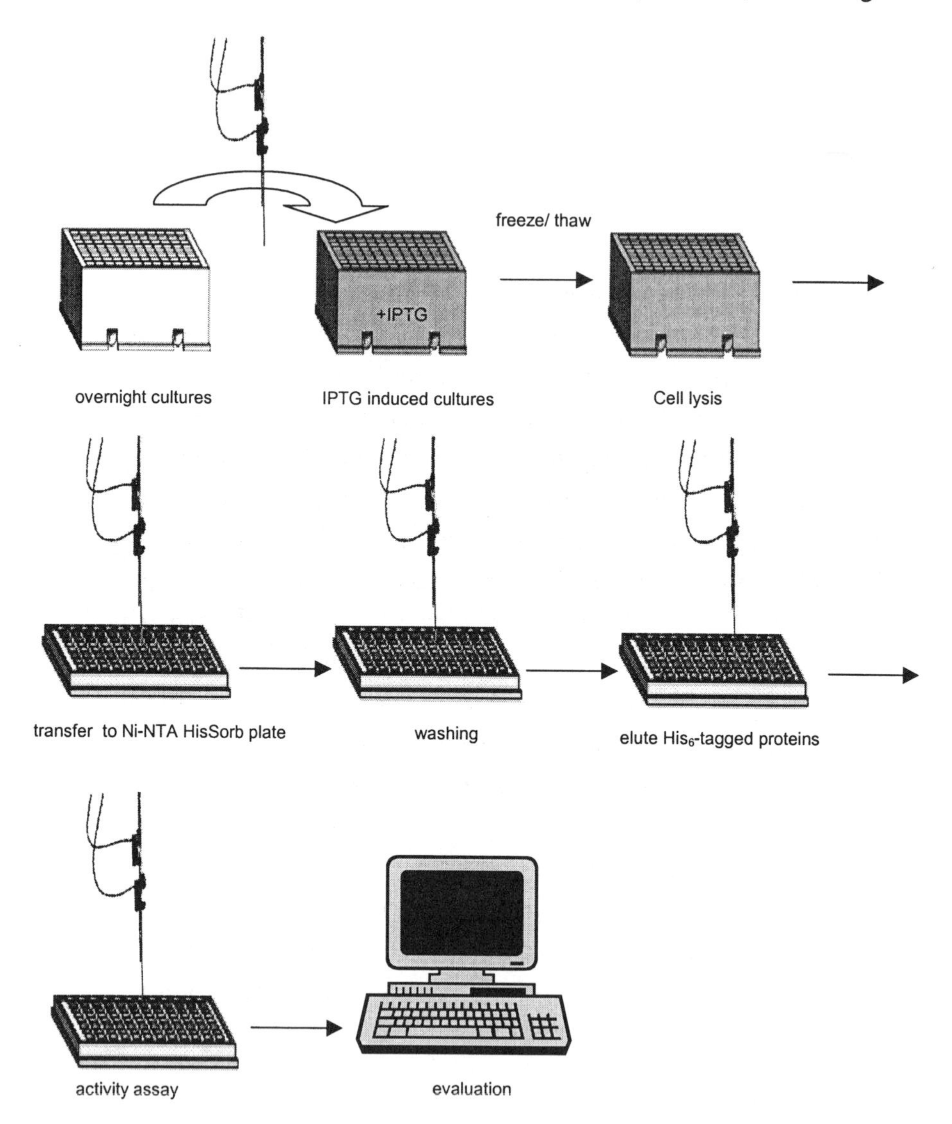

Fig. 1. Schematic overview of the high-throughput protein purification method. Cells are grown, harvested, and lysed in 1-mL square well blocks. The His_6-tagged variants are isolated by transferring the lysate to Ni-NTA coated microplates from which they are eluted after washing. A 1.5-mL cell culture typically yields 5–10 pmol recombinant protein.

that the feature of interest can be examined with a His_6-tagged protein (*see* **Note 1**). The protocol makes use of a pipetting robot (*see* **Note 2**) and 96-well microplates and blocks and yields per variant 5–10 pmol of native protein, which is sufficient for most enzymatic assays or binding studies (*see* **Note 3**).

2. Materials

2.1. Protein Expression

2.1.1. Equipment

1. 96-deep well and 96-square well blocks (NUNC, Wiesbaden, Germany).
2. Automated pipetting system (QIAGEN, Hilden, Germany).
3. Microplate carriers for centrifugation (Beckmann Coulter, Munich, Germany).
4. Refrigerated centrifuge (Beckmann Coulter).

2.1.2. Solutions

1. Standard solutions and reagents for bacterial cell culture.

2.2. Protein Purification

2.2.1. Equipment

1. Ni-NTA HisSorb plates (QIAGEN, Hilden, Germany).
2. Automated pipetting system.
3. Refrigerated centrifuge.
4. Vortex.

2.2.2. Solutions

1. Lysis buffer: 30 m*M* potassium-phosphate, pH 7.5, 0.1 m*M* dithioerythritol (DTE), 0.01% (w/v) lubrol (a mild, non-ionic detergent to stabilize proteins during purification), 500 m*M* NaCl, 20 m*M* imidazole, 100 μg/mL bovine serum albumin, and 1 mg/mL lysozyme.
2. Elution buffer: 30 m*M* potassium-phosphate, pH 7.5, 0.1 m*M* DTE, 0.01% (w/v) lubrol, 200 m*M* imidazole (*see* **Note 1**).

3. Methods

3.1. Protein Expression

In order to facilitate parallel purification, cell culture and expression of the proteins should be carried out in 96-well format (*see* **Note 2**). For that purpose, bacteria are grown in a 96-deep well block in 1 mL of a suitable medium for an appropriate period of time. A fraction of this preculture (e.g., 100 μL) is used to inoculate 1.5 mL fresh medium in a 96-square well block and expression of the proteins is induced by addition of isopropyl-β-D-galactopyranoside (IPTG) or whatever is needed for induction, depending on the expression system employed. The remaining precultures, representing the reference culture for additional characterization of selected variants, are harvested by centrifugation and stored at –20°C.

1. Autoclave a suitable volume of medium, add the required antibiotics and transfer 1 mL into each well of the 96-deep well block, and 1.5 mL into each well of the 96-square well block.

2. Inoculate the 96-deep wells using toothpicks and incubate the block overnight at 37°C (for most *E. coli* strains).
3. Transfer 200 µL cell culture from each deep well to the corresponding square well using an automated pipetting system or a multi-channel pipet.
4. Harvest the remainder of the precultures by centrifugation (typically for 20 min at 2000*g*), seal the deep well block and store it at –20°C. This will allow recovery of the DNA of interesting samples later by PCR or by a plasmid DNA preparation.
5. Incubate the square well blocks containing the expression cultures for several hours and induce expression of the protein of interest by addition of IPTG (final concentration of 1 m*M*) or any other inducer, depending on the expression system.
6. After adequate incubation time (typically 3–5 h) harvest the cells by centrifugation and store the pellets at –70°C.

3.2. Protein Purification

The protein purification scheme is based on the interaction between polyHis-tagged proteins and Ni-NTA coated microplates (*see* **Notes 3** and **4**). The membrane of the bacteria containing the His-tagged proteins is disintegrated by lysozyme treatment and the lysates are transferred without further treatment directly to the Ni-NTA coated microplates. After incubation, the plates are washed several times with lysis buffer and the His-tagged proteins are eluted in the final step of the procedure with a buffer containing high concentration of imidazole or ethylenediamine tetraacetic acid (EDTA).

1. Thaw the cell pellets in the square well block at ambient temperature.
2. Resuspend the pellets in 250 µL lysis buffer (*see* **Notes 5** and **6**).
3. Vortex at 500 rpm for 15 min at ambient temperature (*see* **Note 7**).
4. Transfer 200 µL of the lysate to each well of a Ni-NTA HisSorb plate and incubate at least for 1 h at ambient temperature with vortexing at 500 rpm.
5. Wash the plate twice with 200 µL lysis buffer.
6. Add 40 µL elution buffer to elute the His-tagged protein.

4. Notes

1. The imidazole concentration of the elution buffer (>100 m*M*) must not inhibit the enzyme properties to be examined. Imidazole may be supplemented with or replaced by EDTA (>1 m*M*). The elution buffer may require modification to meet the requirements of the subsequent activity assay.
2. The method described was developed to run on an automated pipetting system. Nevertheless, it can be carried out by hand using a multi-channel pipet. Automated protein purification may be combined with a colony picking robot (BioRobotics, Cambridge, UK), leading to a fully automated clone management *(7)*.
3. If Ni-NTA Superflow™ (QIAGEN, Hilden, Germany) is used, higher amounts of protein can be obtained with a special automated pipetting system.
4. The cultures to be examined can be prepared in advance and stored at –70°C to allow for continuous purification and characterization.

5. For complete cell lysis, the cell pellets should be kept frozen at –70°C for several hours, preferentially over night.
6. Though phosphate buffers are recommended for Ni-NTA handling, other buffers may be used. BSA in the lysis buffer turned out to be of major importance.
7. Cell lysis and protein binding to Ni-NTA plates can also be performed at 4°C, if protein degradation is a problem. In this case the incubation time should be extended.

Acknowledgments

This work has been supported by grants from the Deutsche Forschungsgemeinschaft, the Bundesministerium für Bildung, Wissenschaft, Forschung und Technologie, the European Union and the Fonds der Chemischen Industrie.

References

1. Janknecht, R., de Martynoff, G., Lou, J., Hipskind, R. A., Nordheim, A., and Stunnenberg, H. G. (1991) Rapid and efficient purification of native histidine-tagged protein expressed by recombinant vaccinia virus. *Proc. Natl. Acad. Sci. USA* **88,** 8972–8976.
2. Arnold, F. H. and Volkov, A. A. (1999) Directed evolution of biocatalysts. *Curr. Opin. Chem. Biol.* **3,** 54–59.
3. Benner, S. A. (1993) Catalysis: design versus selection. *Science* **261,** 1402–1403.
4. Moore, J. C. and Arnold, F. H. (1996) Directed evolution of a para-nitrobenzyl esterase for aqueous-organic solvents. *Nat. Biotechnol.* **14,** 458–467.
5. Crameri, A., Raillard, S. A., Bermudez, E., and Stemmer, W. P. (1998) DNA shuffling of a family of genes from diverse species accelerates directed evolution. *Nature* **391,** 288–291.
6. Lanio, T., Jeltsch, A., and Pingoud, A. (2000) Automated purification of His_6-tagged proteins allows exhaustive screening of libraries generated by random mutagenesis. *Biotechniques* **29,** 338–342.
7. Meier-Ewert, S., Maier, E., Ahmadi, A., Curtis, J., and Lehrach, H. (1993) An automated approach to generating expressed sequence catalogues. *Nature* **361,** 375–376.

13

Co-Expression of Proteins in *E. coli* Using Dual Expression Vectors

Karen Johnston and Ronen Marmorstein

1. Introduction

Detailed biophysical, biochemical, and structural studies rely on the preparation of milligram amounts of pure recombinant proteins. Many useful overexpression systems have been developed for this purpose *(1–3)*, and of these, bacterial overexpression is still the most convenient and simplest to use *(4–6)*. Because many proteins are only active as heteromeric complexes, there has been much recent interest in preparing such complexes using recombinant technology *(7–10)*. The preparation of complexes generally relies on the ability to reconstitute such protein complexes from individually prepared recombinant proteins *(10–13)*, a process that often involves refolding *(14,15)*. This is generally not ideal, and in cases where one protein is unstable without the other *(16)*, impossible.

This chapter describes a system that uses two T7 promoter-based vectors *(17)* to co-express binary protein complexes in bacteria. Subsequent purification of the heteromeric protein complex depends on a relatively tight association between the components (>low μ*M*) so that the components remain associated during purification. The two vectors employed for the co-expression have different origins of replication to allow the cell to simultaneously support both expression vectors, and different antibiotic resistances allowing for the selection of cells containing both of the plasmids. An added convenience of having each protein encoded on a separate expression plasmid is the ability to mix and match different protein constructs, greatly simplifying the optimization of different heteromeric protein combinations.

The first protein overexpression plasmid is pRSET *(18)*, commercially available from Invitrogen™ (*see* **Fig. 1A**). This plasmid contains an ampicillin

From: *Methods in Molecular Biology, vol. 205, E. coli Gene Expression Protocols*
Edited by: P. E. Vaillancourt © Humana Press Inc., Totowa, NJ

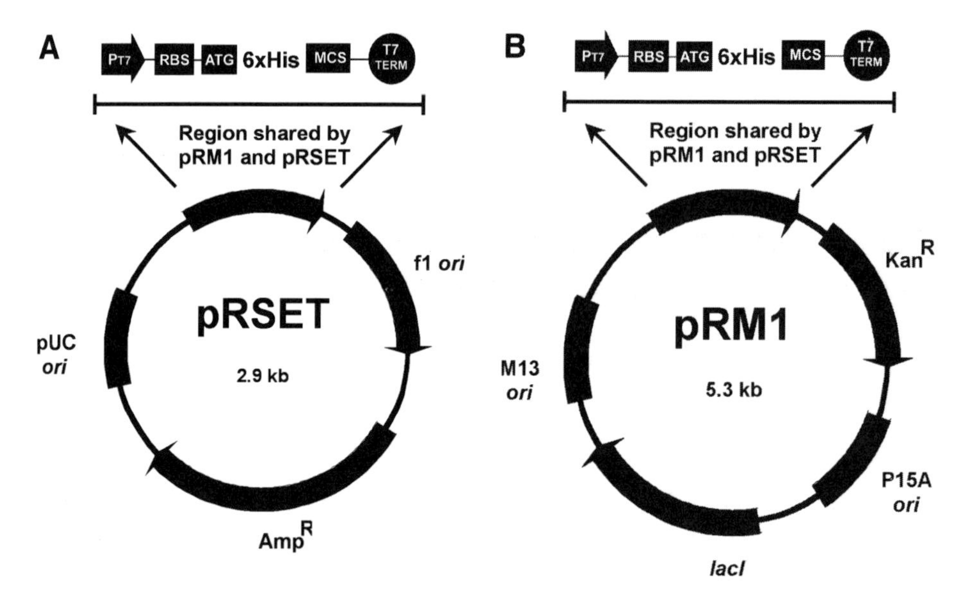

Fig. 1. Schematic representation of the two plasmids pRSET and pRM1 used in this overexpression system. The region common to the two plasmids includes the T7 polymerase promoter (P_{T7}), ribosomal binding site (RBS), 6xHis fusion tag (6xHis), multiple cloning site (MCS), and T7 termination sequence (T7 TERM). **(A)** pRSET contains the pUC origin of replication (PUC ori) and an ampicillin resistance gene (Amp^R). **(B)** pRM1 contains the P15A origin of replication (P15A ori) and a kanamycin resistance gene (Kan^R).

resistance gene and a pUC origin of replication (*see* **Note 1**). Because it is a very high copy number plasmid, each component of the complex being studied is first cloned into this vector. Within the multiple cloning site (MCS) is a 6xHis tag that can be added to one or more components of the complex. Affinity chromatography using the 6xHis tag greatly simplifies the purification of the complex, so one component should be designed to utilize the 6xHis tag. Initial expression, solubility, and purification studies of each of the protein constructs of interest can be carried out using this vector.

The second protein overexpression vector is pRM1, which was developed at The Wistar Institute specifically for use as a co-expression vector with pRSET (*see* **Fig. 1B**; **ref.** ***16***). It is based on pMR103 ***(19)*** and contains an M13 origin of replication and a kanamycin resistance gene. This is a low copy number plasmid and consequently more tedious to clone into directly. However, it also contains the T7 promoter, ribosomal binding site, 6xHis fusion tag, multiple cloning site, and T7 termination sequence derived from pRSET, making it simple to shuttle any sequence cloned into pRSET directly into pRM1 (*see*

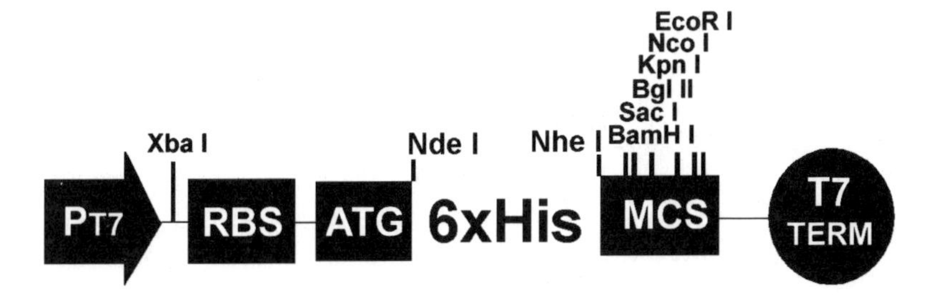

Fig 2. Enlargement of region common to both the pRSET and pRM1 vectors showing the unique restriction sites available for cloning.

Fig. 2). If there is an observable difference in the expression levels between the two protein components over expressed from the pRSET vector, it is recommended that the better expressing protein be transferred into pRM1, as expression levels are generally lower from this vector than from pRSET ***(16)***. Once the pRM1 based protein overexpression construct has been prepared (and correctness of the sequence confirmed) , bacteria containing the pRM1 vector are grown and made competent so that they can subsequently be transformed with the pRSET vector for co-expression studies (*see* **Note 2**).

Once proteins are co-expressed using the pRSET/pRM1 dual vector system, the purification of each protein complex is tailored to the particular protein system under study. If one of the protein components contains the 6xHis tag, then an initial purification step should include Ni-chelate affinity chromatography. A more detailed discussion of other purification strategies is beyond the scope of this chapter. Therefore, here we will describe the protocols for preparing the co-expression vectors and co-expressing the proteins and describe some strategies for using Ni-chelate affinity chromatography to obtain an initial purification of the heteromeric protein complex of interest. In general, if Ni-chelate affinity chromatography is used, complexes should be pure enough to require only one more purification step such as ion exchange or size exclusion chromatography. Specific protocols for using Ni^{2+}-NTA (nickel-nitrilotriacetic acid) to purify 6xHis tagged proteins are well described in product literature as well as in *Current Protocols in Protein Science* ***(20)***, therefore general guidelines for purifying proteins using Ni^{2+}-NTA will not be repeated here.

2. Materials

2.1. Cloning of Protein Constructs into pRSET

1. pRSET (Invitrogen).
2. Gene template for proteins of interest.

3. Synthetically synthesized primers.
4. Polymerase and appropriate buffer.
5. Appropriate restriction enzymes and buffers.
6. Agarose gel and loading buffer.
7. QIAX gel extraction kit (QIAGEN) or equivalent.
8. Ligase and appropriate buffer.
9. DH5α competent cells.
10. Luria-Bertani (LB)-agar plates containing 100 μg/mL ampicillin.

2.2. Test for Protein Overexpression on Small Scale

1. BL21 competent cells.
2. LB-agar plates containing 100 μg/μL ampicillin.
3. Sterile culture tubes.
4. LB media.
5. Ampicillin stock: 100 mg/mL.
6. IPTG (isopropylthio-β-D-galactoside) stock: 200 m*M*.
7. Urea buffer: 50 m*M* Hepes, pH 7.5, 6 *M* urea, 100 m*M* NaCl, 10% glycerol.

2.3. Cloning of Protein Construct from pRSET into pRM1

1. Purified pRSET containing cloned DNA sequence to be moved into RPM I.
2. Purified pRM1 (*see* **Note 3**).
3. Appropriate restriction enzymes and buffers.
4. Agarose gel and loading buffer.
5. QIAX gel extraction kit (QIAGEN) or equivalent.
6. Ligase and appropriate buffer.
7. LB-agar plates containing 50 μg/mL kanamycin.

2.4. Test Protein Expression from pRM1

1. BL21 competent cells.
2. LB-agar plates containing 50 μg/mL kanamycin.
3. Sterile culture tubes.
4. LB media.
5. Kanamycin stock: 50 mg/mL.
6. IPTG stock: 200 m*M*.
7. Urea buffer: 50 m*M* Hepes, pH 7.5, 6 *M* urea, 100 m*M* NaCl, 10% glycerol.

2.5. Prepare Competent BL21 Cells Harboring pRM1 Containing the Cloned Protein Sequence of Interest (see Note 4)

1. Sterile 250-mL centrifuge bottle and lid.
2. Sterile 1.5-mL microfuge tubes (~100).
3. Sterile pipets and pipet tips.
4. LB-agar plate with fresh BL21 cells containing pRM1/protein of interest.
5. LB: 100 mL in sterile Erlenmeyer flask.
6. LB: 5 mL in sterile culture tube.
7. Kanamycin stock: 50 mg/mL.

8. Sterile filtered transforming buffer (TFB) I: 30 m*M* potassium acetate (KOAc), 100 m*M* RbCl (may be replaced with 100 m*M* KCl), 10 m*M* $CaCl_2$, 50 m*M* $MnCl_2$, 15% glycerol, adjusted to pH 5.8 with acetic acid.
9. Sterile filtered TFB II: 10 m*M* MOPS (3-[*N*-morpholino]propanesulfonic acid) (or PIPES (piperazine-*N,N'*-bis(ethanesulfonic acid)), 15% glycerol, 10 m*M* RbCl (may be replaced with 100 m*M* KCl), 75 m*M* $CaCl_2$.

2.6. Co-Expression of Proteins

1. Vector (pRSET) containing coding sequence 1.
2. BL21competent cells harboring pRM1 containing coding sequence 2.
3. LB agar plates containing 25 μg/μL kanamycin and 50 mg/mL ampicillin.
4. LB media (6 L).
5. Kanamycin stock: 50 mg/mL.
6. Ampicillin stock: 100 mg/mL.
7. IPTG stock: 200 m*M*.
8. Low salt (LS) buffer (*see* **Note 5**):100 m*M* Hepes, pH 7.5, 200 m*M* NaCl, 35 m*M* imidizole (if Ni-chelate affinity chromatography used in purifying the complex), 20 m*M* β-mercaptoethanol,
9. Urea buffer (*see* **Subheading 2.4.**, **item7.**)

3. Methods

3.1. Cloning Into pRSET

1. Clone each component of the complex into the MCS of pRSET according to manufacturer's instructions. When designing primers, consider which component of the complex should have the affinity tag (*see* **Note 6**).
2. Sequence the inserts before proceeding.

3.2. Test for Protein Expression on Small Scale (see Note 7)

1. Transform each plasmid into an expression strain such as BL21 and grow on LB agar plates containing 100 μg/mL ampicillin (*see* **Note 8**).
2. Incubate plates approx 12–16 h at 37°C until visible colonies are observed.
3. Inoculate 3 mL cultures containing LB supplemented with 100 μg/mL ampicillin. Grow with agitation at 37°C until cultures appear turbid (~0.4–0.7 OD_{595}).
4. Remove 1 mL of the culture. Harvest pellet by centrifugation and label as "uninduced" sample.
5. Induce expression in the remaining 2 mL by addition of IPTG to a final concentration of 0.5–1.0 m*M*.
6. Grow with agitation for an additional 3 h at 37°C.
7. Remove 1 mL of culture, and harvest pellet by centrifugation. Label the culture as "induced" sample.
8. Dissolve each of the cell pellets in 100 μL urea buffer.
9. Analyze 10 μL of each sample by sodium clodecylsulfate-polyacrylamide gel electrophoresis (SDS-PAGE) to determine the approximate levels of recombinant protein expression.
10. Choose the higher expressing protein for insertion into pRPM1.

3.3. Transfer of the High-Expressing Component into pRM1

1. Digest the protein sequence-cloned-pRSET and pRM1 with appropriate restriction enzymes (*see* **Note 6**).
2. Purify both the digested vector and insert, using agarose gel electrophoresis.
3. Ligate fragment and pRM1 according to manufacturer's instructions.
4. Transform ligation reaction into DH5α competent cells. Grow on LB agar plates containing 50 μg/mL kanamycin.
5. Incubate plates approximately 12–15 h at 37°C until colonies are visible.
6. Grow cultures from single colonies. Purify protein-sequence-cloned pRM1 from these and verify sequence (*see* **Note 4**).

3.4. Test Expression from pRM1

1. Test for expression on a small scale as described in **Subheading 3.2.**
2. Choose the best expressing protein-sequence cloned pRM1 for the preparation of BL21 transformed competent cells (*see* **Note 2**).

3.5. Prepare Chemically Competent BL21 Cells Containing pRM1 (see Note 4)

1. Add 50 μg/mL kanamycin to 5 mL LB and seed with one colony from plate.
2. Grow at 37°C with agitation until cloudy. (~1–2 h)
3. Add the 5 mL growth solution to 100 mL LB containing 50 μg/mL kanamycin.
4. Grow at 37°C with agitation to a concentration of 0.5–0.7 OD_{595}.
5. Chill cells on ice for 5 min.
6. Harvest cells by centrifugation at 5000*g* for 10 min.
7. Discard the supernatant and re-suspend cells in 33 mL TFB I (*see* **Note 9**).
8. Chill on ice for 5 min.
9. Harvest cells by centrifugation at 5000*g* for 10 min.
10. Discard supernatant and resuspend in 4 mL TfB II (*see* **Note 9**).
11. Chill on ice for 15 min.
12. Aliquot 50 μL competent cell resuspension into sterile microfuge tubes.
13. Freeze immediately on dry ice and store at –70°C.
14. Test viability of cells by streaking three LB agar plates containing no antibiotic, 50 μg/mL kanamycin, and 100 μg/mL ampicillin. Healthy cells containing the pRM1 plasmid will grow to a lawn on the LB plate and to single colonies on the LB/kanamycin plate. No colonies should grow on the ampicillin plate.

3.6. Co-Expression of Proteins

1. Transform the pRSET cloned vector into pRM1-cloned competent cells. Grow on LB agar plates containing 25 μg/mL kanamycin and 50 mg/mL ampicillin (*see* **Note 10**).
2. Grow 6 L culture of dually transformed cells at 37°C (*see* **Note 11**) in LB media containing 25 μg/mL kanamycin and 50 μg/mL ampicillin (*see* **Note 10**).

3. When the cell turbidity reaches 0.4–0.7 OD_{595}, remove 1 mL to be used as "uninduced" sample and add 0.5–1 m*M* IPTG. (Centrifuge the uninduced sample and discard supernatant. Suspend pellet in 100 µL urea buffer.
4. Grow the remaining solution an additional 3 h (*see* **Note 11**).
5. Remove 1 mL to be used as "induced" sample. Centrifuge the induced sample, discard the supernatant and suspend pellet in 100 µL urea buffer.
6. Harvest the remaining cells by centrifugation at >3000*g* and discard the supernatant. Re-suspend cell pellet in 100 mL LS buffer (*see* **Note 5**). Keep sample on ice from this point on.
7. Lyse cells by sonication being careful not to over heat sample.
8. Remove 100 µL to test solubility.
   a. Spin 100 µL sample at >10,000*g* for 10 min.
   b. Remove soluble portion of sample to second tube, being careful not to disturb pellet.
   c. Rinse pellet gently with 100 µL LS buffer. Spin briefly and remove wash.
   d. Dissolve pellet in 100 µL urea buffer
   e. Run 10 µL of each sample on SDS-PAGE to confirm solubility of complex. (Also include 10 µL of each of the induced and uninduced samples for comparison.)
9. Separate soluble and insoluble portions of the remaining cell lysate by centrifugation at >35,000*g* for 20 min.
10. Proceed with purification.

4. Notes

1. Previous publications from Invitrogen™ have stated that pRSET contains a ColE1 origin of replication. This information is incorrect . (The sequences of ColE1 and pUC differ by only a few nucleotides.) As of their 2001 catalog, the error has been corrected. The plasmid is as it has always been so this change makes no difference to the co-expression system.
2. Although it is possible to transform cells with both plasmids at once, it has been found that transforming cells with one plasmid, preparing them as competent, and then transforming with the second produces a higher percentage of dually transformed cells ***(16)***. Having prepared stocks of competent cells containing one plasmid also simplifies the testing of different possible protein partners.
3. pRM1 is a low copy number plasmid and should be purified accordingly.
4. There are many protocols for preparing chemically competent cells and any should be appropriate. The protocol described here has been used successfully with this system.
5. Composition of buffer will be protein specific. Consult product literature for QIAGEN Ni-NTA resin for details on the principles and limitations of the resin. A few basic concepts should be kept in mind. It is recommended that the minimal salt concentration during binding and washing steps be 300 m*M* NaCl. If this ionic strength will destabilize the complex, lower salt levels may be used if imidazole is present at low levels to minimize nonspecific binding to the resin. Optimization of imidazole levels during binding can greatly increase yield and

purity. Ni-NTA matrices cannot be used with strong chelating agents such as ethylenediaminetetraacetic acid (EDTA) or strong reducing agents such as dithiothreitol (DTT). In most cases, β-mercaptoethanol can be used up to 20 m*M*.

6. The pRSET vector is designed to add the 6xHis fusion tag to the protein of interest when it is cloned using an N-terminal *Nhe*I site (*see* **Fig. 1A**). Four restriction sites that are unique in pRSET are known not to be unique in pRM1: *Hind*III, *Nhe*I, *Pvu*II, and *Pst*I. Therefore, if the 6xHis tagged protein is to be transferred into pRM1, it cannot be done using the *Nhe*I site. However, an *Xba*I site, common to both vectors and upstream of the 6xHis tag, can easily be used instead to shuttle the 6xHis tagged construct from pRSET into pRM1 (*see* **Fig. 2**).
7. Expression studies on this small scale give a quick, crude estimate of expression levels. However, for very low expressing proteins, it is certainly not definitive; ambiguous results should not be taken as negative ones. Protein bands that correspond to the cloned protein, but are not readily visible in small-scale inductions can often be isolated following Ni-chelate affinity chromatography .
8. Strains such as BL21/LysS that already contain plasmid DNA are not recommended as it is difficult for the cells to maintain 3 separate plasmids.
9. Be sure all cells are resuspended and no lumps remain. Cells on the inside of these clumps will not be exposed to the chemicals, decreasing the transformation efficiency of the entire batch.
10. Decreased cell growth is observed when antibiotics are present at the higher levels used for expressing from individual vectors. A final concentration of 25 μg/mL kanamycin and 50 μg/mL ampicillin is adequate for selection of cells containing both plasmids.
11. Optimizing growth temperature can greatly increase the solubility of complexes. It is useful at this step to grow cultures at several temperatures for comparison (i.e., 30°C, 25°C, 15°C). Keep in mind that it is the temperature at which the cells are growing while producing the protein that matters. It is possible, for example, to grow a culture at 37°C up to an OD_{595} of ~0.15 and then lower the temperature to 15°C. By the time the concentration has reached a point for induction (0.4–0.7 OD_{595}), the media has cooled to 15°C and cell growth has slowed accordingly. When expressing proteins at this temperature, it is best to continue growth 15–20 h after induction.

References

1. Makrides, S. C. (1996) Strategies for achieving high-level expression of genes in *Escherichia coli. Microbiol. Rev.* **60,** 512–538.
2. Roy, P. and Jones, I. (1996) Assembly of macromolecular complexes in bacterial and baclovirus expression systems. *Curr. Opin. Struc. Biol.* **6,** 157–161.
3. Geisse, S., Gram, H., Kleuser, B., and Kocher, H. P. (1996) Eukaryotic expression systems: A comparison. *Protein Expr. Purif.* **8,** 271–282.
4. Morgan, D. O. (1995) Principles of CDK regulation. *Nature* **374,** 131–134.
5. Shaw, P. E. (1992) Ternary complex formation over the c-fos serum response element: $p62^{TCF}$ exhibits dual component specificity with contacts to DNA and and extended structure in the DNA binding domain of $p67^{SRF}$. *EMBO J.* **11,** 3011–3019.

6. Treisman, R. and Ammerer, G. (1992) The SRF and MCM1 transcription factors. *Curr. Opin. Gen. & Dev.* **2,** 221–226.
7. Li, T., Stark, M. R., Johnson, A. D., and Wolberger, C. (1995) Crystal structure of the MATa1/MAT alpha2 homeodomain heterodimer bound to DNA. *Science* **270,** 262–269.
8. Jeffrey, P. D.,Russo, A. A., Polyak, K., et al. (1995) Mechanism of CDK activation revealed by the structure of a cyclinA-CDK2 complex. *Nature* **376,** 313–320.
9. Russo, A. A., Jeffrey, P. D., Patten, A. K., Massague, J., and Pavletich, N. P. (1996) Crystal structure of the p27Kip1 cyclin-dependent-kinase inhibitor bound to the cyclin A-CDK2 complex. *Nature* **382,** 325–331.
10. Tan, S., Hunziker, Y., Sargent, D. F., and Richmond, T. J. (1996) Crystal structure of a yeast TFIIA/TBP/DNA complex. *Nature* **381,** 127–134.
11. Garboczi, D. N., Gosh, P., Utz, U., Fan, Q. R., Biddison, W. E., and Wiley, D. C. (1996) Structure of the complex between human T-cell receptor, viral peptide and HLA-A2. *Nature* **384,** 134–141.
12. Garcia, K. C., Degano, M., Stanfield, R. L. et al. (1996) An alpha beta T cell receptor structure at 2.5 angstrom and its orientation in the TCR-MHC complex. *Science* **274,** 209–219.
13. Geiger, J. H., Hahn, S., Lee, S., and Sigler, P. B. (1996) The crystal structure of the yeast TFIIA/TBP/DNA complex. *Science* **272,** 830–836.
14. Guise, A. D., West, S. M., and Chaudhuri, J. B. (1996) Protein folding in vivo and renaturation of recombinant proteins from inclusion bodies. *Mol. Biotechnol.* **6,** 53–64.
15. Middelberg, A. P. J. (1996) Large-scale recovery of recombinant protein inclusion bodies expressed in *Escherichia coli. J. Microbiol. Biotechn.* **6,** 225–231.
16. Johnston, K., Clements, A., Venkataramani, R., Treivel, R., and Marmorstein, R. (2000) Co-expression of Proteins in Bacteria Using T7-Based Expression Plasmids: Expression of Heteromeric Cell-Cycle and Transcriptional Regulatory Complexes. *Protein Expr. Purif.* **20,** 435–443.
17. Studier, F. W., and Moffatt, B. A. (1986) Use of bacteriophage T7 RNA polymerase to direct selective high-level expression of cloned genes. *J. Mol. Biol.* **189,** 113–130.
18. Schoepfer, R. (1993) The pRSET family of T7 promoter expression vectors for *Escherichia coli. Gene* **124,** 83–85.
19. Munson, M., Predki, P. F., and Regan, L. (1994) ColE1-compatible vectors for high-level expression of cloned DNAs from the T7 promoter. *Gene* **144,** 59–62.
20. Springer, T. A. (1996) In: *Current Protocols in Protein Science,* John Wiley & Sons, New York, NY, pp. 9.4.1.–9.4.16.

14

Small-Molecule Affinity-Based Matrices for Rapid Protein Purification

Karin A. Hughes and Jean P. Wiley

1. Introduction

Affinity chromatography, the method of purifying target proteins from complex mixtures using immobilized affinity ligands on chromatographic supports, is perhaps the most common of all affinity techniques *(1–3)*. Many affinity chromatography systems are comprised of activated supports requiring direct ligand-coupling procedures that can be complex, time-consuming, and result in low- or variable-capacity columns supporting immobilized ligands with poor activity *(2–5)*. We describe here a method that improves this technology by using a small-molecule based chemical affinity technology *(6)* to quickly and easily prepare high-capacity affinity columns supporting functionally active capture ligands for purifying proteins from crude mixtures *(7,8)*. This innovation is based on the specific interaction between two, non-biological, small molecules, phenyl-diboronic acid (PDBA) and salicylhydroxamic acid (SHA) (*see* **Fig. 1**).

In order to prepare an affinity column, PDBA is first covalently attached to an affinity ligand, such as a protein, through the use of any one of three PDBA derivatives: N-hydroxysuccinimidyl (NHS) ester, hydrazide, or maleimide. The PDBA derivative should be selected to modify solvent-accessible functional groups present on the ligand while avoiding the active site. Conjugation of proteins with PDBA occurs in solution under mild reaction conditions separate from immobilization, resulting in high retention of protein activity relative to other methods. After conjugation, and without the need for purification, the PDBA-modified ligand is immobilized on an SHA-modified chromatographic support such as cross-linked agarose. The immobilization step (PDBA:SHA complex formation) is compatible with a wide variety of reaction conditions (*see* **Fig. 2**) and affords a high capacity column supporting functionally active

From: *Methods in Molecular Biology, vol. 205, E. coli Gene Expression Protocols*
Edited by: P. E. Vaillancourt © Humana Press Inc., Totowa, NJ

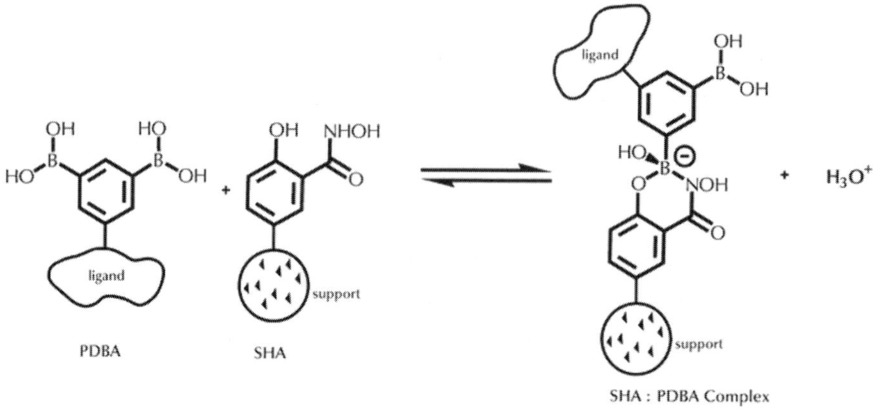

Fig. 1. PDBA:SHA Chemical Affinity System. The reaction of phenyldiboronic acid (PDBA) with salicylhydroxamic acid (SHA). The SHA is covalently anchored to the surface of crosslinked agarose chromatography media and a PDBA derivative is conjugated to a protein of choice.

ligands. Next, a crude mixture containing the protein target is applied to the column. The immobilized ligand captures and retains the target. The column is washed to remove any impurities, and the purified target recovered using elution conditions appropriate to breaking the affinity ligand:target complex while leaving the capture ligand attached to the solid support.

2. Materials

2.1. Conjugation of Protein Capture Ligands

2.1.1. Conjugation of Lysine Residues

1. Protein capture ligand containing solvent-accessible lysines.
2. PDBA-X-NHS (FW = 500.11); (Prolinx Inc., cat no. VER5000-1) store desiccated at –20°C.
3. 0.1 *M* Sodium bicarbonate, pH 8.0.
4. N, N-dimethylformamide (DMF), anhydrous.
5. Dialysis tubing or size exclusion column.

2.1.2. Conjugation of Glycoproteins

1. Glycoprotein capture ligand.
2. PDBA-X-hydrazide (FW = 373.41); (Prolinx Inc, cat. no. VER5100-50) store desiccated at –20°C.
3. Sodium periodate (350 m*M*): prepare fresh, light sensitive.
4. Hydrazide reaction buffer: 0.1 *M* sodium acetate, 0.1 *M* sodium chloride, adjusted to pH 5.5 with 6 *M* HCl.

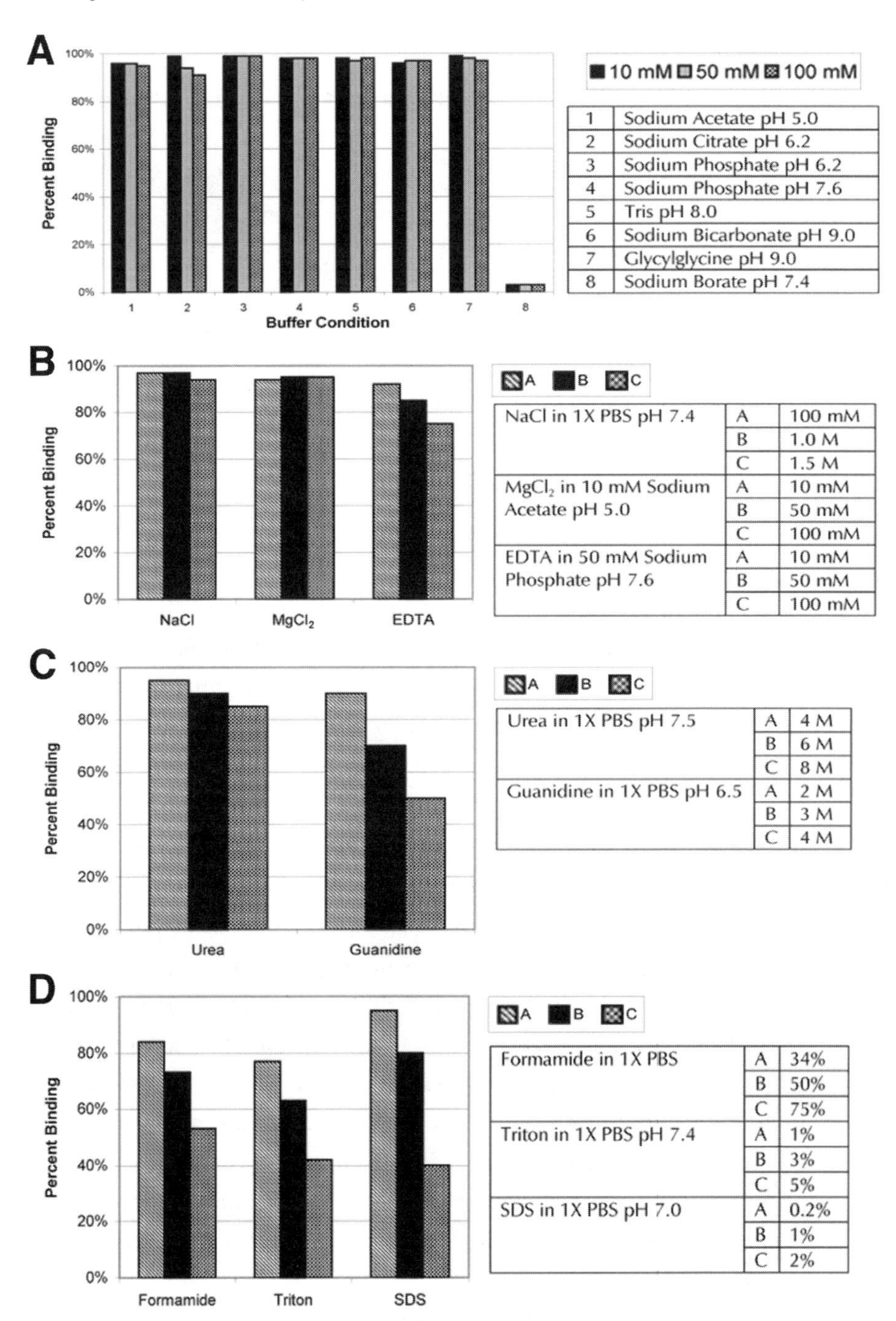

Fig. 2. Effects of pH and common buffer components **(A)**, ionic strength **(B)**, denaturants **(C)**, and detergents **(D)** on PDBA:SHA complex formation.

5. 0.1 *M* Sodium bicarbonate, pH 8.0.
6. 400 m*M* Sodium sulfite.
7. Methyl sulfoxide.
8. Dialysis tubing or size exclusion column.

2.1.3. Conjugation of Sulfhydryl Residues

1. Sulfhydryl-containing capture ligand.
2. PDBA-X-maleimide (MW = 459.07); (Prolinx Inc, cat. no. VER5050-50) store desiccated at –20°C.
3. Dithiothreitol (DTT), molecular biology grade.
4. N, N-dimethylformamide (DMF), anhydrous.
5. 0.1 *M* Sodium phosphate buffer, pH 7.0.
6. Dialysis tubing or size exclusion column.

2.2. Preparing a 0.5 mL SHA Column

1. Empty column (e.g., 0.5 cm diameter column is suitable for 0.5 mL column).
2. SHA agarose, 4% crosslinking (Prolinx Inc, cat. no. VER1000-10).
3. 0.1 *M* Sodium bicarbonate, pH 8.0.

2.3. Immobilizing the PDBA-Modified Capture Ligand on an SHA Column

1. 0.1 *M* Sodium bicarbonate, pH 8.0.

2.4. Loading the Crude Protein Mixture

1. Crude protein mixture containing target protein.
2. 0.1 *M* sodium bicarbonate, pH 8.0.

2.5. Eluting the Purified Protein

1. Desired elution buffer (e.g., 50 m*M* phosphate buffer, pH 11.2, or 100 m*M* glycine hydrochloride, pH 2.5).

3. Methods

3.1. Conjugation of Protein Capture Ligands

In order to immobilize a capture ligand on an SHA support, it is necessary to first conjugate the protein ligand with PDBA. In order to choose which PDBA-derivative to use, it is useful to know the active site of the protein ligand responsible for binding the target protein, and to avoid modifying that active site during the conjugation reaction. If the active site is unknown, a small amount of the capture ligand may be conjugated using each of three PDBA derivatives at varying molar input ratios, and the resulting conjugates tested to determine which gives optimal capture of the protein target.

3.1.1. Conjugation of Lysine Residues

1. Prepare a 5 mg/mL protein ligand solution (*see* **Note 1**) in 0.1 *M* sodium bicarbonate buffer, pH 8 (*see* **Note 2**). If the protein ligand is already in solution, dialyze the protein against this buffer.
2. Measure the UV absorbance of the protein ligand solution. Using the literature value for the absorbtivity and molecular weight of the protein ligand, calculate the concentration of the stock solution (in micromolar units) and the micromoles of protein ligand to be conjugated.
3. Prepare 1 mL of 100 m*M* PDBA-X-NHS in anhydrous DMF. Vortex to dissolve.
4. Add 10 mole equivalents of 100 m*M* PDBA-X-NHS solution to the protein ligand solution (*see* **Note 3**), keeping the final concentration of DMF below 10% (v/v) (*see* **Note 4**).
5. Incubate the reaction on wet ice for 1 h.
6. (Optional, *see* **Note 5**) Purify the protein conjugate of unwanted by-products and reactants by using a size exclusion column (such as Sephadex G-25) or by dialysis.
7. If desired, the concentration of the protein conjugate may be estimated as described in **Note 6**.

3.1.2. Conjugation of Glycoproteins

1. Dissolve the glycosylated protein ligand into 10 mL hydrazide reaction buffer to afford a 10 mg/mL protein solution. If the protein is already in solution, dialyze the protein against this buffer.
2. Measure the UV absorbance of the protein ligand solution. Using the literature value for the absorbtivity and molecular weight of the protein ligand, calculate the concentration of the stock solution (in micromolar units) and the micromoles of protein ligand to be conjugated.
3. Freshly prepare a 350 m*M* solution of sodium periodate in water. Protect this solution from the light.
4. Add 280 µL of 350 m*M* sodium periodate solution to the protein ligand solution (10 mL) to afford a final concentration of 10 m*M* sodium periodate.
5. React on wet ice, in the dark, for 30 min.
6. Quench the reaction by adding 0.5 mL of freshly prepared 0.4 *M* sodium sulfite solution (final concentration 20 m*M* sodium sulfite). Vortex to mix. The quenching reaction should occur immediately.
7. Dissolve 10 mole equivalents (based on the number of micromoles of protein ligand to be conjugated determined in **step 2**) of PDBA-X-hydrazide in 0.8 mL methyl sulfoxide. Add 0.8 mL of hydrazide reaction buffer. Vortex to mix.
8. Add the entire contents of the PDBA-X-hydrazide solution to the glycoprotein solution (*see* **Note 3**) and incubate on wet ice for 4 h.
9. (Optional, *see* **Note 5**) Purify the protein conjugate of unwanted by-products and reactants by using a size exclusion column (such as Sephadex G-25) or by dialysis.
10. If desired, the concentration of the protein conjugate may be estimated as described in **Note 6**.

3.1.3. Conjugation of Sulfhydryl Residues

1. Dissolve the sulfhydryl-containing protein ligand in 2 mL of 0.1 *M* sodium phosphate buffer, pH 7 at a concentration of at least 0.5 mg/mL (*see* **Note 7**). If the protein ligand is already in solution, dialyze the protein ligand against this buffer.
2. Measure the UV absorbance of the protein ligand solution. Using the literature value for the absorbtivity and molecular weight of the protein, calculate the concentration of the stock solution (in micromolar units) and the micromoles of protein ligand to be conjugated.
3. Warm the protein solution to 37°C in a water bath.
4. Add 2 µL of 0.5 *M* dithiothreitol (DTT) solution to the warmed protein solution to afford a final concentration of 0.5 m*M* DTT (*see* **Note 8**). Incubate the solution for an additional 10 min at 37°C.
5. Remove the reducing agent by passing the protein solution through a size exclusion column (e.g., Sephadex G-25). Monitor the fractions by UV and pool all those fractions containing proteinaceous material.
6. Prepare 1 mL of 100 m*M* PDBA-X-maleimide in anhydrous N,N-dimethylformamide. Vortex to mix.
7. Add 5 mole equivalents of the 100 m*M* PDBA-X-maleimide solution to the protein solution (*see* **Note 7**), keeping the final concentration of DMF below 10% (v/v) (*see* **Note 4**). Vortex to mix.
8. Warm the reaction mixture at 37°C for 30 min.
9. (Optional, *see* **Note 5**) Purify the protein conjugate of unwanted by-products and reactants by using a size exclusion column (such as Sephadex G-25) or by dialysis.
10. If desired, the concentration of the protein conjugate may be estimated as described in **Note 6**.

3.2. Preparing a 0.5 mL SHA Column

1. Attach an empty 0.5-cm column to a stand (*see* **Note 9**).
2. Thoroughly resuspend the SHA-agarose slurry by inverting the bottle several times.
3. Remove approximately 1 mL of the slurry to the column by applying it to the sides of the column (*see* **Notes 10** and **11**). Allow the storage solution to drain from the column.
4. Equilibrate the column by washing with 20 column volumes of 0.1 *M* sodium bicarbonate, pH 8 (*see* **Note 12**). Do not allow the column to run dry.

The column is now ready for immobilization of the PDBA-modified capture ligand. At this point, columns may be capped and stored in 0.1 *M* sodium bicarbonate, pH 8 containing 0.02% azide and kept at 4°C for about one month.

3.3. Immobilizing the PDBA-Modified Capture Ligand on an SHA Column

1. Add the PDBA-modified capture ligand (prepared in **Subheading 3.1.**) to the top of the SHA-agarose column and charge the column by gravity or batch loading (*see* **Note 13**).
2. After loading, wash the column with 20 column volumes (e.g., 10 mL for a 0.5 mL column) of 0.1 *M* sodium bicarbonate, pH 8 (*see* **Note 12**).

3.4. Loading the Crude Target Protein Mixture

1. Add the crude target protein mixture to the top of column, being careful not to overload the column (*see* **Note 14**), and charge the column by gravity or batch loading (*see* **Note 13**).
2. Collect the flow through. (You may wish to analyze the flow-through to ensure that the column was not overloaded).
3. Wash the column with 20 column volumes (e.g., 10 mL for a 0.5 mL column) of a low ionic strength buffer (e.g., 0.1 *M* $NaHCO_3$, pH 8).

3.5. Eluting the Purified Target Protein

The target protein of interest may now be recovered by elution. Before beginning the elution step, you will need to determine the optimal elution conditions for your target protein. A target protein may be eluted using a variety of solutions such as high pH (>10) or low pH (2.5) buffers, denaturants, substrates, competitive inhibitors and peptide mimics. First verify that the activity of the target protein is not significantly diminished by the eluant.

4. Notes

1. To maximize the modification of lysines and minimize the effects of hydrolysis of the NHS ester, it is important to maintain a high concentration of protein in the reaction.
2. Although sodium bicarbonate is suggested as the buffer of choice, other buffers may be used in the pH range of 7–9. Exceptions include those buffers that contain free amine or sulfhydryl groups such as Tris, glycine, β-mercaptoethanol, dithiothreitol (DTT) or dithioerythritol (DTE). The mechanism of this reaction is based on nucleophilic attack of the deprotonated amine groups (i.e., the ∈-amine groups of lysine in proteins) on the NHS ester. The pH is best kept below 9 to minimize the competing hydrolysis of the NHS ester.
3. The input ratio of PDBA-conjugation reagent to protein may need to be optimized for a particular ligand. By adjusting the molar ratio of PDBA-conjugation reagent to the target protein, the level of modification may be controlled to create a PDBA-ligand with optimal activity.
4. Above concentrations of 10% (v/v) DMF, protein ligands may begin to denature.

5. The PDBA-SHA system demonstrates modest 1:1 affinity but high avidity through the use of multiple labels. As a result, purification of PBDA-conjugates is not necessary for most applications since excess unincorporated PDBA reagent does not compete with multiply-modified PDBA-proteins for SHA sites and is efficiently removed by washing. However, if you wish to estimate the moles of PDBA per mole of protein, you will need to purify the conjugate to remove free PDBA.
6. The actual moles of PDBA per mole of protein can be estimated by quantitatively comparing the absorbance at 260 nm of unmodified protein (P) versus modified protein (PDBA-P). An increase in absorbance at 260 nm is due to PDBA addition. The molar absorbtivity of PDBA at 260 nm is 4000. Dilute (as necessary) an aliquot of the PDBA-modified protein with an appropriate buffer. Dilute (as necessary) an aliquot of the unmodified protein with the same buffer. Measure A_{280} and A_{260} of both solutions. Use the following equations to estimate the degree of modification. DF refers to the dilution factor used to determine A_{260}.

$$A_{260}{}^{PDBA} = (DF)(A_{260}{}^{PDBA\text{-}P}) - ((DF)(A_{280}{}^{PDBA\text{-}P}) \times (A_{260}P/A_{280}{}^{P}))$$
$$[PDBA, \mu M] = A_{260}{}^{PDBA} * 10^6/4000$$
$$PDBA{:}P = [PDBA, \mu M]/\ [P, \mu M]$$

7. The pH of this reaction needs to be controlled to minimize reaction with free amines on the protein ligand. Maleimides react specifically with sulfhydryl groups in the pH range of 6.5–7.5. At more basic pH, however, maleimides show cross reactivity with amines such as those found on lysine residues. The input concentration of PDBA-X-maleimide may need to be optimized in order to minimize reaction with free amines on the protein ligand.
8. DTT serves to reduce disulfide bonds generating free sulfhydryl groups, that can react with the maleimide. Alternatives to DTT include tris(2-carboxyethyl) phosphine hydrochloride (TCEP), and mercaptoethylamine hydrochloride (MEA). If the protein ligand already has free sulfhydryl groups, this step may be eliminated.
9. The desired bed volume of the SHA column depends on the amount of conjugated ligand to be immobilized. The amount of PDBA-ligand that can be bound to the column is specific to the characteristics of the ligand (e.g., molecular weight, number and distribution of PDBA moieties). For example, more than 20 mg of PDBA-conjugated bovine serum albumin (MW = 68,000 Da) or up to 5 mg of PDBA-conjugated IgG (MW = 150,000 Da) can be bound to a milliliter of settled bed resin.
10. 1 mL of the SHA-agarose slurry will yield a settled bed volume of approximately 0.5 mL.
11. The column may be poured and used at room temperature or 4°C, depending upon the stability of the protein ligand and target.
12. Other buffers in the pH range of 5–9 may be used. A list of buffers, salts, detergents and denaturants that are compatible with the PDBA-SHA system is described in **Fig. 2**. If in doubt, test for leaching of the ligand by analyzing the column flow-through for by absorbance before purifying the protein of interest.
13. Application of the PDBA-ligand by gravity loading will result in over 85% of the PDBA-ligand being bound. However, batch loading of the PDBA-ligand by cap-

ping the column and rotating the column end over end for at least 30 min (at room temperature or 4°C, depending upon ligand stability) will ensure that greater than 95% of the PDBA-ligand will bind. For convenience, the reaction can go as long as overnight.

14. The amount of crude target protein solution that may be loaded onto the column will vary depending on the solution composition, for example, pH, viscosity, and so on. Testing the flow through for the presence of the target protein is one way of determining when the column is overloaded.

References

1. Burton, S. J. (1996) Affinity chromatography, in *Downstream processing of natural products* (Verrall, M. S., ed.), John Wiley and Sons, New York, pp. 193–207.
2. Carlsson, J., Janson, J.-C., and Sparrman, M. (1998) Affinity chromatography, in *Protein purification: Principles, high-resolution, methods, and applications, second edition* (Christer, J.-C. and Ryden, L., eds.), Wiley-Liss, New York, pp. 375–442.
3. Hermanson, G. T. (1996) *Bioconjugate Techniques*. Academic, New York, New York.
4. Liapis, A. I. and Unger, K. K. (1994) The chemistry and engineering of affinity chromatography, in *Highly Selective Separations in Biotechnology* (Street, G., ed.), Blackie, Glasgow, UK, pp. 121–162.
5. Pepper, D. S. (1994) Some alternative coupling chemistries for affinity chromatography. *Mol. Biotechnol.* **2,** 157–178.
6. Stolowitz, M. L., Ahlem, C., Hughes, K. A., et al. (2001) Phenylboronic acid-salicylhydroxamic acid bioconjugates 1: A novel boronic acid complex for protein immobilization. *Bioconj. Chem.* **12,** 229–239.
7. Wiley, J. P., Hughes, K. A., Kaiser, R .J., Kesicki, E. A., Lund, K. P., and Stolowitz, M. L., (2001) Phenylboronic acid-salicylhydroxamic acid bioconjugates 2: Polyvalent immobilization of protein ligands for affinity chromatography. *Bioconj. Chem.* **12,** 240–250.
8. Bergseid, M., Baytan, A. R., Wiley, J. P., et al. (2000) Small-molecule based chemical affinity system for the purification of proteins. *Biotechniques* **29,** 1126–1133.

15

Use of tRNA-Supplemented Host Strains for Expression of Heterologous Genes in *E. coli*

Carsten-Peter Carstens

1. Introduction

Though widely used, expression of heterologous genes in *E. coli* can be cumbersome and often fails to yield significant amounts of the desired protein. There are a variety of reasons why a particular heterologous gene fails to yield significant amount of protein, including susceptibility of the protein to degradation or presence of sequences that act as transcriptional pause sites. However, probably the most commonly encountered problem is a mismatch of the codon preference observed in the heterologous gene from the codon usage of the *E. coli* host. The general problem arises form the fact that due to the redundancy of the genetic code one amino acid can be encoded by more than one codon and different organisms do not utilize each codon at the same frequency (for codon frequencies in *E. coli* and other organisms, *see* **Table 1**). Since *E. coli* contains 46 different tRNA genes (some of them exist in multiple copies), most codons (but not all) have a corresponding cognate tRNA *(1)*. The relative expression of each tRNA gene is typically matched to the frequency of the corresponding codon in the translated RNA species. Forced, high-level expression of heterologous genes containing codons rarely utilized in *E. coli* can lead to depletion of the corresponding tRNA pools and subsequently to stalling of the translation process and degradation of the translated mRNA *(2–4)*. The apparent effect is the failure of product formation, and the potential accumulation of aborted translation products that can have the appearance of degradation products. Other less obvious effects of mismatched codon usage are translational frameshifts *(5)*, skipping of a particular codon *(6)*, or the incorporation of wrong amino acids *(7)*. Misincorporation rates of lysine for arginine encoded by a rare codon of up to 40% have been observed at certain

From: *Methods in Molecular Biology, vol. 205, E. coli Gene Expression Protocols*
Edited by: P. E. Vaillancourt © Humana Press Inc., Totowa, NJ

Table 1
Codon Usage of Rare *E. coli* Codons in Other Organisms[a]

Codon (cognate tRNA gene)	AGG/AGA (*argU*)	CGA	CUA (*leuW*)	AUA (*ileX/ileY*)	CCC (*proL*)
Encoded amino acid	Arginine	Arginine	Leucine	Isoleucine	Proline
E. coli K12	1.2/2.1	3.6	3.9	4.4	5.5
Homo sapiens	11.5/11.3	6.3	7.0	7.2	20.0
Drosphila melanogaster	6.4/5.1	8.5	8.3	9.3	17.9
S. cerivisiae	9.3/21.3	3.0	13.4	17.8	6.8
Plasmodium falciparium	4.0/16.7	2.4	5.3	44.3	2.6
Pyrococcus furiosus	20.5/29.5	0.6	16.2	34.8	8.8
Thermus aquaticus	14.5/1.4	1.6	3.6	1.4	39.3
Arabidopsis thaliana	10.9/19.0	6.3	10.1	13.1	5.2
Triticum aestivum	11.8/6.5	3.2	6.3	6.1	14.4

[a]Codon frequencies are presented as codons encountered in 1000 codons of coding sequence. This table was compiled from the codon usage web site (www.kazusa.or.jp/codon). If every codon would be evenly presented, a codon frequency of 15 would be expected.

positions of some proteins *(8)*. However, most users will not realize these effects since they only become apparent in detailed analysis of the products.

The effects of poor codon usage can be alleviated either by synonymous replacement of rare *E. coli* codons in the heterologous gene through site directed mutagenesis or by co-expression of extra copies of the rare tRNA genes along with the desired product *(1,4,9–12)*. Since the former method, though effective, is very tedious and time consuming, the preferable option is the use of *E. coli* host containing the genes for the rare tRNAs on a compatible multicopy plasmid. In order to provide a generic host that allows for adjustment of the codon bias of heterologous genes, we have constructed pACYC-based vectors that contain copies either of the *argU*, *ileY,* and *leuW* or the *argU* and *proL* tRNA genes. *ArgU*, *IleY*, *leuW*, and *proL* encode tRNAs specific for the arginine codons, AGA/AGG; the isoleucine codon, AUA; the leucine codon, CGA; and the proline codon, CCC. The choice of tRNA genes was determined by the rarity of the cognate codons in *E. coli* and their effect on protein expression (*see* **Table 1**). Other codons may be rarer then the ones selected, but often their presence does not significantly affect expression level. This arises because other tRNA genes for the same amino acid may show sufficient wobble in the third nucleotide of the recognition codon to substitute during translation. Some codons reported in the literature as affecting protein expression will also only do so if they are arranged in a stretch of 8 or more consecutive rare codons *(13)*, a situation unlikely to be encountered in wild type genes. Ideally, all tRNA genes would be placed on the same vector to provide a general expression host. However, we have observed a functional incompatibility of the *proL* and the *ileY* tRNA genes, leading to a loss of functional *ileY* tRNA formation. Therefore, two vectors were constructed with the tRNA genes assembled by their relative GC-richness of their cognate codons. The AGA (*argU*), AUA (*ileY*) and CUA (*leuW*) codons are more likely to be encountered in AT rich organisms, whereas the AGG (*argU*) and CCC (*proL*) codons are more likely to be encountered in GC-rich organisms. The tRNA expression vectors were introduced into the BL21-gold cells (Stratagene), that offer the advantage over conventional BL21 cells that their transformation efficiency is about 100-fold higher and that they are *endA*⁻. The resulting cells are available from Stratagene under the trade name BL21-CodonPlus™.

There are no parameters that will predict with certainty whether the expression of a given heterologous gene in *E. coli* will be affected by the codon usage. However, there are some useful rules that can be used as predictors: (1) It is well documented in the literature that the effect of rare codons is more pronounced if they occur closer to the N-terminus of the protein ***(14)***; (2) There is no minimum frequency for rare codons to have an effect on expression. However, most naturally occurring proteins that are affected in their expression in *E. coli* will have frequencies of at least one rare codon type of 3% of all codons; (3) The most important factor predicting the susceptibility of expression to codon usage problems is the occurrence of consecutive rare codons. The rare codons don't have to be recognized by the same rare tRNA to significantly affect expression (e.g., an AGG or AGA codon followed by an AUA codon will reduce expression). Also, three consecutive rare codons placed near the N-terminus of an otherwise well-expressed protein will effectively abolish expression even if no other rare codons are present in the gene of question. In cases where two genes have the same number and frequency of rare codons, but only in one do gene clusters of consecutive rare codons occur, only the gene with the consecutive rare codons will be affected in its expression level by the codon usage. Generally, the likelihood of genes from specific organisms to be affected by codon usage in expression in *E. coli* can be predicted by its codon preferences (*see* **Table 1**). However, each construct should be evaluated individually. It should be kept in mind that even *E. coli* contains genes with rare codons. It should also be kept in mind that some genes have more than one feature preventing their expression in *E. coli*.

tRNA-supplemented host strains such as BL21-CodonPlus do not behave significantly different from conventional BL21 or other corresponding expression strains. Therefore, all protocols used for other expression systems should be easily adaptable to tRNA supplemented host strain variants. The protocols are designed to allow quick assessment of the potential benefits of tRNA supplemented strains for expression of the gene of interest. Detection of expression by an antibody specific for the product of interest is generally preferable since it will also allow assessment of protein stability, but detection of protein expression with non-specific dyes like Coomassie or Silver staining is sufficient in a large number of cases. Since the detection method will vary for each specific construct, no detailed protocols are given for the detection steps.

2. Materials

2.1. Transformation and Induction of Protein Expression

1. Transform competent host cells such as BL21-CodonPlus(DE3)-RIL or BL21-CodonPlus-RP (*see* **Note 1**). Store chemically or electrocompetent cells at –80°C.
2. Rich media such as 1X NZY: 5 g NaCl, 2 g $MgSO_4 \cdot 7H_2O$, 5 g yeast extract, and 10 g NZ amine (casein hydrolysate). Add deionized H_2O to a final volume of 1 L and adjust the pH to 7.5 with NaOH. Autoclave and store at room temperature.
3. Chloramphenicol stock solution (50 mg/mL in ethanol). Store at –20°C.
4. Stock solution of the selection drug used for the expression construct (usually ampicillin or kanamycin).
5. Isopropyl-β-D-galactopyranoside (IPTG) stock solution of 500 m*M* in water. Store at –20°C.

2.2. Induction of Expression of Toxic Genes

1. LE392 host cells. Available as glycerol stocks from several vendors.
2. λCE6 phage. Commercially available stocks are typically provided at 10^{10} plaque forming units (pfu) in glycerol. Store at –80°C.
3. SM solution: 5 g NaCl, 2 g $MgSO_4 \cdot 7H_2O$, 50 mL 1 *M* Tris-HCl, pH 7.5, 5 mL 2% gelatin. Add deionized H_2O to a final volume of 1 L. Adjust the pH to 7.5. Autoclave and store at room temperature.
4. Top agar: 0.6% agar in deionized water. Autoclave and store at room temperature
5. 20% glucose in water. Filter sterilize and store at room temperature.

3. Methods

3.1. Transformation of BL21-CodonPlus Cells

1. Thaw the competent cells on ice (*see* **Note 2**).
2. Gently mix the competent cells. Aliquot 100 µL of the competent cells into the appropriate number of pre-chilled 15-mL Falcon 2059 polypropylene tubes. Aliquot 100 µL of competent cells into an additional pre-chilled 15-mL Falcon 2059 polypropylene tube for use as a transformation control.
3. Add 1–50 ng of expression plasmid DNA containing the gene of interest to the competent cells and swirl gently.
4. Incubate the reactions on ice for 30 min.
5. Preheat 1X NZY medium (*see* **Subheading 2.1.**) in a 42°C water bath for use in **step 8**.

6. Heat-pulse each transformation reaction in a 42°C water bath for 20 s (*see* **Note 3**).
7. Incubate the reactions on ice for 2 min.
8. Add 0.9 mL of preheated (42°C) NZY medium to each transformation reaction and incubate the reactions at 37°C for 1 h with shaking at 225–250 rpm.
9. Use a sterile spreader to plate ≤200 µL of the cells transformed with the experimental DNA directly onto Luria-Bertani (LB) agar plates that contain the appropriate antibiotic (*see* **Note 4**).
10. Transformants will appear as colonies following overnight incubation at 37°C (*see* **Note 5**).

3.2. Induction of the Gene of Interest

The actual conditions for optimal production of the protein of interest, such as induction method, induction time, and growth conditions, needs to be optimized for each specific protein. Since extra copies of the tRNA genes will only effect the translation efficiency of the heterologous gene, any *E. coli* expression system can be used in conjunction with CodonPlus cells. However, the pET system appears to be the most commonly used *E. coli* expression platform. Therefore an induction protocol for a pET-based system using small-scale cultures is given below. The actual conditions need to be optimized for each construct. The below protocol provides reasonable starting conditions.

1. Inoculate 1 mL aliquots of NZY or LB broth (containing 50 µg/mL of chloramphenicol and the appropriate antibiotic) with single colonies from the transformation. Shake at 220–250 rpm at 37°C overnight (*see* **Note 6**).
2. The next morning, pipet 50 µL of each culture into fresh 1-mL aliquots of LB broth containing no selection antibiotics. Incubate these cultures with shaking at 220–250 rpm at 37°C for 2 h.
3. Split each sample into two 500-µL aliquots.
4. To one of the 500-µL aliquots, add IPTG to a final concentration of 1 m*M*. Incubate with shaking at 220–250 rpm at 37°C for 2 h.
5. After the end of the induction period, place the cultures on ice.
6. Pipet 30 µL of each of the cultures into clean microcentrifuge tubes. Add 30 µL of 2X sodium dodecylsulfate (SDS) gel sample buffer to each sample and denature by boiling for 2 min.
7. Heat all tubes to 95°C for 5 min and analyze the samples by Coomassie® brilliant blue staining of an SDS-PAGE gel, loading 30 µL of associated non-induced/induced samples in adjacent lanes (*see* **Note 7**).

3.3. Induction of Toxic Proteins Using the pET System

Since the expression of the T7-polymerase in cells containing the DE3 lysogen is leaky, expression of toxic genes in the pET system requires measures to prevent premature transcription triggered by the T7 RNA polymerase. The tightest control is achieved by introducing T7 RNA polymerase by infection with a λ-phage (called CE6) that carries the gene for the T7 RNA polymerase. When using this approach, do not use host strains that already carry the DE3 lysogen. The below protocol describes both phage production and the infection protocol for a small scale culture.

3.3.1. Production of CE6 Phage

1. Inoculate 5 mL of NZY broth with a single colony of LE392 host cells. Shake overnight at 37°C at 220–250 rpm (*see* **Note 8**).
2. Centrifuge the overnight culture for 15 min at 1700–2000*g* at 4°C. Resuspend the cells in 10 m*M* $MgSO_4$ to a final OD_{600} of 0.5.
3. Combine 250 μL of cells (at OD_{600} = 0.5) with 1×10^6 pfu of CE6 in Falcon® 2059 polypropylene tubes in triplicate. Incubate at 37°C for 15 min.
4. Add 3 mL of melted NZY top agar (equilibrated to 48°C prior to addition) to each cell suspension and plate on warm agarose plates. Incubate the plates overnight at 37°C.
5. Flood each plate with 5 mL of SM solution and rock the plates for 2 h at room temperature.
6. Remove the SM solution (which contains the lambda CE6) from each plate and pool the volumes in a 50-mL conical tube.
7. Centrifuge the SM solution at 1700–2000*g* for 15 min at 4°C.
8. Remove the supernatant and determine the titer of the solution.
9. Store the lambda CE6 stock at 4°C (*see* **Note 9**).

3.3.2. Induction of Expression

1. Inoculate 5 mL of NZY broth containing 50 μg/mL chloramphenicol and the antibiotic required to maintain the expression plasmid with a single colony of BL21 cells (not the DE3 lysogen) harboring the expression plasmid. Shake overnight at 37°C at 200–250 rpm.
2. In the morning, centrifuge 1.0 mL of the overnight culture, resuspend the cells in 1.0 mL of fresh NZY broth, and pipet the resuspended cells into a flask containing 50 mL of fresh NZY broth (no selection antibiotics).
3. Record the A_{600} of the diluted culture. It should be ≤ 0.1. If the A_{600} is > 0.1, use more fresh NZY broth to dilute the culture to $A_{600} \leq 0.1$. If the A_{600} is < 0.1, the time required to reach an A_{600} of 0.3 (in **step 4**) will be extended.
4. Grow the culture to an A_{600} of 0.3, and add glucose to a final concentration of 4 mg/mL (e.g., 1.0 mL of a 20% glucose solution to the 50-mL culture).
5. Grow the culture to an A_{600} of 0.6–1.0 and add $MgSO_4$ to a final concentration of 10 m*M* (e.g., 500 μL of a 1.0 *M* solution of $MgSO_4$ to the 50-mL culture).
6. Remove a portion of the culture to serve as the uninduced control and infect the rest with bacteriophage lambda CE6 at a multiplicity of infection (MOI) of 5–10 particles per cell (*see* **Note 10**). To optimize induction, cultures may be split into 3 or 4 aliquots and infected with varying dilutions of bacteriophage lambda CE6. The subsequent induction can be monitored by SDS-PAGE or by a functional assay, if available.

7. Grow the culture for 2–3 h.
8. Remove 5–20 μL of the culture for determination by SDS-PAGE, and harvest the remaining culture by centrifugation. Store the pellets at –70°C (*see* **Note 11**).

4. Notes

1. Any *E. coli* host can be used provided it contains the tRNA expression plasmid. In most cases, BL21-derived hosts are preferable since they are naturally *ompT* deficient. The use of BL21gold-based strains offers the additional advantage that these strains are *endA*– and contain a mutation that allows for ≈100-fold higher transformation efficiency. Cells can be rendered transformation competent using any of the standard protocols
2. Store the competent cells on ice at all times while aliquoting. It is essential that the Falcon 2059 polypropylene tubes are placed on ice before the competent cells are thawed, and that 100 μL of competent cells are aliquoted directly into each prechilled polypropylene tube. Do not pass the frozen competent cells through more than one freeze-thaw cycle.
3. The length of the heat pulse is optimized for the use of Falcon 2059 tubes and the volume of competent cells used. Changing the conditions will affect the optimal length of the heat pulse. If in doubt, it is generally safer to extend the duration of the heat pulse.
4. The cells may be concentrated by centrifuging at 200*g* for 3–5 min at 4°C if desired. Resuspend the pellet in 200 μL of 1X NZY broth.
5. Assuming a transformation efficiency of 1×10^7 colony forming units/μg (cfu/μg) (which is the efficiency of commercially available BL21-CodonPlus cells), plating 100 μL ***(10%)*** of a transformation with 1 ng plasmid should yield ≈100 colonies
6. Generally, pACYC based plasmids such as the tRNA expression plasmids pACYC-RIL and pACYC-RP are maintained stable in the absence of selection. However, if the expression of the gene of interest is dependent on the tRNA function and is also toxic to the host cell, the tRNA expression plasmid may be selected against in the absence of chloramphenicol. It is therefore advisable to add chloramphenicol in addition to the selection marker for the expression plasmid to the overnight cultures.
7. When analyzing a cell extract by gel electrophoresis, chloramphenicol acetyl transferase, the protein that provides chloramphenicol resistance, will be observed at ~25,660 Da.
8. The bacteriophage λCE6 requires a *supF* host such as LE392 for replication. CE6 is not replication competent in non-supF host such as BL21.
9. If the titer drops over time, or if more phage are needed, grow up LE392 cells in 10 mL of medium and add bacteriophage lambda CE6 at a multiplicity of infection of 1:1000 (CE6-to-cell ratio). Continue growing the culture at 37°C for 5–6 h and spin down the cellular debris. Titer of the supernatant should be $\geq 5.0 \times 10^9$ pfu/mL.
10. An A_{600} of 0.3 corresponds to $\approx 1.2 \times 10^8$ cells/mL.

11. Infection with CE6 will lead to also lead to production of phage specific proteins. If induction will be monitored using Coomassie stain, silver stain, or another nonspecific protein stain, it is advisable to run a control of CE6-infected BL21 cells harboring the plasmid without a cloned insert.

References

1. Riley, M. and Labedan, B. (1996) *Escherichia coli* gene products: physiological functions and common ancestries, in *Escherichia coli* and *Salmonella: cellular and molecular biology,* 2nd Edition (Neidhardt, F. C., ed.), ASM Press, Washington, DC, pp. 2118–2202.
2. Kane, J. F. (1995) Effects of rare codon clusters on high-level expression of heterologous proteins in *Escherichia coli. Curr. Opin. Biotechnol.* **6,** 494–500.
3. Bonekamp, F., Andersen, H. D., Christensen, T., and Jensen, K. F. (1985) Codon-defined ribosomal pausing in *Escherichia coli* detected by using the *pyrE* attenuator to probe the coupling between transcription and translation. *Nucleic Acids Res.* **13,** 4113–4123.
4. Deana, A,. Ehrlich, R., and Reiss, C. (1998) Silent mutations in the *Escherichia coli ompA* leader peptide region strongly affect transcription and translation in vivo. *Nucleic Acids Res.* **26,** 4778–4782.
5. Spanjaard, R. A., Chen, K., Walker, J. R., and van Duin, J. (1990) Frameshift suppression at tandem AGA and AGG codons by cloned tRNA genes: assigning a codon to *argU* tRNA and T4 tRNA(Arg). *Nucleic Acids Res.* **18,** 5031–5036.
6. Kane J. F., Violand, B. N., Curran, D. F., Staten, N. R., Duffin, K. L., and Bogosian, G. (1992) Novel in-frame two codon translational hop during synthesis of bovine placental lactogen in a recombinant strain of *Escherichia coli. Nucleic Acids Res.* **20,** 6707–6712.
7. Calderone, T. L., Stevens, R. D., and Oas, T. G. (1996) High-level misincorporation of lysine for arginine at AGA codons in a fusion protein expressed in *Escherichia coli. J. Mol. Biol.* **262,** 407–412.
8. Forman, M. D., Stack, R. F., Masters, P. S., Hauer, C. R., and Baxter, S. M. (1998) High level, context dependent misincorporation of lysine for arginine in Saccharomyces cerevisiae a1 homeodomain expressed in *Escherichia coli. Protein Sci.* **7,** 500–503.
9. Hu, X., Shi, Q., Yang, T., and Jackowski, G. (1996) Specific replacement of consecutive AGG codons results in high-level expression of human cardiac troponin T in *Escherichia coli. Protein Expr. Purif.* **7,** 289–293.
10. Del Tito, B. J. Jr., Ward, J. M., Hodgson, J., et al. (1995) Effects of a minor isoleucyl tRNA on heterologous protein translation in *Escherichia coli. J. Bacteriol.* **177,** 7086–7091.
11. Saxena, P. and Walker, J. R. (1992) Expression of *argU*, the *Escherichia coli* gene coding for a rare arginine tRNA. *J. Bacteriol.* **174,** 1956–1964.
12. Garcia, G. M., Mar, P. K., Mullin, D. A., Walker, J. R., and Prather, N. E. (1986) The *E. coli dnaY* gene encodes an arginine transfer RNA. *Cell* **45,** 453–459.

13. Goldman, E., Rosenberg, A. H., Zubay, G., and Studier, F. W. (1995) Consecutive low-usage leucine codons block translation only when near the 5' end of a message in *Escherichia coli. J. Mol. Biol.* **245,** 467–473.
14. Rosenberg, A. H., Goldman, E., Dunn, J. J., Studier, F. W., and Zubay, G. (1993) Effects of consecutive AGG codons on translation in *Escherichia coli*, demonstrated with a versatile codon test system. *J. Bacteriol.* **175,** 716–722.

16

Screening Peptide/Protein Libraries Fused to the λ Repressor DNA-Binding Domain in *E. coli* Cells

Leonardo Mariño-Ramírez, Lisa Campbell, and James C. Hu

1. Introduction

The use of λ repressor fusions to study protein-protein interactions in *E. coli* was first described by Hu and others *(1)*. Since then, the repressor system has been employed by several laboratories to screen genomic *(2–5)* and cDNA libraries *(6)* for homotypic or heterotypic interactions. λ repressor consists of distinct and separable domains: the N-terminal domain which has DNA binding activity and the C-terminal domain which mediates dimerization. The repressor fusion system is based on reconstituting the activity of the repressor by replacing the C-terminal domain with a heterologous oligomerization domain. The interaction is detected when the C-terminal domain forms a dimer (or higher order oligomer) with itself (homotypic interaction) or with a different domain from other fusion (heterotypic interaction) (*see* **Fig. 1**).

Repressor fusions are usually expressed from multicopy plasmids; for a detailed discussion of repressor fusion plasmids available from our laboratory *see* **ref.** *7*. Similar plasmids have been constructed by other groups *(5,8–10)* with a variety of modifications. In all cases, unique restriction sites are available for cloning a desired insert in-frame with the N-terminal domain of repressor. **Table 1** lists the features of several of the repressor plasmid vectors in the literature.

The identification and characterization of homotypic or heterotypic interactions is done by fusing a target DNA (fragments from a specific gene of interest, or a genomic, cDNA, randomized, or rationally designed library) to the λ repressor DNA binding domain. Repressor fusion libraries are made by using appropriate vectors with standard cloning methods. Library construction is not discussed further in this chapter (*see* **Note 1**). Here, we focus on the evaluation

From: *Methods in Molecular Biology, vol. 205, E. coli Gene Expression Protocols*
Edited by: P. E. Vaillancourt © Humana Press Inc., Totowa, NJ

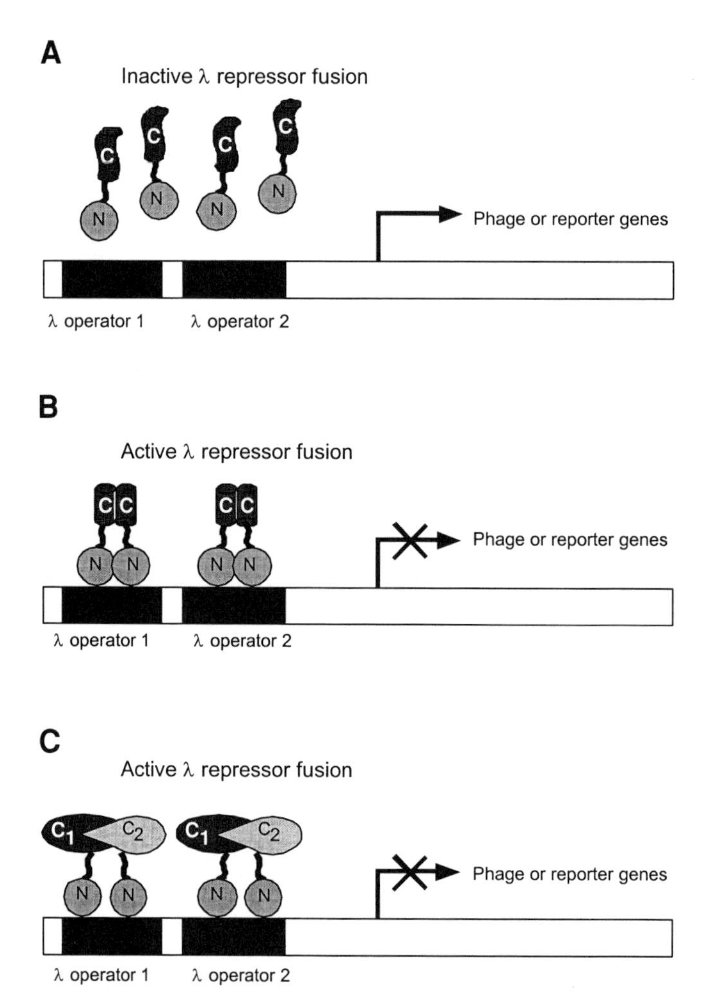

Fig. 1. The rationale of λ repressor fusions. Repressor fusions are used to detect protein-protein interactions in vivo. Protein or peptide targets are fused to the λ repressor DNA binding domain; these fusions can be evaluated for repressor activity using direct selection with λ phage, or a variety of reporter genes suitable for library screening. **(A)** Inactive repressor fusions are unable to bind its target DNA sequences (λ operators in promoters regulating phage or reporter genes). The expression of phage or reporter genes remains unaffected. In this case the fused peptide/protein is monomeric in vivo. **(B)** Active repressor fusions can be reconstituted when a dimeric peptide/protein is placed at the C terminus. The fusions are able to bind λ operators in the promoter and the reporter or phage genes are repressed. In this example the fusion is dimeric but a higher order oligomer can also reconstitute the activity of the repressor. **(C)** Heterodimers can also reconstitute the activity of the λ repressor. In this example, a target peptide (C_1) is encoded in a first plasmid and a peptide library is introduced in the cell by transformation. One of the library encoded peptides (C_2) is able to form a heterodimer with the target peptide reconstituting the activity of λ repressor.

of the resultant repressor fusions for repressor activity using either immunity to phage infection (*see* **Subheading 3.1.**) or a variety of reporters under λ repressor control (*see* **Table 2** and **Subheadings 3.2–3.4.**). Further screening is useful to ensure that the repressor activity of the fusion protein is dependent on the insert, especially when evaluating clones isolated by selection. A simple a high-throughput screening strategy based on nonsense suppression is described in **Subheading 3.5.**

2. Materials

Different subsets of the materials listed below are needed for the different protocols

2.1. General Use Media, Antibiotics, and Materials

1. Luria-Bertani (LB) broth and agar: Premixed LB broth (DIFCO, cat. no. 244620) and agar (DIFCO, cat. no. 244620) are prepared according to the vendors instructions.
2. 2XYT broth per L: 16 g tryptone, 10 g yeast extract, 10 g NaCl. Dissolve in 1 L distilled H_2O. Autoclave.
3. Antibiotics: Ampicillin 200 mg/mL in H_2O (1000X stock, use at a final concentration of 200 μg/mL); kanamycin 20 mg/mL in H_2O (1000X stock, use at a final concentration of 20 μg/mL).
4. Sterile 96-well microplates (clinical "V bottom").
5. Microplate replicator 96 pin (Boekel Model 140500).
6. Multichannel pipetter (8 or 12-channel) to handle volumes from 5–200 μL.
7. Sterile toothpicks.

2.2. Strains

Strains used are listed in **Table 3**. Different strains are used for each of the screening approaches described below.

2.3. For Phage Immunity Selections and Screens

1. AG1688 and JH787 (*see* **Note 2**).
2. λKH54 and λKH54h80 phage stocks at 10^9–10^{10} plaque forming units (pfu)/mL (*see* **Note 3**).
3. Tryptone broth per L: 10 g Tryptone, 5 g NaCl. Dissolve in 1 L H_2O. Autoclave.
4. Tryptone agar: 13 g Bacto-Agar/L of tryptone broth before autoclaving.
5. Tris-Magnesium (TM) buffer: 10 m*M* Tris-HCl, pH 8.0, 10 m*M* $MgSO_4$. Autoclave.
6. Tryptone top agar: 0.7 g Bacto agar/100 mL of tryptone broth before autoclaving.
7. Chloroform.
8. 15-cm LB plates containing ampicillin and kanamycin (*see* **Note 4**).
9. 100-mm LB Amp Kan plates containing 25 m*M* sodium citrate, added from a sterile 1M stock solution.

Table 1
Repressor Fusion Vectors Used for Peptide/Protein Library Screening

Name (size)	Promoter	Cloning sites/Comments	Ref.
pJH370	lacUV5	*Sal*I *Nde*I *Sac*I GCG GAG AGA TGG GTG TCG ACA CAT ATG AAA CAG CTG GAA GAC AAA GTT GAA GAG CTC TCT CTC TCT ACC CAC AGC TGT GTA TAC TTT GTC GAC CTT CTG TTT CAA CTT CTC GAG a e r w v s t H M K Q L E D K V E E L *Xho*I *Spe*I CTG TCT AAA AAC TAC CAC CTC GAG AAC GAA GTT GCG CGC CTG AAA AAA CTA GTT GGT GAC AGA TTT TTG ATG GTG GAG CTC TTG CTT CAA CGC GCG GAC TTT TTT GAT CAA CCA L S K N Y H L E N E V A R L K K L V G *Bam*H1 GAA CGT TGA GGA TCC CTT GCA ACT CCT AGG E R Opa Original CI-GCN4 fusion construct. Also contains the ind1 *Hind*III site at position 117 of the linker between the N and C terminal domains. In principle, this could also be used to generate fusions with a shorter linker.	*(1)*

pJH391
7 kb
	lacUV5	<pre> SalI SacI
		GCG GAG AGA TGG GTG TCG AC GGATCGATCCC GTCCG TTT GAG CTC
		CGC CTC TCT ACC CAC AGC TG CTC GAG
		A E R W V S E L
		cI 130 ... lacZ... GCN4
		
		                        XhoI SpeI
		CTG TCT AAA AAC TAC CAC CTC GAG AAC GAA GTT GCG CGC CTG AAA AAA CTA GTT GGT
		GAC AGA TTT TTG ATG GTG GAG CTC TTG CTT CAA CGC GCG GAC TTT TTT GAT CAA CCA
		L S K N Y H L E N E V A R L K K L V G
		
		            BamH1
		GAA CGT TGA GGATCC GGCTG CTAAC AAAGC CCGAA AGGAA GCTGA GTTGG CTGCT GCCAC
		CTT GCA ACT CCTAGG CCGTC GATTG TTTCG GGCTT TCCTT CGACT CAACC GACGA CGGTG
		E R Opa T7 terminator</pre>pJH370 + a "stuffer fragment that allows easier purification of backbone DNA cut with *Sal*I and *Bam*HI from singly cut vector DNA.
			(17)

pJH391s 7 kB	lacUV5	*Bam*HI Contains an S10 epitope tag to allow the identification of fusion proteins.	*(5)*

(continued)

Table 1 ***(continued)***

Name (size)	Promoter	Cloning sites/Comments	Ref.

pLM99
3.4 kB
	7107	*SalI* *SmaI* *SphI* *BstBI* *BglII* *BamHI*
		`G TCG ACC CGG GCA TGC TTC GAA GAT CTT AAT TAA TTAAGGATCC`
		`C AGC TGG GCC CGT ACG AAG CTT CTA GAA TTA ATT AATTCCTAGG`
		`  S   T   R   A   C   F   E   D   L   N   Ocr`
		
		*Sal*I, *Sma*I, *Sph*I, *Bst*BI, *Bgl*II, *Bam*HI
		pLM99 (GenBank Acc. No. AF308739) contains a triple mutation in the cI DNA binding domain that makes the repressor a better activator at the P_{RM} promoter ***(20)*** without a detectable effect in DNA binding, an amber mutation at position 103 of the cI DBD and a FLAG epitope tag in the linker to allow the identification of fusion proteins. Expression of the fusion proteins is from the weak constitutive promoter 7107 ***(19)***.
			(7)

pLM100
3.4 kb
	7107	*SalI* *SmaI* *SphI* *BstBI* *BglII* *BamHI*
		`G TCG ACG CCC GGG CAT GCT TCG AAG ATC TTA ATT AAT TAA GGATCC`
		`C AGC TGC GGG CCC GTA CGA AGC TTC TAG AAT TAA TTA ATT CCTAGG`
		`  S   T   P   G   H   A   S   K   I   L   I   N   Ocr`
		
		*Sal*I, *Sma*I, *Sph*I, *Bst*BI, *Bgl*II, *Bam*HI
		pLM100 (GenBank Acc. No. AF308740) is identical to pLM99 except for a frameshift at position 7 of the linker.
			(7)

pLM101 3.4 kB	7107	*Sal*I *Sma*I *Sph*I *Bst*BI *Bgl*II *Bam*HI G TCG ACC CCG GGC ATG CTT CGA AGA TCT TAA TTAATTAAGGATCC C AGC TGG GGC CCG TAC GAA GCT TCT AGA ATT AATTAATTCCTAGG S T P G M L R R S Ocr *Sal*I, *Sma*I, *Sph*I, *Bst*BI, *Bgl*II, *Bam*HI pLM101 (GenBank Acc. No. AF308741) is identical to pLM99 except for a frameshift at position 7 of the linker.	*(7)*
pME10 2.8 kB	lacUV5	Multiple cloning site from pSP72 (Promega, Madison, WI) Contains the cI DBD amino acids 1-92.	*(10)*
pAC117	434 repressor	Contains the cI DBD amino acids 1-101.	*(9)*

Table 2
Reporters Available for Library Screening Using Repressor Fusions

Name	Reporter	Principle	Ref.
λ200	$O_R^+P_R$-*lacZ*	An active repressor fusion binds to the PR promoter, down-regulating the lacZ gene.	*(23)*
λ202	$O_R2^-P_R$-*lacZ*	An active repressor fusion binds to a single operator in the PR promoter, down-regulating the lacZ gene.	*(1)*
λ112O_sP_s	$O_s1^+O_s2^+P_s$-*cat-lacZ*	An active repressor fusion binds to two synthetic operators in a promoter, down-regulating the lacZ gene. Reporter used testing cooperative DNA binding of for repressor fuions to operator sites.	*(24)*
λXZ970	$O_s1^-O_s2^+P_s$-*cat-lacZ*	An active repressor fusion binds to a single synthetic operator in a promoter, down-regualting the *lacZ* gene. Reporter used for testing cooperative DNA binding of repressor fusions.	*(18)*
λLS100	$O_{434}^-O_s2^+P_s$-*cat-lacZ*	Same as above.	*(25)*
λLM58	$O_L^+P_L$-*cat-lacZ*	An active repressor fusion binds to the O_L1 and O_L2 operator in the P_L promoter, down-regulating the *cat* and *lacZ* genes.	*(7)*
λLM25	P_L-*GFP*	An active repressor fusion binds to the $O_L1$and O_L2 operator in the P_L promoter, down-regulating the GFPmut2 gene.	L. Mariño-Ramírez, unpublished.
λO_LP_L—amb sup tRNA in Q537	P_L-amber suppressor tRNA	An active repressor fusion down-regulates the *lacZ* amber gene indirectly by repressing the transcription of an amber suppressor tRNA.	*(8)*

Table 3
***E. coli* Strains Used for Peptide/Protein Library Selection and Screening**

Strain	Genotype	Uses	Ref.
AG1688	[F'128 *lacIq lacZ::Tn5*] *araD139*, Δ(*ara-leu*)7697, Δ(*lac*)X74, *galE15*, *galK16*, *rpsL*(Str^R), *hsdR2*, *mcrA*, *mcrB1*	Host for libraries made with repressor fusion vectors lacking an amber mutation. Allows M13-mediated transduction.	*(26)*
JH371	AG1688 [λ200]	Same as AG1688. Allows screening with the *PR-lacZ* reporter (*see* **Table 2**).	*(1)*
JH372	AG1688 [λ202]	Same as AG1688. Allows screening with the *PR-lacZ* reporter (*see* **Table 2**).	*(1)*
JH787	AG1688 [φ80 Su-3]	Host for libraries made with repressor fusion vectors containing an amber mutation.	*(7)*
Q537	F^- *mcrA*, *mcrB*, *r⁻k m+k*, *i*, *lac amU281*, *arg*Eam, *gal*, *rif*, *nal*, *sup0*	Allows screening with the P_L-amber suppressor tRNA reporter.	*(4)*
LM58[a]	JH787 [λLM58] [φ80 Su-3]	Allows screening with the P_L-*cat-lacZ* reporter. Allows amber suppression.	*(7)*
LM59[a]	AG1688 [λLM58]	Allows screening with the P_L-*cat-lacZ* reporter.	*(7)*
LM25	JH787 [λLM-GFP]	Allows screening with the P_L-*GFP* reporter.	L. Mariño-Ramírez, unpublished.

2.4. For Screening with lacZ*-Based Reporters*

Materials for β-galactosidase assay of choice ***(11)***.

2.5. For Screening with Cat-Based Reporters

1. LM58 and/or LM59 (*see* **Note 5**).
2. Chloramphenicol 25 mg/mL in 100% ethanol (1000X stock, use at a final concentration of 25 μg/mL).
3. 15-cm LB plates containing ampicillin.
4. 15-cm LB plates containing ampicillin and chloramphenicol.

2.6. For Screening with Green Fluorescent Protein (GFP) Reporters

1. Repressor fusion libraries in LM25 (*see* **Note 6**).
2. 9-cm LB plates containing ampicillin and kanamycin.
3. LB-ampicillin-kanamycin broth.
4. Disposable analytical filter unit (NALGENE Cat. No. 140–4045).
5. Multiple-fluorophore purple/yellow low intensity beads (Spherotech Cat. No. FL-2060-2) (Working solution is 5 μL beads in 5 mL H_2O supplemented with 0.02% Sodium azide).
6. Flow cytometer FACSCalibur (Becton Dickinson).

2.7. Transfer of Plasmids by M13-Mediated Transduction

1. M13 rv-1 1×10^{11} pfu/mL (*see* **Note 7**).
2. 2XYT broth supplemented with ampicillin, kanamycin and 25 m*M* sodium citrate (if using colonies from phage selections).

3. Methods

Preparation of vector DNA, construction of libraries in repressor fusion vectors and transformation of competent cells can be done by a variety of standard molecular biology methods. The protocols below assume that you are starting with a freshly transformed or amplified library containing the desired inserts.

3.1. Selection or Screening for Phage Immunity

Cells expressing repressor activity are immune to λ infection. This provides a simple selection for active repressor fusions. Cells containing plasmids of interest are spread onto plates pre-seeded with phage. Any cells that lack repressor activity will be killed, and only the survivors need to be studied further.

Selection for active repressor fusions is done in the presence of two λ phage derivatives with different receptor specificities. λKH54 uses the LamB porin as the receptor for infection, whereas λKH54h80 is a ϕ80 hybrid phage that uses the TonB protein as the receptor. We estimate that double mutations resulting in simultaneous loss of both receptors occur at a frequency of around 10^{-9}, while the single mutations in each receptor occur at around 10^{-4}. Because the power of phage selection lies in its ability to process on the order of 10^7 clones/plate, the use of both phages is important to minimize the background of survivors due to host mutations.

Note that in freshly transformed cells, the intracellular concentration of repressor will be zero at the moment the plasmid is introduced, and the steady-state level of repressor will not be achieved for several generations after transformation. Thus, while plating a transformation directly on phage reduces the numbers of siblings recovered, there is a trade-off in a reduction in the recovery of active fusions.

1. Preseed plates by spreading approximately 10^8 phage each of λKH54 and λKH54h80. Allow the plates to dry briefly.
2. Plate cells from amplified or unamplified libraries onto plates containing λ phage. We have plated up to 10^7 cells from an amplified library on a single 150-mm plate. Allow plates to dry.
3. Incubate at 37°C overnight. Immune survivors should show up as single colonies the next day.
4. Pick colonies onto plates or into liquid cultures in microtiter plates containing sodium citrate (*see* **Note 8**).

3.2. Screening with lacZ Reporters

Repressor activity can also be evaluated using reporter constructs that place a screenable or selectable marker under the control of λ operators. Several reporters are available that use natural or artificial promoter-operators to drive *lacZ* expression under λ repressor control. However, these are generally based on strong promoters, and the repressed level of β-galactosidase is still high enough to give blue colonies on X-gal plates. Thus, it is necessary to screen transformants by enzyme assays. The protocol below is based on using the reporters λ200, λ202, λ112O_sP_s, λXZ970, or λLS100. The specialized uses of these reporters are described in **Table 2**.

1. Select transformants on LB Amp Kan plates.
2. Grow individual cultures of each transformant.
3. Assay for β-galactosidase activity using any of a variety of standard assays *(11)*.

3.3. Screening with Chloramphenicol Acetyl Transferase (cat) Reporter

λLM58 carries a chloramphenicol reporter under the control of the P_L promoter, which can be down-regulated by an active repressor fusion (*see* **Table 2**). This allows simple screening on plates.

1. Select transformants on LB Amp Kan plates.
2. Replica plate or pick onto parallel LB Amp Kan plates in the presence and absence of 25 μg/mL chloramphenicol. Active fusions will be sensitive to chloramphenicol while inactive fusions will be resistant.

3.4. Green Fluorescent Protein (GFP) Reporter for the Screening of Active Repressor Fusions

λLM25 carries a GFPmut2 reporter is under the control of the P_L promoter, which can be repressed by an active repressor fusion (*see* **Table 1** and **Note 6**). The activity of a fluorescent reporter can be monitored by fluorescence-activated cell sorting (FACS); additionally FACS can be used to isolate a subpopulation of cells where the reporter has been repressed (*see* **Fig. 2**). For recent

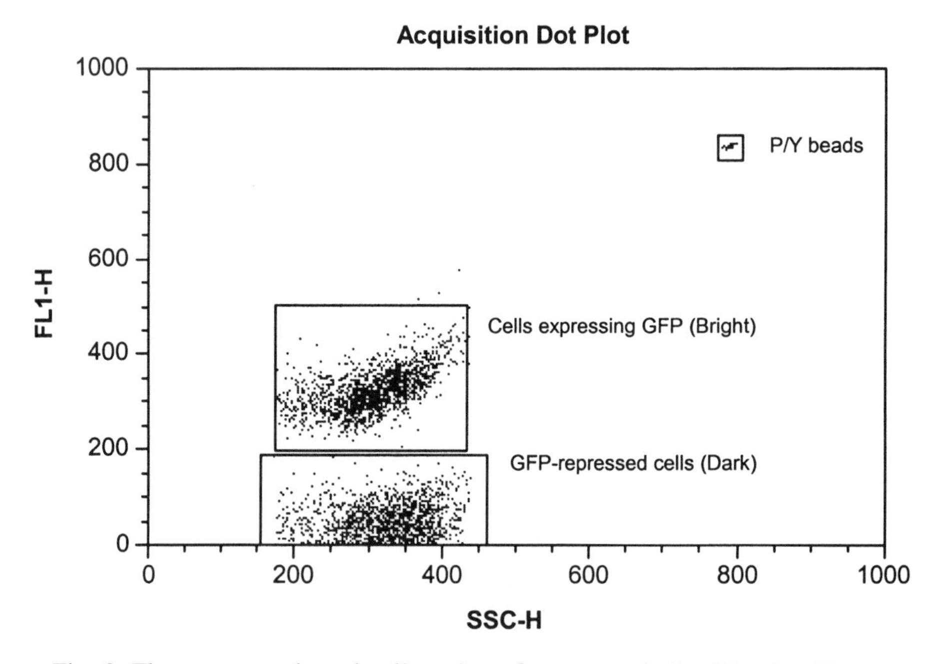

Fig. 2. Fluorescent-activated cell sorting of repressor fusion libraries. Repressor fusion librarires containing yeast genomic DNA were introduced into LM25 cells by electroporation and the libraries sorted as described in **Subheading 3.4.** The cells corresponding to the box labeled as GFP-repressed cells were collected, concentrated and plated as described in the text. A total of 81 cfu's were recovered and transduced into AG1688 (sup^0) and LM25 (supF). Forty three of these clones displayed an immune phenotype dependent on the insert; this fraction is similar to what is observed from this library when clones are isolated by phage selection.

reviews about the application of flow cytometry to various biological systems, *see* **ref. *12,13***. The expression level of the GFP reporter in the cell population is highly homogeneous, as detected by FACS. The homogeneous expression of the GFP reporter is due to the single copy lysogen carrying the reporter. This is important because multi-copy GFP reporters have great variations in the expression of reporters in a cell population.

1. Inoculate 3 mL LB-ampicillin-kanamycin broth with 1/100 vol of an amplified or unamplified library. Incubate at 37°C for 14 h.
2. Prepare 1 mL samples by diluting cells 10,000 fold with deionized water sterilized by filtration through a 0.2 µm filter.
3. Add purple/yellow low intensity beads (10 µL/mL of sample) as fluorescence control.
4. Sterilize the cell sorter by running 70% ethanol for 20 min followed by a wash with MilliQ water for 20 min. Perform cell sorting at a rate of less than 300 events/s

(collect light-scatter and green fluorescence data). Sort at least 50,000 events. Sort the fraction of cells with no detectable green fluorescence. Filtered MilliQ water was used as a sheath into which the cells were sorted.

5. Concentrate the sorted cells by filtration using a disposable analytical filter unit. Place the filter onto a 9-cm LB-ampicillin-kanamycin plate. Incubate at 37°C for 16 h (*see* **Note 9**).
6. Confirm immunity status of positive clones by transducing them into an appropriate background for evaluation by either phage or β-galactosidase assays.

3.5. Nonsense Suppression to Evaluate Insert-Dependence

It is important to check that the repressor activity expressed from a recombinant plasmid is actually due to the fusion of a self-assembly domain rather than some other plasmid mutation that increases expression of the N-terminal DNA-binding domain. Although this can be done by subcloning, conditional expression of the insert can be achieved by nonsense suppression when vectors pLM99-101 are used. These each contain an amber mutation at position 103 of the cI gene. Screening for repressor activity must be done in a host containing an amber suppressor, such as JH787 or LM58. These strains are paired with isogenic strains that are unable to suppress nonsense mutations, AG1688 and JH787, respectively.

1. Pick single colonies from one of the selections or screens above using sterile toothpicks and inoculate 150 μL of 2XYT-ampicillin-kanamycin broth + 25 m*M* sodium citrate (necessary if cells are from phage selection, *see* **Note 8**) in sterile 96-well microplates. Incubate at 37°C and grow for 16 h (*see* **Note 10**).
2. Mix 5 μL M13 rv-1 and 5 μL of each overnight culture. Incubate at 37°C for 10 min to allow phage to adsorb. Add 0.15 mL 2XYT+ 25 m*M* sodium citrate in sterile 96-well microplates broth. Grow for 6 h at 37°C.
3. Heat at 65°C for 20 min to kill *E. coli*. Spin the plates at 1000*g* for 15 min. Store the plate, which contains the M13 transducing phage stocks at 4°C.
4. Transfer the plasmid DNA containing the repressor fusions to an isogenic pair of strains, either AG1688 (Sup^0) and JH787 (SupF) or LM58 (SupF) and LM59(Sup^0) by M13 transduction. Mix 5 μL M13 transducing phage and 50 μL overnight culture from the SupF and Sup^0 strains. Incubate at 37°C for 30 min. Use the microplate replicator to transfer the transductions to LB-ampicillin plates. Incubate at 37°C overnight.
5. Screen the colonies for repressor activity by the appropriate method described above (phage immunity for AG1688 and JH787 or chloramphenicol sensitivity for LM58 and LM59).

4. Notes

1. Highly representative repressor fusion libraries are critical for a successful screening. In addition to methods described in popular cloning manuals ***(14,15)***,

construction of repressor fusion libraries have been described *(3–5)*. Note that genomic libraries require higher coverage than is needed for genome sequencing because large numbers of fusion joints within every gene are needed for library saturation. Vectors pLM99-101 contain polylinkers that allow compatible ligation with a variety of blunt and sticky ends *(16)*. For the generation of blunt ended fragments from the yeast genome, we have used DNA partially digested with *Cvi*TI (Megabase Research).

2. AG1688 *(17)* and JH787 (*see* Table 3) are both sensitive to λKH54 and λKH54h80. JH787, which contains an amber suppressor, should be used when the plasmid vector used for library construction contains an amber mutation, i.e., pLM99-101, between the cI DNA binding domain and the insert *(7)* to allow expression of the full-length fusions.
3. The KH54 deletion removes the cI gene, which is required for establishment and maintenance of lysogens. This is important because lysogens will pass as false positives in a library screen. The h80 substitution replaces λ genes with those of ϕ80. for this use, the relevant change replaces the receptor specificity of λ, which uses the LamB protein, with that of ϕ80, which uses the TonB protein. A mixture of phage is used to eliminate background due to spontaneous receptor mutants. Thus, for phage selection using this mixture of phage to be effective, the starting strain must contain wt alleles for both *lamB* and *tonB*.
4. Ampicillin selects for the plasmid vectors. Kanamycin selects for the F' episome in strains derived from AG1688. This F' carries the *lacI*^{q} allele needed to repress the expression of the fusion proteins expressed from the *lacUV5* promoter in pJH370 and pJH391. In addition, F functions are needed for M13-mediated transduction of the plasmids containing M13 origins (*see* **Subheading 3.5.**).
5. LM58 and LM59 are isogenic strains containing the chloramphenicol reporter carried by λLM58 (*see* **Table 2**). As with AG1688 and JH787, one strain (LM58) contains the SupF amber suppressor, while the other (LM59) is a nonsuppressor strain. The suppressor strain should be used for repressor fusion vectors that contain an amber mutation at position 103 in the cI DNA binding domain.
6. LM25 (JH787 [λLM-GFP]). λLM-GFP is λimm^{21} P_L-*GFP*. Constructed by recombination between λXZ1 *(18)* and Plasmid pLM10 (GenBank Acc. No. AF108217). This strain contains the GFPmut2 allele, which has been optimized for use with fluorescence-activated cell sorting (FACS) *(19)*. GFPmut2 was cloned from pDS439 *(20)* under the control of the P_L promoter from phage λ. The P_L-*GFP* reporter is present in *E. coli* JH787 (*see* **Table 3**) as a single copy lysogen.
7. M-13 rv-1 *(21)* is used to transduce plasmids that contain an M13 ssDNA replication origin and M13 packaging signals *(22)*. Phage stocks are prepared in the same manner as that used to prepare transducing stocks (*see* **Subheading 3.5.**) using a plasmid-free strain as the host. Mix 5 μL M13 rv-1 and 50 μL of a fresh overnight culture in a sterile test tube. Incubate at 37°C to preadsorb the phage. Add 5 mL 2XYT broth, incubate with aeration at 37°C for 6–8 h or overnight. Pellet cells by centrifugation. Save the supernatant. Pasteurize the phage stock by heating to 65°C for 20 min. Store at 4°C.

8. Sodium citrate chelates magnesium ions needed for phage infection. Citrate in the plates prevents reinfection by λ phage carried over from the selection plates.
9. Cells with reduced expression of GFP should contain active repressor fusions. The filter should have about 100 colonies. Adjust cell density to obtain isolated colonies if necessary.
10. Cultures in 96-well plates have a tendency to dry, to avoid this we incubate them for no longer than 16 h. Additionally, we incubate the culture plates on top of two plates that have been filled with distilled water and we keep a 500-mL beaker with distilled water in the incubator to increase humidity.

References

1. Hu, J. C., O'Shea, E. K., Kim, P. S., and Sauer, R. T. (1990) Sequence requirements for coiled-coils: analysis with λ repressor-GCN4 leucine zipper fusions. *Science* **250,** 1400–1403.
2. Park, S. H. and Raines, R. T. (2000) Genetic selection for dissociative inhibitors of designated protein- protein interactions. *Nat. Biotechnol.* **18,** 847–851.
3. Zhang, Z., Murphy, A., Hu, J. C., and Kodadek, T. (1999) Genetic selection of short peptides that support protein oligomerization in vivo. *Curr. Biol.* **9,** 417–420.
4. Jappelli, R. and Brenner, S. (1999) A genetic screen to identify sequences that mediate protein oligomerization in *Escherichia coli. Biochem. Biophys. Res. Commun.* **266,** 243–247.
5. Zhang, Z., Zhu, W., and Kodadek, T. (2000) Selection and application of peptide-binding peptides. *Nat. Biotechnol.* **18,** 71–74.
6. Bunker, C. A. and Kingston, R. E. (1995) Identification of a cDNA for SSRP1, an HMG-box protein, by interaction with the c-Myc oncoprotein in a novel bacterial expression screen. *Nucleic Acids Res.* **23,** 269–276.
7. Mariño-Ramírez, L. and Hu, J. C. (2001) Using λ repressor fusions to isolate and characterize self-assembling domains, in *Protein-Protein Interactions: A Laboratory Manual*, (Golemis, E. and Serebriiskii, I., ed.), Cold Spring Harbor Laboratory, Cold Spring Harbor, NY, pp. 375–393.
8. Cairns, M., Green, A., White, P., Johnston, P., and Brenner, S. (1997) A novel bacterial vector system for monitoring protein-protein interactions in the cAMP-dependent protein kinase complex. *Gene* **185,** 5–9.
9. Jappelli, R. and Brenner, S. (1998) Changes in the periplasmic linker and in the expression level affect the activity of ToxR and λ-ToxR fusion proteins in *Escherichia coli. FEBS Lett.* **423,** 371–375.
10. Edgerton, M. D. and Jones, A. M. (1992) Localization of protein-protein interactions between subunits of phytochrome. *The Plant Cell* **4,** 161–171.
11. Miller, J. H. (1972) Experiments in molecular genetics, Cold Spring Harbor Laboratory, Cold Spring Harbor, NY.
12. Jaroszeski, M. J. and Radcliff, G. (1999) Fundamentals of flow cytometry. *Mol. Biotechnol.* **11,** 37–53.
13. Radcliff, G. and Jaroszeski, M. J. (1998) Basics of flow cytometry. *Methods Mol. Biol.* **91,** 1–24.

14. Sambrook, J., Fritsch, E. F., and Maniatis, T. (1989) *Molecular Cloning, a laboratory manual* 2nd Ed., Cold Spring Harbor Laboratory, Cold Spring Harbor, NY.
15. Cowell, I. G. and Austin, C. A., eds. (1996) *Methods in Molecular Biology.* Vol. 69: cDNA Library Protocols. Humana Press, Totowa, NJ.
16. James, P., Halladay, J., and Craig, E. A. (1996) Genomic libraries and a host strain designed for highly efficient two- hybrid selection in yeast. *Genetics* **144,** 1425–1436.
17. Hu, J., Newell, N., Tidor, B., and Sauer, R. (1993) Probing the roles of residues at the e and g positions of the GCN4 leucine zipper by combinatorial mutagenesis. *Protein Science* **2,** 1072–1084.
18. Zeng, X. and Hu, J. C. (1997) Detection of tetramerization domains in vivo by cooperative DNA binding to tandem lambda operator sites. *Gene* **185,** 245–249.
19. Cormack, B. P., Valdivia, R. H., and Falkow, S. (1996) FACS-optimized mutants of the green fluorescent protein (GFP). *Gene* **173,** 33–38.
20. Siegele, D. A., Campbell, L., and Hu, J. C. (2000) Green fluorescent protein as a reporter of transcriptional activity in a prokaryotic system. *Methods Enzymol.* **305,** 499–513.
21. Zagursky, R. J. and Berman, M. L. (1984) Cloning vectors that yield high levels of single-stranded DNA for rapid DNA sequencing. *Gene* **27,** 183–191.
22. Vershon, A. K., Bowie, J. U., Karplus, T. M., and Sauer, R. T. (1986) Isolation and analysis of Arc repressor mutants: evidence for an unusual mechanism of DNA binding. *Proteins: Structure Function and Genetics* **1,** 302–311.
23. Meyer, B. J., Maurer, R., and Ptashne, M. (1980) Gene regulation at the right operator (OR) of bacteriophage lambda. II. OR1, OR2, and OR3: their roles in mediating the effects of repressor and cro. *J. Mol. Biol.* **139,** 163–194.
24. Beckett, D., Burz, D. S., Ackers, G. K., and Sauer, R. T. (1993) Isolation of lambda repressor mutants with defects in cooperative operator binding. *Biochemistry* **32,** 9073–9079.
25. Hays, L. B., Chen, Y. S., and Hu, J. C. (2000) Two-hybrid system for characterization of protein-protein interactions in *E. coli. Biotechniques* **29,** 288–290, 292–294, 296.
26. Hu, J. C. and Gross, C. A. (1988) Mutations in rpoD that increase expression of genes in the mal regulon of *Escherichia coli* K-12. *J. Mol. Biol.* **203,** 15–27.

17

Studying Protein–Protein Interactions Using a Bacterial Two-Hybrid System

Simon L. Dove

1. Introduction

Two-hybrid systems are powerful genetic assays that allow the interaction between two proteins to be detected in vivo. Although originally described in yeast *(1,2)*, several bacterial two-hybrid systems have recently been developed (reviewed in *3–6*). This chapter will describe the use of one such bacterial system: a transcriptional activation-based two-hybrid system for the analysis of protein-protein interactions in *Escherichia coli*. This system, like the classic yeast two-hybrid system, involves the synthesis of two fusion proteins within the cell whose interaction stimulates expression of a suitable reporter gene. This bacterial system has been used successfully to detect and analyze the interactions between a number of different proteins from both prokaryotes and eukaryotes *(7–10)*, including a phosphorylation-dependent protein-protein interaction between two mammalian transcription factors *(11)*, and the interaction between a peptide aptamer and its intracellular target *(12)*. The use of selectable reporter genes with this system should facilitate the selection of interacting proteins from complex protein libraries *(11,13)*.

The principle behind the bacterial two-hybrid system detailed here is that any sufficiently strong interaction between two proteins can activate transcription in *E. coli* provided one of the interacting proteins is tethered to the DNA by being fused to a DNA-binding protein, and the other is fused to a subunit of RNA polymerase *(7,8)*. In particular, we have shown that interaction between a protein domain X, fused to the amino terminal domain of the α subunit of RNA polymerase (α-NTD), or to the ω subunit, and a second domain Y fused to a suitable DNA-binding protein can mediate transcriptional activation from a suitably designed test promoter that contains a cognate recognition site for

From: *Methods in Molecular Biology, vol. 205, E. coli Gene Expression Protocols*
Edited by: P. E. Vaillancourt © Humana Press Inc., Totowa, NJ

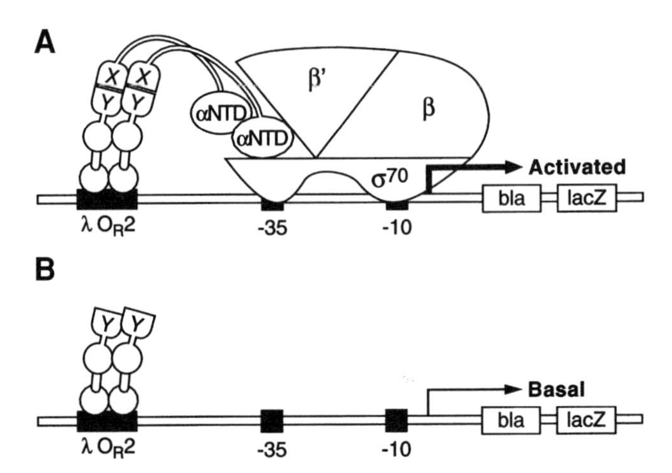

Fig. 1. Principle of bacterial two-hybrid system. **(A)** Contact between protein domains X and Y fused, respectively, to the α-NTD and to λcI activates transcription from the test promoter. The illustrated test promoter p*lac*O_R2–62 contains a λ operator (O_R2) positioned 62 base pairs upstream from the transcriptional start site of the *lac* core promoter. The –10 and –35 elements that constitute the *lac* core promoter, and to which RNAP binds, are depicted as small black boxes. In reporter strain US3F'3.1 and the BacterioMatch reporter strain the test promoter drives the expression of both the *bla* and *lacZ* reporter genes. **(B)** In the absence of any interaction between protein domains X and Y, the binding of RNAP to the test promoter is not stabilized and transcription is not activated.

the DNA-binding protein (*see* **Fig. 1**; **refs.** *7,8*). The protein-protein interaction between domains X and Y presumably serves to stabilize the binding of RNA polymerase (RNAP) to the promoter, with the strength of the protein-protein interaction between domains X and Y determining the magnitude of the activation *(7)*.

In order to use this two-hybrid system to test whether two proteins can interact, one of the proteins under investigation (the bait) is fused to the end of the bacteriophage λcI protein (a sequence-specific DNA-binding protein that binds its recognition site as a dimer), while the other protein (the prey) is fused to the α-NTD via the long flexible linker that α naturally contains (*see* **Note 1**; **refs.** ***14,15***). Alternative DNA-binding proteins, such as derivatives of the monomeric zinc-finger protein Zif268 ***(13)***, and alternative subunits of RNA polymerase, such as the ω subunit ***(3,8)***, can also be used in this system (*see* **Note 2**). For the purposes of this chapter, only the version of this bacterial two-hybrid system that makes use of fusions to λcI and the α-NTD will be detailed. Compatible plasmids directing the synthesis of the λcI fusion protein and the α fusion protein are transformed into a suitable *E. coli* reporter strain. The reporter strain typically contains a test promoter driving expression of a linked reporter gene such as *lacZ,* which encodes β-galactosidase. The test promoter in the reporter strain consists of the *lac* core promoter with a binding site for λcI (O_R2) positioned upstream. If the two fused protein domains can interact, the λcI fusion protein will stabilize the binding of RNAP (containing the α fusion protein) to the test promoter, thereby stimulating expression of the reporter gene. The level of *lacZ* reporter gene expression, which in part reflects the strength of the protein-protein interaction under investigation, can be assayed either quantitatively, using a liquid β-galactosidase assay, or qualitatively by examining colony color on indicator medium containing the chromogenic substrate X-Gal (5-bromo-4-chloro-3-indolyl-β-D-thiogalactopyranoside). Proteins that interact with a protein of interest can be identified from suitable libraries using a modified version of this system in which the test promoter directs transcription of both a selectable gene (the *bla* gene, which encodes β-lactamase) and the *lacZ* gene ***(11)*** (*see* **Fig. 1**). In this case, the activation of reporter gene transcription results in increased expression of the *bla* gene, rendering cells resistant to β-lactams such as carbenicillin. Any carbenicillin resistant clone can then be assayed qualitatively (with indicator medium) and quantitatively (by liquid β-gal assay) for *lacZ* expression.

2. Materials

2.1. Making Fusions to λcI

1. Plasmids pBT and pACλcI32 can be used for making fusions to the end of λcI (*see* **Table 1**).
2. Competent cells (*see* **Note 3** about making chemically competent cells): Strains XL1-Blue, XL1-Blue MRF' Kan, and JM109 can be used for propagating plasmids pBT and pACλcI32 (*see* **Table 2**).
3. Antibiotic stock solutions (1000X): Carbenicillin (100 mg/mL in distilled water, filter sterilize), Chloramphenicol (25 mg/mL in methanol), Kanamycin (50 mg/mL in distilled water, filter sterilize), Tetracycline (10 mg/mL in methanol).
4. Luria-Bertani (LB)-agar plates (containing appropriate antibiotics): 10 g bacto tryptone, 5 g yeast extract, 10 g NaCl and 15 g bacto agar in 1 L distilled water. Autoclave LB-agar to sterilize. Antibiotics should be added once the media has cooled to below 50°C. LB-agar plates should be stored at 4°C.
5. LB broth: 10 g bacto tryptone, 5 g yeast extract, and 10 g NaCl in 1 L distilled water. Autoclave to sterilize. Store at room temperature.

2.2. Making Fusions to the α Subunit of E. coli *RNA Polymerase*

1. Plasmids pTRG, pBRSTAR and pBRαLN can be used for making fusions to the α subunit of *E. coli* RNAP (*see* **Table 1**).

Table 1
Plasmids for Use with Bacterial Two-hybrid System

Plasmid	Description	Antibiotic Resistance	Useful cloning sites	Source/ Ref.
pBT	Bait plasmid, encodes λcI	Chloramphenicol (25 μg/mL)	*Not*I, *Eco*RI, *Sma*I, *Bam*HI, *Xho*I, *Bgl*II	Stratagene
pBT-LGF2	Positive control plasmid, encodes λcI-Gal4 fusion	Chloramphenicol (25 μg/mL)		Stratagene
pACλcI32	Bait plasmid, encodes λcI	Chloramphenicol (25 μg/mL)	*Not*I, *Bgl*II (*Bst*YI), *Asc*I, *Bst*YI	*(3)*
pTRG	Prey plasmid, encodes α-NTD	Tetracycline (10 μg/mL)	*Bam*HI, *Not*I, *Eco*RI *Xho*I, *Spe*I	Stratagene
pTRG-Gal11P	Positive control plasmid, encodes α-Gal11P fusion	Tetracycline (10 μg/mL)		Stratagene
pBRSTAR	Prey plasmid, encodes α-NTD	Tetracycline (10 μg/mL)	*Not*I, *Bam*HI	*(11)*
pBRαLN	Prey plasmid, encodes α-NTD	Carbenicillin (50 μg/mL)	*Not*I, *Bam*HI	*(3)*
pBRα-Gal11P	Positive control plasmid, encodes α-Gal11P fusion	Carbenicillin (50 μg/mL)		*(8)*

Table 2
Bacterial Strains for Use with Bacterial Two-hybrid System

Strain	Relevant genotype	Antibiotic resistance	Reporter genes	Source/ Ref.
KS1	F' *lacI*^{q}	Kanamycin (50 μg/mL)	*lacZ*	*(7)*
BacterioMatch Reporter[a]	*recA*, F' *lacI*^{q}	Kanamycin (50 μg/mL)	*bla, lacZ*	Stratagene
US3F'3.1	*recA*, F' *lacI*^{q}	Kanamycin (50 μg/mL)	*bla, lacZ*	***(11)***
XL1-Blue[b]	*recA*, F' lacI^q	Tetracycline (10 μg/mL)		Stratagene
XL1-Blue MRF' Kan[b]	*recA*, F' lacI^q	Kanamycin (50 μg/mL)		Stratagene
JM109[b]	*recA*, F' lacI^q			***(18)***

[a]Higher transformation efficiencies can be achieved with this strain than can be achieved with strain US3F'3.1.
[b]Strains used for propagation of plasmids used with the bacterial two-hybrid system.

2. Competent cells (*see* **Note 3** about making chemically competent cells). Strains XL1-Blue MRF' Kan and JM109 can be used for manipulating plasmids pTRG and pBRSTAR, whereas strains XL1-Blue, XL1-Blue MRF' Kan, and JM109 can be used to manipulate plasmid pBRαLN (*see* **Table 2**).
3. Antibiotic stock solutions (*see* **Subheading 2.1.**).
4. LB-agar plates containing appropriate antibiotics (*see* **Subheading 2.1.**).
5. LB broth (*see* **Subheading 2.1.**).

2.3. Testing Whether Two Proteins Can Interact

2.3.1. Rapid Cotransformation of Reporter Strains with Bait and Prey Plasmids

1. Competent cells of reporter strain KS1 or the BacterioMatch reporter strain (*see* **Note 3**).
2. LB-agar plates containing appropriate antibiotics (*see* **Subheading 2.1.**).
3. Antibiotic stock solutions (*see* **Subheading 2.1.**).

2.3.2. Quantitative Assay of Reporter Gene Expression Using Liquid Cultures: The β-Galactosidase Assay

1. LB broth (*see* **Subheading 2.1.**).
2. Antibiotic stock solutions (*see* **Subheading 2.1.**).
3. IPTG stock solution (100 m*M*). Dissolve 0.238 g isopropyl-β-D-thiogalactoside (IPTG) in 10 mL distilled water, filter sterilize, aliquot and store at –20°C.
4. Z-buffer: Mix 16.1 g $Na_2HPO_4 \cdot 7H_2O$, 5.5 g $NaH_2PO_4 \cdot H_2O$, 0.75 g KCl, 0.246 g $MgSO_4 \cdot 7H_2O$, 2.7 mL β-mercaptoethanol, and distilled water to 1 L. Adjust pH to 7.0. Do not autoclave.
5. 0.1% Sodium dodecylsulfate (SDS).
6. Chloroform.
7. ONPG solution. O-Nitrophenyl-β-D-galactoside (ONPG) 4 mg/mL in distilled water or Z-buffer. Store at –20°C.
8. 1 *M* Na_2CO_3.

2.4. Screening Libraries in E. coli for Protein Interaction Partners

2.4.1. Library Screening

1. Antibiotic stock solutions (*see* **Subheading 2.1.**).
2. X-Gal stock solution. X-Gal (10 mg/mL) in dimethylformamide (DMF).
3. 150-mm LB-agar plates containing kanamycin (50 μg/mL), tetracycline (10 μg/mL), chloramphenicol (25 μg/mL), IPTG (20 μ*M*), carbenicillin (500–2000 μg/mL).
4. 100-mm LB-agar plates containing kanamycin (50 μg/mL), tetracycline (10 μg/mL), chloramphenicol (25 μg/mL), IPTG (20 μ*M*), carbenicillin (500–2000 μg/mL).
5. LB broth (*see* **Subheading 2.1.**).
6. IPTG (*see* **Subheading 2.3.2.**).

3. Methods

3.1. Making Fusions to λcI

Plasmids pBT and pACλcI32 are derived from the low copy number vector pACYC184; they confer resistance to chloramphenicol and harbor the *cI* gene under the control of the IPTG-inducible *lac*UV5 promoter. Plasmids pBT and pACλcI32 differ from one another with respect to the multiple cloning sites that have been introduced at the end of the *cI* gene to facilitate the fusion of bait proteins to the end of λcI (*see* **Table 1**). Typically the bait is fused to the end of λcI by cloning a suitably designed PCR product into pBT or pACλcI32 such that the gene encoding the bait is in-frame with the *cI* coding region (*see* **Note 4**). Because plasmids pBT and pACλcI32 contain the *lac*UV5 promoter, they should only be propagated in strains of *E .coli* that contain *lacI^q^* such as XL1-Blue, XL1-Blue MRF' Kan, and JM109 (*see* **Note 5**). In addition, because of the relatively low copy number of these plasmids, larger culture volumes should be used when isolating plasmid DNA. Bait plasmids should be constructed using standard molecular biology procedures and the materials listed in **Subheading 2.1.**

3.2. Making Fusions to the α Subunit of E. coli *RNA Polymerase*

Plasmids pTRG, pBRSTAR, and pBRαLN are derived from the low-medium copy number plasmid pBR322 and harbor a truncated version of the *rpoA* gene (encoding amino acids 1–248 of the α subunit of RNAP) under the control of tandem *lpp* (a strong constitutive promoter) and *lac*UV5 promoters. Plasmids pTRG and pBRSTAR confer resistance to tetracycline, whereas plasmid pBRαLN confers resistance to carbenicillin (or ampicillin). Plasmids pTRG and pBRSTAR differ from one another with respect to the multiple cloning sites that have been introduced at the end of the truncated *rpoA* gene to facilitate the fusion of prey proteins to the end of the α linker (*see* **Table 1**). Typically the prey is fused to the end of the α linker by cloning a suitably designed PCR product into these plasmids such that the gene encoding the prey is in-frame with the *rpoA* coding region (*see* **Note 4**). Because plasmids pTRG, pBRSTAR and pBRαLN contain the *lac*UV5 promoter they should only be propagated in strains of *E .coli* that contain *lacI^q^*. Note that plasmids pTRG and pBRSTAR cannot be propagated in strain XL1-Blue because this strain carries a resistance determinant for tetracycline on an F' episome. Prey plasmids should be constructed using standard molecular biology procedures and the materials listed in **Subheading 2.2.**

3.3. Testing Whether Two Proteins Can Interact

In order to test whether two proteins can interact with one another the bait and prey plasmids directing the synthesis of the desired λcI fusion protein and the desired α fusion protein, respectively, are co-transformed into a suitable reporter strain of *E. coli* together with the appropriate controls. Transformants are then assayed for β-galactosidase activity. Two proteins are said to interact with one another when the bait stimulates production of β-galactosidase only in the presence of the prey. A suitable positive control is provided by co-transforming plasmids directing the synthesis of λcI-Gal4 and α-Gal11P fusion proteins (*see* **Table 1**), which have been shown previously to interact with one another in *E. coli* and strongly stimulate production of β-galactosidase in an appropriate reporter strain *(8)*. Suitable negative controls comprise the bait plasmid co-transformed with the parent vector of the prey plasmid (i.e., pTRG, pBRSTAR, or pBRαLN), and the prey plasmid co-transformed with the parent vector of the bait plasmid (i.e., pBT or pACλcI32). Alternative negative controls comprise the bait plasmid co-transformed with a plasmid that directs the synthesis of the α-Gal11P fusion protein (i.e., pTRG-Gal11P or pBRα-Gal11P), and the prey plasmid co-transformed with a plasmid directing the synthesis of the λcI-Gal4 fusion protein (i.e., pBT-LGF2).

Several different reporter strains have been designed for use with this bacterial two-hybrid system (*see* **Table 2**). The author routinely uses reporter strain KS1 which is a derivative of *E. coli* strain MC1000 that harbors an F' episome that confers resistance to kanamycin and carries *lacI^q* *(7)*. KS1 also harbors an imm 21 prophage on the chromosome that bears a *lacZ* reporter gene linked to the test promoter p*lac*O_R2–62 depicted in **Fig. 1**. This test promoter consists of the *lac* core promoter together with a λ operator (O_R2) positioned 62 bp upstream from the *lac* transcriptional start site. Reporter strain US3F'3.1 and the BacterioMatch reporter strain can also be used. These reporter strains contain the same F' episome carrying both *bla* and *lacZ* reporter genes under the control of the p*lac*O_R2–62 test promoter (as depicted in **Fig. 1**). In these strains the F' episome also bears *lacI^q* and a kanamycin resistance cassette. The US3F'3.1 and BacterioMatch reporter strains have lower apparent basal or unstimulated levels of *lacZ* expression (~15 Miller Units of β-galactosidase activity) compared to that of reporter strain KS1 (~60 Miller Units of β-galactosidase activity). Furthermore, it should be noted that higher transformation efficiencies can be achieved with the BacterioMatch reporter strain than with strain US3F'3.1, presumably as a result of differences in the genotypes of these two strains.

3.3.1. Rapid Co-transformation of Reporter Strains with Bait and Prey Plasmids

1. Thaw chemically competent reporter strain cells on ice (*see* **Note 3** about making chemically competent cells).
2. Add approx 10 ng each of the bait and prey plasmids or control plasmids to a sterile 1.5-mL microcentrifuge tube and place on ice for 5 min.

3. Add chemically competent reporter strain cells (25 µL) directly to the DNA in each tube.
4. Incubate tubes on ice for 10 min, heat shock at 42°C for 2 min and then incubate on ice for a further 2 min.
5. Spread each transformation mix directly onto an LB-agar plate containing the appropriate antibiotics (*see* **Note 6**).
6. Incubate plates at 37°C overnight.

3.3.2. Quantitative Assay of Reporter Gene Expression Using Liquid Cultures: The β-Galactosidase Assay

This protocol is derived from that originally described by Miller ***(16)***.

1. Individual colonies from the transformation plates (*see* **Subheading 3.3.1.**) are inoculated into 3 mL of LB containing the appropriate antibiotics (*see* **Note 7**) and IPTG at a concentration of between 0 and 200 µ*M* (*see* **Note 8**).
2. Incubate cultures overnight (~16 h) at 37°C with aeration.
3. Inoculate 30 µL of each overnight culture into 3 mL of LB supplemented with antibiotics and IPTG at the same concentration used in the overnight culture (0–200 µ*M*).
4. Incubate cultures at 37°C with aeration until cultures reach an OD_{600} of 0.3–0.7 (*see* **Note 9**).
5. Place cultures on ice for 20 min. This stops further protein synthesis and growth of the bacteria.
6. Transfer 1 mL of bacterial culture from each tube to a cuvette and record the OD_{600}. At this point, an aliquot of the bacterial culture may also be taken for future Western blot analysis if desired (*see* **Note 10**).
7. Transfer 200 µL of each culture to a small glass test tube containing 800 µL of Z buffer (the assay tube). This should be performed in duplicate for each culture. Duplicate tubes that will serve as the blank should similarly be prepared using 200 µL of LB.
8. Add 30 µL 0.1% SDS to each assay tube.
9. Add 60 µL chloroform to each assay tube (*see* **Note 11**).
10. Vortex each pair of duplicate assay tubes for 6 s. The tubes are now ready to be assayed for β-galactosidase activity. Assays can be performed straight away or tubes can be kept at 4°C for several hours.
11. Place tubes in a 28°C waterbath and allow them to equilibrate to 28°C by incubation for 10 min.
12. Start the assay by adding 200 µL of ONPG (4 mg/mL) (equilibrated to 28°C) to each assay tube and record the time at which each assay was started using a timer.
13. Mix the contents of the assay tubes by gentle agitation or vortexing.
14. Stop the reaction when the tubes become sufficiently yellow (*see* **Note 12**) by adding 500 µL of 1 *M* Na_2CO_3 and record the time at which each assay was stopped.
15. Vortex the tubes briefly at a low setting, and let sit at room temperature or 4°C until the remaining reactions are complete.

16. Transfer 1 mL from each assay tube to a cuvette (*see* **Note 13**) and determine the optical density at 420 nm and 550 nm for each assay using the control assay as the blank.
17. Determine β-galactosidase activity (in Miller Units) for each assay using the following equation:

$$\text{Units} = 1000 \times [OD_{420} - (1.75 \times OD_{550})/t \text{ x } v \times OD_{600}]$$

Note that "t" is the time of the reaction in minutes, "v" is the volume of the culture used in the assay in mL (i.e., 0.2) and the OD_{600} is that determined for the culture used in each assay (*see* **ref. *16***).

3.3.3. Interpreting Results of β-galactosidase Assays

An interaction between the bait and prey proteins results in at least a several fold increase in β-galactosidase activity above that seen with the appropriate negative controls. It is important to note that a certain amount of β-galactosidase activity will be apparent for the negative controls, reflecting basal transcription from the test promoter that drives expression of the *lacZ* reporter gene (*see* **refs. *7–11*** for examples).

3.4. Screening Libraries in E. coli for Protein Interaction Partners

The bacterial two-hybrid system described here can be used to identify proteins from a complex library that interact with a given protein of interest ***(11)***. Crucially, interaction between bait and prey proteins in reporter strain cells containing the *bla* gene linked to the activateable p*lac*O_R2–62 test promoter results in cells that are resistant to higher concentrations of carbenicillin. This permits the selection of cells containing interacting bait and prey proteins from large populations of cells in which relatively few of the combined bait and prey proteins interact.

The bait should be made by fusing the protein of interest (or domain thereof) to the end of the λcI protein as described in **Subheading 3.1.** Prior to screening a library for proteins that can interact with the bait, it is perhaps advisable to confirm that the λcI fusion protein to be used as the bait can bind to a λ operator. This can be done using reporter strain S11-LAM1 ***(17)*** which carries an F' episome harboring a λ operator positioned between the –10 and –35 elements of a relatively strong test promoter. The test promoter in this reporter strain drives expression of a linked *lacZ* reporter gene and is repressed when the λ operator is occupied (*see* **Note 14**). The degree to which the test promoter is repressed in strain S11-LAM1 is therefore a measure of how efficiently a particular protein can bind the λ operator. Any λcI fusion protein that binds the λ operator poorly, or not at all, is unsuitable as a bait.

Libraries of α fusion proteins (prey libraries) can be made by cloning genomic DNA, cDNA, or PCR products into plasmids pTRG or pBRSTAR. However, the construction of these libraries is beyond the scope of this Chapter. Certain cDNA libraries made using pTRG are commercially available from Stratagene. What follows is one protocol for screening an α fusion library, made in pTRG or pBRSTAR, for proteins that interact with a predetermined bait.

3.4.1. Library Screening

1. Add 1 μg of bait plasmid to a sterile 1.5 mL microcentrifuge tube and incubate on ice for 5 min (*see* **Note 15**).
2. Add 100 μL of chemically competent BacterioMatch reporter strain cells containing the prey library (cloned into pTRG or pBRSTAR) (*see* **Note 16**).
3. Incubate on ice for 30 min. Heat shock for 2 min at 42°C then return tubes to ice for at least 2 min.
4. Add 1 mL LB to the transformation mix and incubate at 37°C for at least 1 h.
5. Add IPTG to the transformation mix to a final concentration of 20 μ*M* and incubate at 37°C for 1 h.
6. Spread sufficient transformation mix to plate ~10^7 transformants onto 150-mm LB-agar plates containing kanamycin (50 μg/mL), tetracycline (10 μg/mL), chloramphenicol (25 μg/mL), IPTG (20 μ*M*; see **Note 8**), carbenicillin (500–2000 μg/mL; *see* **Note 17**) and X-Gal (*see* **Note 18**).
7. Incubate plates overnight at 37°C (*see* **Note 19**). Positive clones will form blue colonies on these selection plates.

3.4.2. Analysis of Putative Interaction Partners

In order to eliminate certain false positives, any putative positive clone can be first restreaked onto 100-mm LB-agar plates containing kanamycin (50 μg/mL), tetracycline (10 μg/mL), chloramphenicol (25 μg/mL), IPTG (20 μ*M*, or concentration used in original selection), carbenicillin (concentration used in original selection) and X-Gal. A true positive should again give rise to blue colonies on these plates. Any putative positive that passes this test should be inoculated into 3 mL LB containing kanamycin (50 μg/mL) and tetracycline (10 μg/mL) and grown overnight with aeration at the appropriate growth temperature (37°C or 30°C). Plasmid DNA is then isolated from the overnight cultures and transformed into *E. coli* strain XL1-Blue MRF' Kan or strain JM109 so that the tetracycline resistant prey plasmid of interest can be isolated from the bait plasmid and purified (transformants should be plated on LB-agar plates containing tetracycline at 10 μg/mL). Once the desired prey plasmid has been isolated, it can be co-transformed with the original bait plasmid, together with the appropriate controls (*see* **Subheading 3.3.**), into a suitable reporter strain. Whether the prey plasmid directs the synthesis of an α fusion protein that can interact with the bait can then be assessed quantitatively by performing β-galactosidase assays on liquid cultures (*see* **Subheading 3.3.2.**), or qualitatively by plating transformants onto selection plates containing carbenicillin and X-Gal (*see*

Subheading 3.4.1.). Any plasmid that encodes a prey capable of interacting with the bait can then be sequenced to determine the identity of the interaction partner (*see* **Note 20**).

4. Notes

1. The bacterial two-hybrid system detailed here exploits the domain structure of the α subunit of *E. coli* RNAP which consists of two independently folded domains separated by a long flexible linker ***(14,15)***. The α subunit of RNAP is an essential protein and is present in dimeric form in the polymerase molecule. Since the reporter strains used in this bacterial two-hybrid system contain wild type α (expressed from the native *rpoA* gene on the chromosome), any cell expressing a particular α fusion protein (or prey) will contain a population of RNAP molecules that contain either 2, 1 or 0 copies of the α fusion protein.
2. Sequence specific DNA-binding proteins other than λcI, such as a derivative of the monomeric zinc-finger protein Zif268, have also been used in this system as DNA-binding domains to which to fuse proteins of interest to ***(13)***. Furthermore, α is not the only subunit of RNA polymerase that can be used as a fusion point for a protein of interest. We have previously shown that the monomeric ω subunit of *E. coli* RNA polymerase can also be used to display heterologous protein domains ***(8)***.
3. A number of protocols for making chemically competent cells exist and one is described here. An isolated colony of the desired strain is inoculated into 3 mL LB containing the appropriate antibiotic. The culture is incubated at 37°C overnight with aeration. 0.5 mL of the overnight culture is then used to inoculate 200 mL LB (containing the appropriate antibiotic where required) in a 1 L conical flask. 3 mL of filter sterilized 1 *M* $MgCl_2$ are also added to the 200 mL culture. The culture is then incubated at 37°C with aeration until an OD_{600} of approx 0.5 is reached. Cells are transferred to a large sterile centrifuge bottle and pelleted by centrifugation at 4°C. The cell pellet is then resuspended in 60 mL of cold solution A and the suspension is incubated on ice for 20 min. (Solution A is made by combining 10 mL 1 *M* $MnCl_2$, 50 mL 1 *M* $CaCl_2$, 200 mL 50 m*M* 2-morpholinoethanesulfonic acid (MES) pH 6.3, and 740 mL distilled water. The solution is then filter sterilized.) Cells are then pelleted by centrifugation at 4°C and resuspended in 12 mL cold solution A containing 15% glycerol. Aliquot competent cells in 0.5–1.0 mL volumes in sterile pre-cooled 1.5-mL microcentrifuge tubes and quick-freeze on dry ice. Cells are then stored at –80°C.
4. In plasmids pBT and pACλcI32 a *Not*I site has been introduced at the end of the *cI* gene such that the eight base pair *Not*I site, together with an additional base pair (i.e., GCGGCCGCA), adds three alanine residues to the end of the *cI* gene, thus providing a short linker to which to attach a protein of interest (the bait) ***(3)***. Useful cloning sites at the end of the *cI* gene in plasmids pBT and pACλcI32 are listed in **Table 1**. In plasmids pTRG, pBRSTAR, and pBRαLN, a *Not*I site has been introduced into the *rpoA* gene after codon 248 such that the *Not*I site (together with an additional base pair) similarly adds three alanine residues to the

end of the α-linker. The *Not*I restriction sites in the bait and prey plasmids provide a convenient means to clone PCR products encoding the bait and prey, respectively. Useful cloning sites at the end of the truncated *rpoA* gene in plasmids pTRG, pBRSTAR, and pBRαLN are also listed in **Table 1**. PCR products to be cloned into the above vectors (except pTRG) typically contain a *Not*I site at one end and a *Bam*HI site preceded by a stop codon at the other.

5. The *lacI^q* allele is required to provide sufficiently high quantities of Lac repressor to keep the *lac*UV5 promoter efficiently repressed. In the absence of *lacI^q*, the strong *lac*UV5 promoter will be close to being fully derepressed and, as a result, undesirable mutations within the expression vectors may be selected.
6. The rapid transformation protocol does not include a step to allow for phenotypic outgrowth of antibiotic resistant determinants and so does not result in particularly high transformation efficiencies.
7. Antibiotics that select for the bait and prey plasmids should be used in conjunction with kanamycin (50 µg/mL) to select for maintenance of the F' in each of the reporter strains.
8. An initial IPTG concentration of 20 µ*M* is recommended. However, the optimal IPTG concentration (0–200 µ*M*) should be determined empirically by testing a range of concentrations.
9. Generally cultures take 1.5–4 h to reach mid-logarithmic phase; the exact time varies depending on the particular bait and target proteins under investigation and the concentration of IPTG being used to induce their synthesis.
10. An antibody against λcI is available commercially from Invitrogen, and can be used to determine the expression levels of λcI fusion proteins.
11. Chloroform can be dispensed accurately using a pipetman provided the chloroform is pipetted up and down 3–4× before trying to dispense any.
12. Ideally, the OD_{420} (the yellow color) of the stopped reaction should be 0.6–0.9 (*see* **ref. *16***). If the reactions are allowed to proceed for too short or too long a time, then the determined β-galactosidase activity is less accurate. Ideally, each reaction should be stopped when each assay reaches the same intensity of yellow, so that the reaction time is the only variable.
13. Care should be taken not to transfer any of the chloroform residing at the bottom of the assay tube to the cuvette as this will interfere with OD readings.
14. Occupancy of the λ operator in reporter strain S11-LAM1 by λcI, or a λcI fusion protein, prevents RNAP from binding the test promoter, resulting in repression ***(17)***. Plasmids pBT and pACλcI32 are suitable positive controls for use in this strain, whereas plasmid pACΔcI *(7)* is a suitable negative control for use in this strain.
15. The number of potential false positives that might occur as a result of mutations in the BacterioMatch reporter strain for example, can be determined by performing a control transformation with 1 µg of the bait parent vector (pTRG or pACλcI32).
16. Chemically competent cells harboring the prey library should ideally have a transformation efficiency of at least 1×10^7/µg of DNA.

17. For an initial screen, a concentration of carbencillin of 1000 μg/mL is suggested. However, ideally a range of carbenicillin concentrations (500–2000 μg/mL) should be tested. Carbenicillin concentrations as low as 250 μg/mL can be used provided the number of transformants that are spread on a plate are limited to approx 10^5. It is also important to note that ampicillin should not be used instead of carbenicillin.
18. X-Gal can be added to the 150-mm LB-agar plate by first adding 200 μL of LB to the center of the plate followed by 160 μL of X-Gal (10 mg/mL). The X-Gal/LB mix is then spread onto the plate. The plate should then be allowed to dry for ~40 min before any transformation mix is added.
19. The selection can be performed at either 37°C or 30°C. In *E. coli*, certain proteins are more soluble, or exhibit less of a tendency to form inclusion bodies when cells are grown at 30°C.
20. An oligonucleotide (sequence 5'-GCAATGAGAGTTGTTCCGTTGTGG-3') that hybridizes to the end of the *cI* gene and reads towards the *Not*I site can be used for sequencing inserts in bait plasmids pBT and pACλcI32. An oligonucleotide (sequence 5'-GGTCATCGAAATGGAAACCAACG-3') that hybridizes to *rpoA* and reads towards the *Not*I site can be used for sequencing inserts in prey plasmids pTRG, pBRSTAR, and pBRαLN.

References

1. Fields, S. and Song, O. (1989) A novel genetic system to detect protein-protein interactions. *Nature* **340,** 245–246.
2. Gyuris, J., Golemis, E. A., Chertkov, H., and Brent, R. (1993) Cdi1, a human G1 and S phase protein phosphatase that associates with Cdk2. *Cell* **75,** 791–803.
3. Hu, J. C., Kornacker, M. G., and Hochschild, A. (2000) *Escherichia coli* one- and two-hybrid systems for the analysis and identification of protein-protein interactions. *Methods* **20,** 80–94.
4. Ladant, D. and Karimova, G. (2000) Genetic systems for analyzing protein-protein interactions in bacteria. *Res. Microbiol.* **151,** 711–720.
5. Legrain, P. and Selig, L. (2000) Genome-wide protein interaction maps using two-hybrid systems. *FEBS Lett.* **480,** 32–36.
6. Hu, J. C. (2001) Model systems: Studying molecular recognition using bacterial n-hybrid systems. *Trends Microbiol.* **9,** 219–222.
7. Dove, S. L., Joung, J. K., and Hochschild, A. (1997) Activation of prokaryotic transcription through arbitrary protein-protein contacts. *Nature* **386,** 627–630.
8. Dove, S. L. and Hochschild, A. (1998) Conversion of the ω subunit of *Escherichia coli* RNA polymerase into a transcriptional activator or an activation target. *Genes Dev.* **12,** 745–754.
9. Dove, S. L., Huang, F. W., and Hochschild, A. (2000) Mechanism for a transcriptional activator that works at the isomerization step. *Proc. Natl. Acad. Sci. USA* **97,** 13,215–13,220.
10. Dove, S. L. and Hochschild, A. (2001) Bacterial two-hybrid analysis of interactions between region 4 of the σ^{70} subunit of RNA polymerase and the transcrip-

tional regulators Rsd from *Escherichia coli* and AlgQ from *Pseudomonas aeruginosa*. *J. Bacteriol.* **183,** 6413–6421.

11. Shaywitz, A. J., Dove, S. L., Kornhauser, J. M., Hochschild, A., and Greenberg, M. E. (2000) Magnitude of the CREB-dependent transcriptional response is determined by the strength of the interaction between the kinase-inducible domain of CREB and the KIX domain of CREB-binding protein. *Mol. Cell. Biol.* **20,** 9409–9422.
12. Blum, J. H., Dove, S. L., Hochschild, A., and Mekalanos, J. J. (2000) Isolation of peptide aptamers that inhibit intracellular processes. *Proc. Natl. Acad. Sci. USA* **97,** 2241–2246.
13. Joung, J. K., Ramm, E. I., and Pabo, C. O. (2000) A bacterial two-hybrid selection system for studying protein-DNA and protein-protein interactions. *Proc. Natl. Acad. Sci. USA* **97,** 7382–7387.
14. Blatter, E. E., Ross, W., Tang, H., Gourse, R. L., and Ebright, R. H. (1994) Domain organization of RNA polymerase α subunit: C-terminal 85 amino acids constitute a domain capable of dimerization and DNA-binding. *Cell* **78,** 889–896.
15. Jeon, Y. H., Yamazaki, T., Otomo, T., Ishihama, A., and Kyogoku, Y. (1997) Flexible linker in the RNA polymerase alpha subunit facilitates the independent motion of the C-terminal activator contact domain. *J. Mol. Biol.* **267,** 953–962.
16. Miller, J. H. (1972) *Experiments in molecular genetics*. Cold Spring Harbor Laboratory, Cold Spring Harbor, NY.
17. Whipple, F. W. (1998) Genetic analysis of prokaryotic and eukaryotic DNA-binding proteins in *Escherichia coli*. *Nucleic Acids Res.* **26,** 3700–3706.
18. Yanisch-Perron, C., Vieira, J., and Messing, J. (1985) Improved M13 phage cloning vectors and host strains: nucleotide sequences of the M13mp18 and pUC19 vectors. *Gene* **33,** 103–119.

18

Using Bio-Panning of FLITRX Peptide Libraries Displayed on *E. coli* Cell Surface to Study Protein–Protein Interactions

Zhijian Lu, Edward R. LaVallie, and John M. McCoy

1. Introduction

The completion of the human genome project has ushered in a new era of life science *(1,2)* in which the new challenge is to understand functions of the entire collection of the gene products, or the proteome. One important feature of biological research in this post-genomics era is the emphasis on understanding how individual components of a proteome interact with one another temporally and spatially to constitute a living organism. Over the past decade, researchers have developed various methods designed to study protein-protein interactions including displaying proteins and peptides on live microorganisms, the most well-known example being the display of random peptide libraries on filamentous phage *(3,4)*. Many people (including the authors) have explored the use of *E. coli* as an alternative organism for protein and peptide display *(5)*. Based on our expertise and experience with both flagellin *(6)* and thioredoxin *(7)*, we developed FLITRX technology, a unique system that displays conformation-constrained random peptides on the bacterial surface as functional fusions between flagellin and thioredoxin *(8)*.

E. coli thioredoxin is a small cytoplasmic protein involved in oxido-reduction in bacteria *(9)*. Its structural features include, in addition to a stable globular tertiary fold, an active-site loop (Cys-Gly-Pro-Cys) that is disulfide-bonded in oxidizing environments *(10)*. Peptide insertions into this active site loop are tethered at both their N- and C-termini to this defined scaffold, and thus likely to be displayed in stable secondary and tertiary structure. The stabilizing effect of thioredoxin as a scaffold in displaying peptides helped Colas et al. *(11)* to

From: *Methods in Molecular Biology, vol. 205, E. coli Gene Expression Protocols*
Edited by: P. E. Vaillancourt © Humana Press Inc., Totowa, NJ

develop a yeast interaction trap utilizing peptide libraries created within the thioredoxin active-site loop. Using this technique, they identified several different 20-amino acid peptides that inhibited the biological function of cyclin-dependent kinase 2 (cdk-2) by physical interaction with the protein. A further advantage of conformation-constrained display of peptides in the thioredoxin active-site scaffold is the ease with which the fusion proteins containing the selected peptide can be produced for biophysical and biochemical studies ***(7,12)***.

In the native state, thioredoxin resides in the cytoplasm of the bacterial cell, so a "bio-panning" selection for random peptide libraries inserted into the thioredoxin active-site loop against extracellular targets is not possible, unless the thioredoxin peptide-fusion libraries can be brought to the cell surface. To achieve this goal, we explored the use of flagellin, the most abundant structural protein of flagella, to carry thioredoxin to the cell surface. The flagellum is the cell surface apparatus that confers motility to microorganisms ***(13,14)***. Each flagellum fiber is an ordered aggregate of thousands of flagellin protein monomers. Structural studies revealed a region in the central part of the flagellin protein that is dispensible for its function. Deletions in this region still resulted in partially functional flagella which assembled at the extracellular surface of the bacterium ***(15)***. By a screening procedure described previously ***(8)***, we identified a region within this dispensible region of *E. coli* flagellin that can be replaced by thioredoxin to form the chimeric protein FLITRX. Expression of FLITRX in non-motile bacteria that have the deletion of the endogenous flagellin gene can partially restore the motility of the microorganism, indicating the formation of functional flagella by the protein chimera. Further, the active-site loop of thioredoxin in FLITRX is solvent-accessible, as evidenced by the observation that FLITRX-expressing bacteria are anchored on glass-slide coated with antibodies against the active-site sequence (Lu and McCoy, unpublished data). Because of the characteristics described above, we were able to construct a random dodecapeptide library in FLITRX (LO-T), display it on the *E. coli* cell surface, and devise a bio-panning procedure for a prototype application—mapping antibody epitopes ***(8)***.

Several groups have explored wider application of the FLITRX display technology in their studies of protein-protein interactions. These include the identification of binding motifs for protein phosphatase-1 ***(16)*** and proliferating cell nuclear antigen ***(17)***, mapping the interaction between the α and β subunits of voltage-gated potassium channel protein ***(18,19)***, and panning the library on live tumor cells ***(20)***. In our own laboratory we applied the technology to protein engineering for the study of the "switch epitope" concept ***(21)***. The purpose of this chapter is to describe the basic protocol in using FLITRX libraries for bio-panning, as well as the procedure which we used to engineer the "switch-epitopes".

2. Materials

2.1. Apparatus and Special Reagents

1. A shaking incubator with a temperature range of 25–37°C.
2. A rotary platform shaker.
3. Polystyrene tissue culture dishes (60 mm in diameter, from Nunc).
4. 96-Well flat bottom tissue culture plates (Costar, 3596).
5. 100-mm and 150-mm Plastic petri dishes (Fisher).
6. 82 mm and 137-mm diameter Nitrocellulose membrane filters (Millipore, HAHY 13750 and HAHY 08250).
7. α-Methyl D-mannoside (methyl α-D-mannopyranoside, Sigma Chemical Co., M6882).
8. ^{125}I-labeled protein A (DuPont NEN, NEX-146).
9. Rabbit anti-mouse IgG polyclonal antibody (Zymed, 616500).
10. Murine anti-human IL-8 monoclonal antibody HIL8-NR7 (Devaron, Inc., 104-12-2).
11. Purified monoclonal antibody with unknown epitope.
12. *E. coli* strain GI724, a healthy non-motile prototroph which may be used as a host cell for pL expression vectors *(7)*. This strain (available from Wyeth/Genetics Institute, Invitrogen or the American Type Culture Collection) is sensitive to both ampicillin and tetracycline.
13. *E. coli* strain GI808, which is wild-type with respect to flagellar synthesis and cell motility *(8)*. This strain is sensitive to both ampicillin and tetracycline.
14. *E. coli* strain GI826, which carries deletions in the *fliC* (flagellin) and *motB* genes *(8)*. This strain is sensitive to ampicillin but resistant to tetracycline.
15. Plasmid pFLITRX, which carries the gene for a functional fusion of flagellin and thioredoxin under the transcriptional control of the pL promoter *(8)*.
16. Plasmid pALTrxA-781, which carries the gene for *E. coli* thioredoxin under the transcriptional control of the pL promoter *(7)*. Please note that items 13–17 are available from Wyeth/Genetics Institute.
17. "LO-T", a frozen stock (10^{11} cells/vial) of GI826 cells transformed with a population of pFLITRX plasmids *(8)*. The plasmids harbor a dodecapeptide library (diversity: 1.8×10^8) inserted into the thioredoxin active site. (available from Wyeth/Genetics Institute, or Invitrogen).

2.2. Stock Solutions

1. Casamino acids (CAA), 2% solution: Dissolve 20 g of casamino acids (Difco Certified grade) in 1 L deionized water. Autoclave the solution.
2. 10X M9 salts: Dissolve 60 g Na_2HPO_4, 30 g KH_2PO_4, 5 g NaCl, and 10 g NH_4Cl in 800 mL deionized water, adjust pH to 7.4 with NaOH, bring the volume up to 1 L. Autoclave the solution.
3. 10X M9 salts with glycerol: dissolve 60 g Na_2HPO_4, 30 g KH_2PO_4, 5 g NaCl, and 10 g NH_4Cl in 700 mL deionized water, add 100 mL glycerol, adjust pH to 7.4 with NaOH, bring the volume up to 1 L. Autoclave the solution.
4. Glucose, 20% solution: dissolve 20 g of glucose in deionized water and bring the final volume up to 100 mL. Sterilize by filtering through a 0.22-μm membrane.

5. Ampicillin (Amp), 10 mg/mL solution: dissolve 1 g of ampicillin (sodium salt, Sigma) in 100 mL of deionized water. Sterilize by filtering through a 0.22-µm membrane.
6. Tetracycline (Tet), 10 mg/mL solution: dissolve 100 mg of tetracycline (Sigma) in 10 mL 75% v/v ethanol/water.
7. L-tryptophan (Trp), 10 mg/mL solution: dissolve 1 g of L-tryptophan (Sigma) in 100 mL 80°C deionized water. Sterilize by filtering through a 0.22-µm membrane.
8. α-Methyl D-mannoside, 20% solution: dissolve 20 g of α-Methyl D-mannoside in deionized water and bring the final volume up to 100 mL. Sterilize by filtering through a 0.22-µm membrane.
9. Other stock solutions: 1 *M* $MgSO_4$ (autoclaved); 1 *M* $CaCl_2$ (autoclaved); 10% w/v NaN_3; 5 *M* NaCl (filter sterilized); and 1 *M* Tris-HCl, pH 7.5.

2.3. Working Solutions and Media

1. Induction media with casamino acids (IMC)/Amp/Tet: mix 100 mL CAA, 100 mL 10X M9 salts, 25 mL 20% glucose, 1 mL 1 *M* $MgSO_4$, 0.1 mL 1 *M* $CaCl_2$, 10 mL 10 mg/mL Amp, 0.5 mL 10 mg/mL Tet, and 770 mL sterile deionized water. Omit the tetracycline solution for making IMC/Amp medium.
2. High-density plasmid growth media (HPM/Amp): Add 10 mL 10X M9 salts with glycerol, 0.1 mL 1 *M* $MgSO_4$, 0.01 mL 1 *M* $CaCl_2$, and 1 mL 10 mg/mL Amp to 89 mL 2% CAA. Add 0.05 mL 10 mg/mL Tet if required.
3. Luria-Bertani Broth (LB)/Tet: Autoclave 10 g tryptone (Difco), 5 g yeast extract (Difco) and 10 g NaCl in 1000 mL deionized water. After cooling add 0.5 mL 10 mg/mL Tet. For LB medium omit the addition of tetracycline solution.
4. CAA/Amp/Tet plates: Autoclave 20 g casamino acids and 15 g agar (Difco) in 870 mL deionized water. After autoclaving, add the following solutions when the medium cools to 60°C: 100 mL 10X M9 salts, 25 mL 20% glucose, 1 mL 1 *M* $MgSO_4$, 0.1 mL 1 *M* $CaCl_2$, 10 mL 10 mg/mL Amp, and 0.5 mL 10 mg/mL Tet. Pour the plates at 55°C. For IMC/Amp plates omit the tetracycline solution.
5. CAA/Amp/Tet/Trp plates: Autoclave 20 g casamino acids and 15 g agar (Difco) in 870 mL deionized water. Add the following solutions after autoclaving and allowing to cool to 60°C: 100 mL 10X M9 salts, 25 mL 20% glucose, 1 mL 1 *M* $MgSO_4$, 0.1 mL 1 *M* $CaCl_2$, 10 mL 10 mg/mL Trp, 10 mL 10 mg/mL Amp, and 0.5 mL 10 mg/mL Tet. Pour the plates at 55°C.
6. LB/Tet plates: Autoclave 10 g tryptone, 5 g yeast extract, 10 g NaCl, and 15 g agar in 1000 mL deionized water. After cooling to 60°C add 0.5 mL 10 mg/mL Tet and pour the plates at 55°C. For LB plates omit the addition of tetracycline solution.
7. 5X Cell binding buffer: Dissolve 5 g powdered skim milk in 60 mL sterile deionized water, add 15 mL sterile 5 *M* NaCl, 25 mL 20% α-methyl D-mannoside.
8. Blocking buffer: Add 20 mL of 5X cell binding buffer to 80 mL IMC/Amp/Tet.
9. Washing media: Add 25 mL 20% α-methylmannoside to 475 mL IMC/Amp/Tet.
10. Cell-lysis buffer: Dissolve 2 g powdered skim milk in 183 mL sterile deionized water, then add 10 mL 1 *M* Tris-HCl, pH 7.5, 6 mL 5 *M* NaCl, 1 mL 1 *M* $MgSO_4$, 0.4 mL 10% NaN_3, 200 µg DNase, and 8 mg lysozyme.

11. Antibody binding buffer: Dissolve 10 g powdered skim milk in 920 mL deionized water, then add 30 mL 5 *M* NaCl and 50 mL 1 *M* Tris-HCl, pH 7.5.
12. Filter washing buffer: Add 30 mL 5 *M* NaCl and 50 mL 1 *M* Tris-HCl, pH 7.5, to 920 mL deionized water.
13. Tris-EDTA (TE) buffer: 10 m*M* Tris-HCl, 1 m*M* EDTA, pH 8.0.

3. Methods

3.1. Maintaining the E. coli Strains and the "LO-T" Library

1. *E. coli* strains GI724 and GI808 can grow on LB plates and GI826 on LB/Tet plates at 37°C. Grow pFLITRX/GI826 on CAA/Amp/Tet plates at 30°C. Grow pALTrxA-781/GI724 on CAA/Amp plates at 30°C (*see* **Note 1**).
2. Pick single colonies of GI808 or GI826 to inoculate LB or LB/Tet media, respectively, grow at 37°C to saturation. Add 1 mL of 50% sterilized glycerol solution to 1 mL of each culture, and store at –80°C in 2-mL Corning cryo-vials (*see* **Note 2**).
3. Pick single colonies of pFLITRX/GI826 to inoculate IMC/Amp/Tet media, grow at 30°C to saturation. Inoculate IMC/Amp media with pALTrxA-781/GI724, grow at 30°C to saturation. Add equal volumes of 50% sterile glycerol to the cultures, and store at –80°C as above (*see* **Note 3**).
4. Maintaining the library "LO-T". Transfer the entire contents of a master vial (100 OD_{550}/vial, 10^{11} cells) into 1 L of IMC/Amp/Tet and incubate at 30°C with shaking at 250 rpm to saturation. To make duplicate master libraries (or working libraries), briefly centrifuge the culture, resuspend the bacteria to a density of 100 OD_{550}/mL (or 10 OD_{550}/mL for working libraries), mix with equal volume of 50% sterilized glycerol and save 2-mL aliquots at –80°C (*see* **Note 4**).
5. To prepare plasmids, grow pGIS-104/GI826 or pFLITRX/GI826 in HPM/Amp/Tet at 30°C for at least 18 h with vigorous shaking (250 rpm). pALTrxA-781/GI724 can be grown in HPM/Amp (no tetracycline). Plasmids may be prepared from these cultures using standard protocols ***(22)***.

3.2. Mapping Monoclonal Antibody Epitopes

1. Transfer the entire content of a working library (10 OD_{550}, 10^{10} cells) into 100 mL IMC/Amp/Tet. Grow the culture at 25°C overnight with shaking (200 rpm).
2. Inoculate 100 mL of IMC/Amp/Tet, containing 100 µg/mL Trp, with 5 mL of the overnight culture of LO-T. Grow the culture at 25°C for 6 h with shaking (200 rpm) (*see* **Note 5**).
3. To prepare antibody-coated panning surface, add 1.5 mL deionized water to each 60-mm tissue culture dish and to this add 20 µg of the antibody. Spread out the antibody solution and keep the dishes gently agitating at room temperature (60 rpm) for 2 h. Keep the dishes covered to maintain sterility and to prevent evaporation.
4. Pour off the antibody solution and rinse the dishes once with 5 mL sterile deionized water. Next, add 10 mL of blocking buffer to each dish (*see* **Note 6**) and shake the dishes at 60 rpm until use.

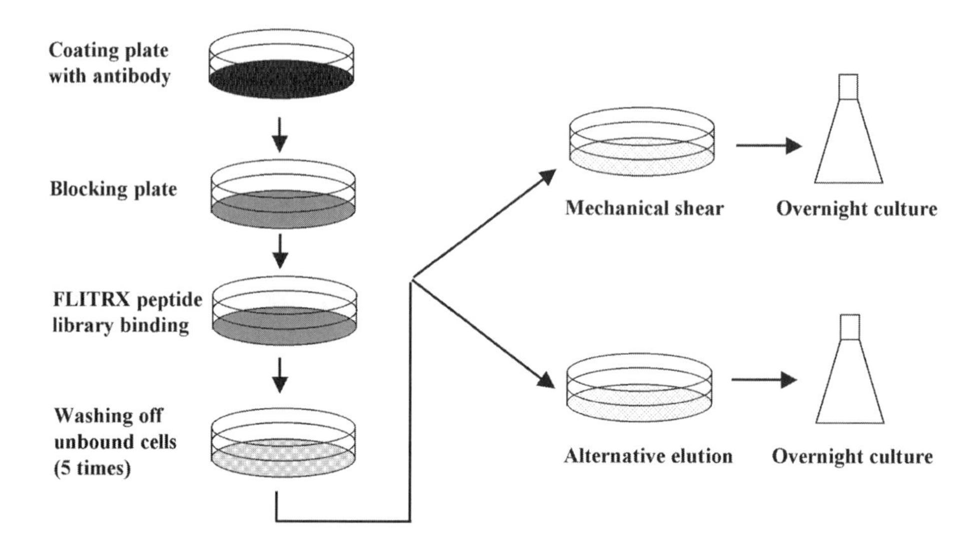

Fig. 1. Flow chart of "bio-panning" of FLITRX peptide library. *See* **steps 3–13** in **Subheading 3.2.** for detail.

5. After 6 h post-induction from step two, measure the OD_{550} (ideally it should lie between 0.8–1.2). Mix 1 part of 5X cell binding buffer with 4 parts of the induced FLITRX library culture and place 10 mL of the mixture into each coated dish.
6. Shake the dishes at 60 rpm for 1 min, then leave them sitting stationary at room temperature for 1 h (*see* **Note 7**).
7. Pour off the bacterial culture, slowly pipet 10 mL of wash buffer into each dish at a marked spot along the rim (*see* **Note 8**), shake the dishes at 60 rpm for 5 min, then remove the wash buffer by pipetting or aspiration.
8. Repeat the washing step 4 more times. Make sure that each time the wash buffer is added slowly to the same marked spot on the dishes.
9. After the fifth wash, leave only 0.4 mL solution in each dish and dissociate the bound bacteria by vortexing the dishes vigorously for 30 s with the lids held tightly on. Next, rinse the dishes with 5 mL IMC/Amp/Tet twice, combining the rinse solutions with 100 mL of fresh, sterile IMC/Amp/Tet media. Also rinse the lids with 1 mL IMC/Amp/Tet, if they appear to be splattered with liquid from the vortexing step, and also add this to the fresh IMC/Amp/Tet media.
10. Alternative elution methods may be used in places of **step 9** based upon physical characteristics of the interaction. For example, in order to obtain zinc-sensitive binders, use a final rinse with the wash buffer containing 2.5 m*M* $ZnCl_2$, instead of mechanical shearing, to collect dissociated bacteria (*see* **Note 9**).
11. Incubate the eluted/dissociated bacteria in IMC/Amp/Tet media in a water-bath shaker at 25°C overnight at 200 rpm (*see* **Note 10**). This completes one round of selection. **Steps 3–11** are illustrated in **Fig. 1**.

12. After overnight incubation, measure the OD_{550} of the culture. Continue incubation, if necessary, until an OD_{550} of at least 0.1 is achieved. If the OD_{550} has already exceeded 0.1, adjust the cell density to 0.1 in a total volume of 100 mL IMC/Amp/Tet and then add Trp to 100 μg/mL. Incubate at 25°C for 6 h with shaking (200 rpm).
13. Repeat **steps 3–8** followed by either **step 9** or **step 10** to perform additional rounds of selection. At the end of the last round, culture the bacteria to saturation in IMC/Amp/Tet media. We originally used 3 rounds of selections followed by a screening method to identify positive clones. However, in many of the applications reported lately ***(16–20)***, more rounds (up to 9) of selections were deployed so that the screening process described in **steps 14–28** can be omitted (*see* **Note 11**).
14. Inoculate 10 mL IMC/Amp/Tet to 0.05 OD_{550}/mL with fresh saturated culture from the final round of selection. Incubate at 30°C with shaking until the OD_{550} reaches 0.6, then make three different dilutions (1:60,000, 1:30,000 and 1:20,000) with IMC/Amp/Tet.
15. Spread 0.2 mL of the culture dilutions onto 150-mm CAA/Amp/Tet plates. Target 4000 non-overlapping colonies per each 150-mm plate, spread 4 plates for each dilution. Leave the plate covers ajar for 30 min at room temperature to allow excess liquid to evaporate. Then incubate the plates upside down at 30°C overnight.
16. After overnight incubation, chill the plates at 4°C for 1 h. The colonies should not be allowed to grow bigger than 0.5 mm in diameter. Meanwhile, pre-incubate an equal number of 150-mm CAA/Amp/Tet/Trp plates at 30°C (*see* **Note 12**).
17. Using a ball-point pen or a soft pencil, individually mark 150-mm nitrocellulose filter membranes at three non-symmetrical intervals along their edges. Center and slowly lay down each filter, with the markings facing down, onto the top of the colonies present on the chilled CAA/Amp/Tet plates (master plates). Let the filters sit on the plates for 5 min to allow for complete wetting and good contact with the colonies. If necessary, gently press out any air bubbles. Trace exactly the positions of the filter numbering and alignment markings onto the outside of each plate with a marker to facilitate later alignment of filters, master plates, and autoradiograms.
18. Gently lift the filters away from the CAA/Amp/Tet plates and lay them on the pre-warmed CAA/Amp/Tet/Trp plates with the bacterial colonies (and pen markings) facing up. Incubate the plates with the filters at 30°C for 5 h (*see* **Note 12**). Reincubate the master plates at 30°C to allow the original colonies to grow back to 0.5–1 mm in diameter. Store these re-grown master plates at 4°C.
19. Take the filters off the CAA/Amp/Tet/Trp plates and place them in the lysis buffer at room temperature with gentle agitation overnight.
20. Wash the filters three times with filter washing buffer, 15 min each wash.
21. Place the filters in antibody binding buffer containing 1 μg/mL (or pre-determined optimal concentration) of the monoclonal antibody used for the selection. Use 20 mL of solution for each 137-mm membrane. Agitate the solution gently at room temperature for 2 h.
22. Wash the filters again three times in filter washing buffer, 15 min each wash, with gentle agitation at room temperature.

23. Place the filters into antibody binding buffer containing the secondary antibody, e.g., rabbit anti-mouse IgG (presorbed with GI808 lysate to remove any *E. coli*-reactive antibodies, *see* **Note 13**). Incubate at room temperature with gentle shaking for 2 h.
24. Wash the filters three times with filter washing buffer, 15 min each wash, with gentle agitation at room temperature.
25. Incubate the filters with ^{125}I labeled protein A solution (a 1:2000 dilution of ^{125}I labeled protein A with the antibody binding buffer) for 2 h at room temperature with gentle agitation.
26. Wash the filters three times with filter washing buffer, 15 min each, with gentle agitation at room temperature. After the final wash, tape one edge of each filter to a piece of Whatman 3MM filter paper and air-dry the filters. Cover the filters and 3MM filter paper with a large piece of plastic film, taping the free edges of the plastic film to the reverse side of the filter paper. Position radiolabeled or luminescent markers to allow for later registration of autoradiograms and filters.
27. Expose to X-ray film using an intesifying screen at –80°C overnight. Develop the film and, if necessary, expose longer or shorter periods of time.
28. Match and align the autoradiograms to the master plates with the help of the three pen marks on each filter and the marks on the plates. Identify and mark the positions of the colonies on the master plates that correspond to the positive signals on the autoradiograms.
29. Pick the positive colonies from the master plates to inoculate 5 mL HPM/Amp/Tet in roller tubes. Grow the bacteria overnight at 30°C to saturation.
30. Perform plasmid mini-preps *(22)*.
31. Sequence the plasmid DNA across the region coding for the inserted peptides. Use an oligonucleotide primer with the sequence: 5'-GACAGTTTTGACACGGATGT-3' for the top strand and 5'-TCAGCGATTTCATCCAGAAT-3' for the bottom strand (*see* **Note 14**).

3.3. Insertion of Peptides into Thioredoxin

As stated in the introduction, once individual clones are selected from the FLITRX library by the epitope mapping procedure, the peptide sequences and their derivatives can then be inserted into the active site-loop of thioredoxin to make Trxloop proteins. The peptide inserts in such TrxLoop fusions are also thought to retain the original conformations adopted as FLITRX inserts. Further, high level production of Trxloop proteins can often be achieved to facilitate subsequent biochemical or structural analyses (*see* **Note 15**). To illustrate this, we describe our work of using TrxLoop proteins to improve the binding affinity of individual peptides identified through FLITRX library selection *(21)*. Many experiments outlined in this section for thioredoxin fusion system have been described by LaVallie et al. (*23,24*; *see* also Chapter 8 in this volume). Further, one may consult existing protocols for some of the standard molecular biology techniques *(22)*.

1. Cleave 10 µg of pALTrxA-781 with endonuclease CspI (Stratagene, La Jolla, CA, an isoschizomer of RsrII) in the manufacturer-supplied buffer. This step linearizes the plasmid by cleavage at the unique CspI site in the region encoding the active-site of thioredoxin (*see* **Note 16**).
2. Phenol-extract and ethanol precipitate the DNA fragment using existing protocols.
3. Redissolve the plasmid DNA precipitate in 50 µL of 50 m*M* Tris-Cl, pH 8.0, 10 m*M* $MgCl_2$ buffer, and dephosphorylate the 5'-ends by incubating with 1 unit of calf intestinal alkaline phosphatase for 5 min. Phenol extract and ethanol precipitate the DNA to inactivate the phosphatase. Gel-purify the linearized plasmid.
4. For saturation mutagenesis two oligonucleotides were synthesized: I) 5'-GACTGACTG*GTCCACAGGTACATCCAAAACACTTCGGTCACGCTC CAATCG*GTCCTCAGTCAGTCAG-3' and II) 5'-CTGACTGACTGAGGACC-3'. Each base at an underlined region was synthesized with approx 86% of the desired specific base and with 14% random N (25% A / 25% C / 25% G / 25% T), with the exception of the first C in boldface, which was synthesized with 14% V (33% A / 33% G / 33% C), to avoid a stop codon (CAG->TAG). The incorporation of mixed nucleotides at the indicated positions resulted in a random population of 7% no mutations, 20% single mutations, 29% double mutations, and 44% with 3 or more mutations. The * symbols denote AvaII restriction cleavage sites on either side of the semi-random region.
5. Anneal oligo II with oligo I and synthesize the second strand with Klenow fragment (New England Biolabs). Phenol-extract and ethanol-precipitate the extended duplex oligonucleotide.
6. Digest the extended duplex oligonucleotide with AvaII restriction enzyme (New England Biolabs). Phenol-extract and ethanol-precipitate the digest.
7. Ligate the products from **step 3** and **step 6**. Transform GI724 with the ligation products by electroporation to generate the *E. coli* library that was estimated to contain approx 75% single oligonucleotide inserts with the rest having more than one inserts, based on sodium dodecylsulfate-polyacrylamide gel electrophoresis (SDS-PAGE) of induced cell lysates of 24 random colonies (*see* **Note 17**).
8. To screen for Trxloop mutants with improved binding affinity to the mAb, grow the bacteria on CAA/Amp plates overnight at 30°C. Then lift the colonies onto nitrocellulose filters and re-grow the plates. Lay the filters on LB/Amp/Trp plates with colony side up and incubate at 30°C for 6 h for Trxloop protein induction.
9. Lyse the cells on the filters by 3 freeze-thaw cycles of –80°C for 10 min followed by 30°C for 30 min (*see* **Note 18**). Block the filters in 1% non-fat milk in filter wash buffer overnight at room temperature. Probe the filters using **steps 20–31** in **Subheading 3.2.** to identify clones with better binding properties by the signal intensity of the colonies on the film. **Steps 10–14** provide instructions for inserting a defined sequence into the active-site loop of thioredoxin.
10. Synthesize oligonucleotide inserts encoding the peptide of interest, with ends compatible with the sticky ends generated by CspI restriction cleavage. For example, the sequences for a pair of DNA oligos coding for a $(Ser)_6$ peptide insertion are:

5'-GT CCA TCA TCA TCA TCA TCA TCA G-3'
3'-GT AGT AGT AGT AGT AGT AGT C CAG-5'

Note that the above design introduces an amino-terminal proline and a carboxyl-terminal glycine in addition to the inserted hexapeptide (*see* **Note 19**).

11. Separately phosphorylate the 5'-ends of 100 pmole of each oligonucleotide with T4 kinase for 30 min, using the manufacturer-supplied buffer in a volume of 20 μL. Heat to 90°C for 10 min to terminate the reaction and then chill on ice.
12. Combine in one tube both phosphorylated oligonucleotides in a final volume of 100 μL of annealing buffer (50 m*M* Tris-HCl, pH 8, 10 m*M* $MgCl_2$). Anneal the complementary strands by heating to 90°C for 5 min, followed by a slow cooling step to <30°C over a period of 1 h.
13. Ligate a 1:1 molar ratio of the phosphorylated, annealed oligonucleotide duplex to linearized and dephosphorylated plasmid pALTrxA-781 (from **step 3**). Incubate overnight at 15°C with T4 DNA ligase in the manufacturer-supplied buffer (*see* **Note 17**).
14. Transform strain GI724 with the ligation mixture by electroporation. Plate the bacteria on CAA/Amp plates and incubate at 30°C (*see* **Note 20**).
15. Pick transformant colonies, or the colonies identified in **step 9**, to inoculate HPM/Amp media to grow overnight at 30°C for plasmid minipreps. Verify the construction by restriction analysis and sequencing.
16. Inoculate 50 mL IMC/Amp media with a fresh overnight culture of a verified candidate clone to 0.05 OD_{550} and grow at 30°C until the OD_{550} reaches 0.5. Add L-Trp to 100 μg/mL and continue growth at 37°C for 4.5 h.
17. Measure the OD_{550} of the resulting culture. Pellet and resuspend the cells to 10 OD_{550}/mL in the lysis buffer (50 m*M* Tris-HCl, pH 8, containing 1 m*M* of *p*-aminobenzamidine and 1 m*M* phenylmethylsulfonyl fluoride). Lyse the cells in a French Pressure Cell and then centrifuge the cell lysate in a Microfuge at 13000 rpm (13,000*g*) for 10 min. Carefully transfer the supernatant into a separate tube and resuspend the pellet in an equal volume of the lysis buffer. Analyze these fractions by SDS-PAGE *(23,24)*.

4. Notes

1. These strains of bacteria are derived from strain GI724 *(24)*. The repression of pL promoter on plasmid by cI repressor, controlled by a Trp repressor/promoter on bacterial chromosome, decreases slightly at temperatures above 30°C. GI826 is tet-resistant because it bears a tetracycline resistance gene close to its *motB* locus.
2. We find that storing *E. coli* in 25% glycerol at –80°C retains viability of the bacteria.
3. Because the expression of TRX gene and FLITRX gene (in both pALTrxA-781 and the pFLITRX) is negatively regulated by the Trp promoter *(24)*, the plasmids and library should be propagated in media that do not contain tryptophan. We recommend Trp-free CAA-based media for plasmid growth, and for the outgrowth prior to induction of protein expression.

4. Always maintain a "master" peptide library, "LO-T", comprising aliquots of 100 OD_{550} (about 10^{11} *E. coli* cells), in order to preserve library diversity. "Working" library aliquots may be prepared and may contain less cells, but always several fold more than the total library diversity (i.e., 1.8×10^8). We recommend that "working" libraries always be prepared from a "master" library and not from another "working" library.
5. Adding L-tryptophan at this step induces protein expression that results in bacteria generating surface flagella. At 25°C the pL promoter is only partially induced, and most bacteria survive the induction process. A full pL induction can kill host *E. coli* cells.
6. Blocking buffer contains α-methyl mannoside to prevent *E. coli* from binding to the oligosaccharides present on glycosylated antibody molecules via interactions with cell-surface fimbriae ***(25–27)***.
7. At this step, the peptide loops present on bacterial flagella bind the mAb coated on the plate. Avoid jarring the plates after this step as mechanical shear will break flagella.
8. Adding the solution at one spot on the plate edge help to minimize loss of bound bacteria and mAb due to mechanical shear forces.
9. In our experiment, we used 2.5 m*M* $ZnCl_2$ solution to elute Zn(II)-sensitive binders after 3 rounds of selection by mechanical shear elution method. One can also experiment elution with buffers at lower or higher pH (within the range that bacteria can survive) to select acid- or base-labile binders.
10. The number of eluted FLITRX/LO-T library members include those that specifically bind to the mAb on the plate, as well as some non-specifically bound bacteria. Relatively few bacteria are left in the elution especially in the early rounds of selections. Thus they must be re-grown overnight.
11. We used 3 rounds of selection followed by a colony detection method (**steps 14–28**) for epitope mapping. Typically, about 1–10% of the eluted bacteria are "hits". However, colonies can be picked directly after more rounds of selection as described by Zhao and Lee ***(16)***, Xu et al. ***(17)***, Lombardi et al., ***(18,19)*** and Brown et al. ***(20)***.
12. When we applied the screening method (**steps 14–28**) to detect "hit" colonies after three rounds of selection, we found the following hints which might be helpful for those who want to practice this procedure: (a.) Chilling the plates to arrest colony growth actually helps to lift colonies onto nitrocellulose. (b.) The growth and induction of *E. coli* in the presence of tryptophan on top of the porous membrane results in some of the expressed FLITRX proteins being immobilized on the membrane. These proteins can be later detected by antibody-based staining techniques. (c.) We use lysozyme to disrupt the *E. coli* cell membranes, and DNase to break up the viscous genomic DNA. (d.) The skim-milk is the blocking reagent of choice to prevent non-specific adsorption of antibodies during the later probing steps. (e.) Because the filters also carry immobilized *E. coli* peroxidase and phosphatase activities, detection of mAb binding with a secondary antibody conjugated with horseradish peroxidase or alkaline phosphatase usually results in all colonies appearing as "hits". A phosphatase enzyme inhibitor such as

levamisole might be useful for AP conjugates, although we have had limited success with this approach. The best signal to noise ratio has been obtained with ^{125}I-labeled protein A, and for this purpose the secondary antibody should be derived from rabbits because of its high affinity to protein A ***(28,29)***

13. To presorb the antibodies with GI808 lysate, resuspend bacteria in cell lysis buffer to an OD_{550} of 100. Lyse the cells by French Pressure cell. Mix the bacterial lysate and the polyclonal antibody preparation (without dilution) at a ratio of 1:1 and incubate on at 4°C for 1 h. Centrifuge the mixture at 13000 rpm (13,000*g*) in a Microfuge for 10 min and then take the supernatant.
14. The sense strand primer lies approx 55 bases upstream from the region encoding the random peptide dodecamer. The reverse strand primer lies approx 25 bases downstream of the region encoding the random dodecapeptide. The usual encoded peptide sequence obtained is: -CGP$(X)_{12}$GPC-, where X is any of the 20 common amino acids. This pair of primers can also be used for sequencing insertion constructs of thioredoxin.
15. Because FLITRX forms flagella and displays peptides in a multi-valent manner, a strong binder identified from the selection may have high affinity, or may have modest to low affinity but bind strongly simply due to the avidity effect, as has been demonstrated in our original report ***(8)***. This phenomenon does not interfere with antibody epitope mapping since peptide sequences from the panning procedure all contribute to building a consensus sequence regardless of whether they have strong or weak affinities. We used monomeric TrxLoop proteins to avoid the avidity complication in our effort to improve binding affinity of zinc-ion dependent binders.
16. The unique RsrII/CspI region in pFLITRX is identical to that in pALTrxA-781 so the considerations for making insertions into the active-site loop in both plasmids are the same (*see* Chapter 8 in this volume). When constructing random peptide libraries, the CspI-digested and dephosphorylated plasmid DNA should be purified by acrylamide gel electrophoresis to remove undigested plasmids in order to increase the insertion rate.
17. It is important to keep a 1:1 ratio between the linearized plasmid and oligo duplex in order to minimize multi-copy insertions (two or more dodecapeptides linked in tandem with the spacer Gly-Pro in between), which in the subsequent affinity screening step tend to give very strong false positive signals.
18. The lysis in this step and in **step 19** of **Subheading 3.2.** is non-denaturing. Therefore, the positive signals are likely due to the detection of epitopes in native proteins.
19. In general, random libraries of DNA oligos for insertion into thioredoxin or FLITRX genes are generated by synthesizing the sense strand chemically, with fixed end sequences containing AvaII sites. The complimentary DNA strands are then synthesized by annealing a primer to the fixed region of the 3' end of the sense strand, followed by PCR or Klenow fragment treatment to generate the random region and the other fixed end region. The resulting mixture of oligo duplexes are then cut with AvaII endonuclease to obtain a pool of double-stranded oligos with ends complimentary to the sticky ends generated by CspI digestion. For constructing the particular dodecapeptide library in pFLITRX (LO-T), we

used the following oligonucleotides: oligo 1, 5'-GACTGACTG*GTCCG $(XNN)_{12}$G*GTCCTCAGTCAGTCAG-3'; oligo 2, 5'-CTGACTGACTGAGGACC-3'. Note that a proline and a glycine are introduced into the N- and C-termini of the peptides, respectively.

20. Electroporation is necessary to transform plasmid libraries into *E. coli* hosts. Because *E. coli* GI826 does not express wild-type flagellin and is defective in flagellar motor mechanism, it is the appropriate strain for hosting pFLITRX plasmid random peptide libraries ***(8)***.

References

1. Venter, J. C., et al. (2001) The sequence of the human genome. *Science* **291,** 1304–1351.
2. Lander, E. S., et al. (2001) Initial sequencing and analysis of the human genome. *Nature* **409,** 860–921.
3. Smith, G. P. (1985) Filamentaous fusion phage: novel expression vectors that display cloned antigens on the virion surface. *Science* **228,** 1315–1317.
4. Scott, J. K. and Smith, G. P. (1990) Searching for peptide ligands with an epitope library. Science, **249,** 386–390.
5. Li, M. (2001) Application of display technology in protein analysis. *Nat. Biotechnol.* **18,** 1251–1256.
6. LaVallie, E. R. and Stahl, M. L. (1989) Cloning of the flagellin gene from *Bacillus subtilis* and complementation studies of an in vitro-derived deletion mutation. *J. Bacteriol.* **171,** 3085–3094.
7. LaVallie, E. R., Diblasio, E. A., Kovacic, S., Grant, K. L., Schendel, P. F., and McCoy, J. M. (1993) A thioredoxin gene fusion expression system that circumvents inclusion body formation in the *E. coli* cytoplasm. *Bio/Technology* **11,** 1187–1193.
8. Lu, Z., Murray, K. S., van Cleave, V., LaVallie, E. R., Stahl, M. L., and McCoy, J. M. (1995) Display of Random Peptide Libraries on the *Escherichia coli* Cell Surface: A System for Exploring Protein-Protein Interactions and its Applications in Epitope Mapping. *Bio/Technology* **13,** 366–372.
9. Holmgren, H. (1989) Thioredoxin and glutaredoxin. *J. Biol. Chem.* **264,** 13,963–13,966.
10. Eklund, H., Gleason, F. K., and Holmgrem, H. (1991) Structural and functional relations among thioredoxins of different species. *Proteins Struct. Funct. Genet.* **11,** 13–28.
11. Colas, P., Cohen, B., Jessen, T., Grishna, I., McCoy, J., and Brent, R. (1996) Genetic selection of peptide aptamers that recognize and inhibit cyclin-dependent kinase 2. *Nature* **380,** 548–550.
12. Smith, P. A., Tripp, B. C., DiBlasio-Smith, E. A., Lu, Z., LaVallie, E. L., and McCoy J. M. (1998) A plasmid expression system for quantitative in vivo biotinylation of thioredoxin fusion proteins in *E. coli. Nucleic Acid Res.* **26,** 1414–1420.
13. Namba, K., Yamashita, I., and Vonderviszt, F. (1989) Structure of the core and central channel of bacterial flagella. *Nature* **342,** 648–654.

14. Wilson, A. R. and Beveridge, T. J. (1993) Bacterial flagellar filaments and their component flagellins. *Can. J. Microbiol.* **39,** 451–472.
15. Kuwajima, G. (1998) Construction of a minimum-size functional flagellin of *Escherichia coli. J. Bacteriol.* **170,** 3305–3309.
16. Zhao, S. and Lee, E. Y. (1997) A protein phosphatase-1-binding motif identified by the panning of a random peptide display library. *J. Biol. Chem.* **272,** 28,368–28,372.
17. Xu, H., Zhang, P., Liu, L., and Lee, M. Y. (2001) A novel PCNA-binding motif identified by the panning of a random peptide display library. *Biochemistry* **40,** 4512–4520.
18. Lombardi, S. J., Truong, A., Spence, P., Rhodes, K. J., and Jones, P. G. (1998) Structure-activity relationships of the Kvbeta1 inactivation domain and its putative receptor probed using peptide analogs of voltage-gated potassium channel alpha- and beta-subunits. *J. Biol. Chem.* **273,** 30,092–30,096.
19. Lombardi, S. J., Truong, A., Spence, P., Rhodes, K. J., and Jones, P. G. (1999) Probing the potassium channel Kv beta 1/Kv1.1 interaction using a random peptide display library. *Ann. NY Acad. Sci.* **868,** 427–430.
20. Brown, C. K., Modzelewski, R. A., Johnson, C. S., and Wong, M. K. (2000) A novel approach for the identification of unique tumor vasculature binding peptides using an *E. coli* peptide display library. *Ann. Surg. Oncol.* **7,** 743–749.
21. Tripp, B. C., Lu, Z., Bourque, K., Sookdeo, H., and McCoy, J. M. (2001) Investigation of the 'switch-epitope' concept with random peptide libraries displayed as thioredoxin loop fusions. *Protein Eng.* **14,** 367–377.
22. Sambrook, J., Fritsch, E. F., and Maniatis, T. (1989), *Molecular cloning: A Laboratory Manual*, 2nd edition, Cold Spring Harbor Laboratory Press, Cold Spring Harbor, NY.
23. McCoy, J. M. and LaVallie, E. R. (1994) Expression and purification of thioredoxin fusion proteins in *Current Protocols in Molecular Biology*. (Janssen, K., ed.) Wiley, NY, Unit 16.8.
24. LaVallie, E. R., Lu, Z., Racie, L., DiBlasio-Smith, E., and McCoy, J. (2000) Thioredoxin as a Fusion Partner for Production of soluble Recombinant Proteins in *E. coli. Methods Enzymol.* **326,** 322–340.
25. Diderichsen, B. (1980) *flu,* a metastable gene controlling surface properties of *Escherichia coli. J. Bacteriol.* **141,** 858–867.
26. Ofek, I. and Beachy, E. H. (1978) Mannose binding and epithelial cell adherence of *Escherichia coli. Infect. Immun.* **22,** 247–254.
27. Ponniah, S., Endres, R. O., Hasty, D. L., and Abraham, S. N. (1991) Fragmentation of *E. coli* type 1 fimbriae exposes cryptic D-mannose-binding sites. *J. Bacteriol.* **173,** 4195–4202.
28. Lu, Z., Tripp, B. C., and McCoy, J. M. (1991) Displaying libraries of conformationally constrained peptides on the surface of *Escherichia coli* as flagellin fusions. In *Methods in Molecular Biology, Combinatorial Peptide Library Protocols,* Vol. 87, Cabilly, S., ed., Humana, Totowa, NJ, pp. 265–280.
29. Harlow E. and Lane, D. (1998) *Antibodies: A laboratory manual.* Cold Spring Harbor Laboratory, Cold Spring Harbor, NY.

19

Use of Inteins for the In Vivo Production of Stable Cyclic Peptide Libraries in *E. coli*

Ernesto Abel-Santos, Charles P. Scott, and Stephen J. Benkovic

1. Introduction

Advances and opportunities in drug discovery and functional genomics have put methods for generating molecular diversity at a premium. Both chemical and biological approaches for the production of compound libraries have been pursued. Combinatorial chemistry has been used to synthesize molecular libraries in vitro, while molecular biology has been exploited to biosynthesize molecular libraries within cells. Unlike synthetic methods, which have largely focused on the production of libraries of small molecules, biosynthetic libraries must contend with the catabolic machinery of the host cell. Thus, variable segments are typically embedded within or fused to large biomolecules *(1)*. The resulting random sequences have been described as peptides, but they display the physical characteristic of the scaffold biopolymer.

Stability against cellular degradation can also be achieved by constraining the ends of a molecule with non-covalent and covalent interactions. Attaching dimerization domains to a random amino acid sequence afforded peptides that were stable when expressed in mammalian cells *(2)*. The protein microdomains formed proved to be sparingly soluble restricting their general use in biological systems. Disulfide bonds provide stability against protease digestion in vitro *(3)*, but are incompatible with the reducing environment of the bacterial cytoplasm.

We have pursued intracellular backbone cyclization as an alternative method to stabilize biosynthesized peptide libraries against catabolism. This procedure, named *s*plit *i*ntein *c*ircular *l*igation *o*f *p*eptides and *p*rotein*s* (SICLOPPS) *(4)*, harnesses the protein ligase activity of inteins (*see* **Fig. 1A**, for review on intein chemistry and applications *see* **ref. 5**). In the SICLOPPS construct, the intein is circularly permuted such that the *C*-terminal domain (I_C) precedes its *N*-terminal

From: *Methods in Molecular Biology, vol. 205, E. coli Gene Expression Protocols*
Edited by: P. E. Vaillancourt © Humana Press Inc., Totowa, NJ

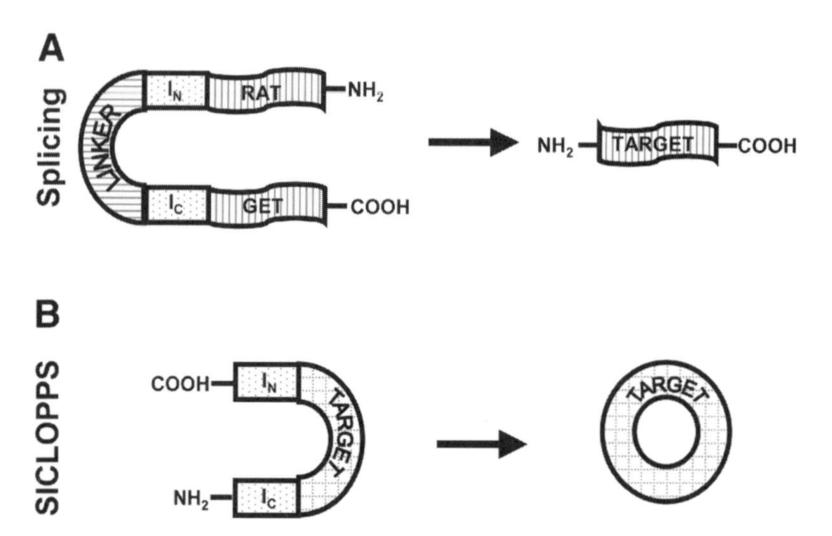

Fig. 1. The concept of SICLOPPS. **(A)** Naturally occurring inteins catalyze a multi-step reaction whereby the peptide backbone is broken in two places (between "TAR" and "I_N", and between "I_C" and "GET") and the flanking polypeptides are ligated together to produce the mature host gene product ("TARGET). The linker domain that separates these two elements is not essential for peptide cleavage or ligation. **(B)** A circularly permuted intein is reengineered to eliminate the linker domain. The resulting construct can catalyze the same multi-step reaction, but will form a peptide bond between the *N*-and *C*-terminus of an internal target, thus liberating cyclic peptides and proteins.

domain (I_N). The sequence to be cyclized serves as the linker between the intein halves (*see* **Fig. 1B**).

The SICLOPPS construct was prepared by cloning I_C upstream of I_N in a tandem configuration. Restriction sites engineered into I_C and I_N ensure that any target can be cloned into the SICLOPPS vector without limitations due to sequence identity (*see* **Fig. 2**). A chitin binding domain was fused *C*-terminal to I_N to aid in precursor protein purification and cyclic peptide characterization *(4,6)*. SICLOPPS was first used to produce the backbone cyclic form of *E. coli* dihydrofolate reductase and the naturally occurring eight-amino acid peptide pseudostellarin F. Cyclic dihydrofolate reductase showed increased thermostability, and pseudostellarin F synthesis was apparent in vivo through the inhibition of melanin production. Subsequently, other groups have used SICLOPPS-like methods to obtain cyclic maltose binding protein *(6)* and green fluorescent protein *(7)*.

Since the increase in stability conferred by cyclization is independent of folded structure or molecular weight, intracellular cyclization offers an attrac-

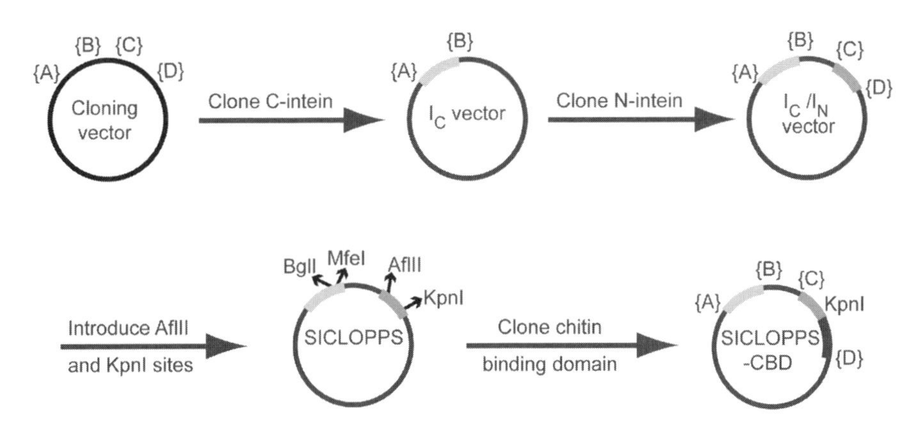

Fig. 2. Construction of SICLOPPS vectors. The I_C domain (light gray) of an intein is cloned between restrictions sites {A} and {B} to yield the I_C vector. The I_N domain (dark gray) is cloned into the I_C vector between {C} and {D}. This vector is further mutagenized to introduce *Afl*II and *Kpn*I sites into I_N yielding the SICLOPPS vector. Target sequences inserted between the 5'-end of I_C and the 3'-end of I_N can be cloned using *Bgl*I/*Afl*II or *Mfe*I/ *Afl*II. An affinity tag (black) is inserted into the SICLOPPS vector using *Kpn*I and {D} restriction sites to form the SICLOPPS-CBD vector.

tive method for the intracellular production of vast, genetically encoded libraries of small molecules. In order to make a library of SICLOPPS peptides, degenerate oligonucleotides had to be introduced between the I_C and I_N genes while keeping the correct reading frame throughout the tripartite construct. Several methods to introduce random sequences into expression vectors have been described. In one approach, the oligonucleotide encoding random positions is annealed with two adapter DNA fragments. The construct is ligated into an appropriate plasmid, leaving the library sequence as single stranded DNA *(8)*. The ligated plasmid can be transformed and the single stranded region is repaired intracellularly. A related procedure anneals a primer to the 5'-end of the library oligonucleotide and a DNA polymerase is used to extend the primer in vitro. The DNA cassette is digested and, after purification, the fragment containing the random region is ligated into an expression vector *(9)*.

These methods proved unsatisfactory for SICLOPPS library construction owing to the small size of the oligonucleotides encoding the six and nine amino acid peptides tested. A PCR-based technique was developed to transform short DNA sequences into longer, more manageable fragments. Primers were prepared by positioning the codons encoding the library between the 3'-end sequence of I_C and the 5'-end sequence of I_N. The library primer was used in conjunction with a reverse primer annealing to the chitin binding domain (CBD) to amplify the I_N-CBD fusion gene. Because of library sequence complexity,

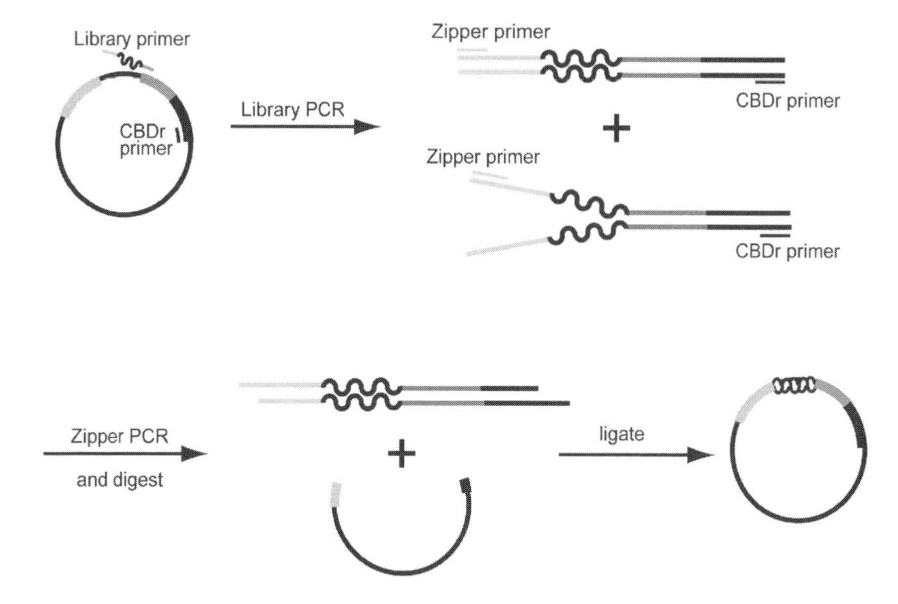

Fig. 3. Construction of library vectors: A library primer encoding the 5'-end of I_C (light gray), a degenerate sequence (wavy lines), and the 3'-end of I_N (dark gray) is used to amplify an I_N-CBD fragment. The resulting PCR product is subjected to a second round of amplification with the zipper primer to eliminate mismatches from the library sequences. The amplified fragments are digested and re-introduced into a similarly digested SICLOPPS-CBD vector.

half of the amplified DNA fragments contained mismatches in the random nucleotide region. A second PCR reaction using a "zipper" primer corresponding to the 3'-end of I_C ensured that all DNA sequences annealed to their complementary strand, creating viable substrates for restriction enzymes (*see* **Fig. 3**). Using this procedure, we demonstrated that SICLOPPS could be used to biosynthesize cyclic peptide libraries containing 10^7 to 10^8 members ***(10)***.

2. Materials

2.1. Preparation of Competent Cells

1. Bacterial strain: ElectroMAX DH5α-E (Invitrogen Corporation, Carlsbad, CA).
2. Bacterial strain: Tuner(DE3), (Novagen, Inc., Madison, WI).
3. Luria-Bertani (LB) broth: Mix 10 g bacto-tryptone, 5 g bacto-yeast extract, (Becton Dickinson Microbiology Systems, Sparks, MD) and 10 g NaCl with 1 L water. Adjust to pH 7.0 with 5 *N* NaOH. Sterilize by autoclaving (*see* **Note 1**).
4. Deionized, autoclaved water.
5. 10% glycerol solution: Mix 10 mL glycerol with 90 mL deionized water. Autoclave.

2.2. Construction of SICLOPPS Vector

2.2.1. Cloning of I_C Gene

1. 10X Thermophilic DNA polymerase buffer, 25 m*M* $MgCl_2$, and Taq DNA polymerase (5 U/µL), (Promega Corporation, Madison, WI).
2. dNTP mix: dATP, dCTP, dGTP, and dTTP each at a concentration of 100 m*M* (Boehringer Mannheim GmBH, Mannheim, Germany).
3. *Synechocystis* sp. PCC6803 genomic DNA (*see* **Note 2**).
4. Expression vector with appropriately located restriction sites (*see* **Note 3**).
5. I_Cf primer: 5'-{A}ATGGTTAAAGTTATCGGTCGTCGTTCCC -3' (*see* **Note 4**).
6. I_Cr primer: 5'-{B}ATTGTGCGCAATCGCCCCAT-3'.
7. Seakem GTG agarose (for DNA fragments >1 kb) and NuSieve GTG agarose (for DNA fragments <1 kb) (Biowhittaker Molecular Applications, Rockland, ME).
8. 1X Tris-Acetate EDTA (TAE) buffer: Prepare a 50X stock solution by mixing 242 g Tris base, 57.1 mL glacial acetic acid, and 100 mL 0.5 *M* ethylenediaminetetraacetic acid (EDTA), pH 8.0, and water to 1 L final volume. Dilute 20 mL to 1 L to obtain a working 1X solution.
9. QiaQuick PCR purification and QiaQuick gel extraction kits (QIAGEN, Valencia, CA).
10. Appropriate 10X restriction enzyme buffer, 100X BSA, and restriction enzymes {A}, {B}, {C}, {D}, *Afl*II, and *Kpn*I (New England BioLabs, Beverly, MA).
11. Shrimp alkaline phosphatase (1 U/µL), (USB Corporation, Cleveland, OH)
12. PhiX174 DNA/HaeIII and Lambda DNA/HinDIII markers (Promega Corp.).
13. 10X Ligation buffer and T4 DNA ligase (3 U/µL), (Promega Corp.).
14. 0.2-cm Gene pulser cuvets (Bio-Rad Laboratories, Hercules, CA).
15. SOC medium: Mix 20 g bacto-tryptone, 5 g bacto-yeast extract, and 0.5 g NaCl. Add 10 mL of 250 m*M* KCl. Adjust to pH 7.0 with 5 *N* NaOH. Dilute to 1 L with distilled water. Sterilize by autoclaving. Before using, add 5 mL sterile 2M $MgCl_2$ and 20 mL sterile 1 *M* glucose.
16. LB agar plates: Add 15 g bacto-agar (Becton Dickinson Microbiology Systems) to 1 L of LB broth. Sterilize by autoclaving. Cool the agar to 60°C and add the appropriate antibiotic. Dispense in 100 × 15 mm Petri dishes.
17. Wizard *Plus* Minipreps DNA purification kit (Promega).
18. QIAGEN Plasmid Midi kit (QIAGEN).

2.2.2. Cloning of I_N Gene

1. I_Nf primer: 5'-{C}GCCTCAGTTTTGGC-3'.
2. I_Nr primer: 5'-{D}TTATTTAATAGTCCCAGCGTC-3'.
3. Quick Change site-directed mutagenesis kit (Stratagene, La Jolla, CA).
4. AFLf primer: 5'-{C}TGCTTAAGTTTTGGCACC-3'.
5. AFLr primer: 5'-GGTGCCAAAACTTAAGCA{C}-3'.
6. KPNf primer: 5'-TACTTGACGCTGGTACCATTAAATAA{D}-3'.
7. KPNr primer: 5'-{D}TTATTTAATGGTACCAGCGTCAAGTA-3'.

2.2.3. Cloning of Chitin Binding Domain

1. CBDf primer: 5'-GGGGTACCATTAAAACGACAAATCCTGGTGTA-3'.
2. CBDr primer: 5'-{D}TCATTGAAGCTGCCACAAGG-3' (*see* **Note 5**).
3. pCYB1 plasmid vector (New England Biolabs).

2.3. SICLOPPS Library Construction

1. Library primers (*see* **Note 6**)
   S + 5: 5'-GGAATTCGCCAATGGGGCGATCGCCCACAATTCCNNSN NSNNSNNSNNSTGCTTAAGTTTTGGC-3'
   Δ4 + 5: 5'-GGAATTCGCCAATGGGGCGATCGCCCACAATTCCGGAN NSNNSNNSNNSNNSCCGCTGTGCTTAAGTTTTGGC-3'
2. Zipper primer: 5'-GGAATTCGCCAATGGGGCGATCGCC-3' (*see* **Note 7**).
3. Pellet Paint ethanol precipitation kit (Novagen).
4. Absolute ethanol chilled to –20°C.
5. 70% ethanol chilled to –20°C.
6. Library agar plates: Add 200 mL melted LB agar with appropriate antibiotic to a 243 × 243 × 18 mm agar diffusion assay dish (Nalge Nunc, Int., Naperville, IL).
7. Recovery media: Mix 16 g bacto-tryptone, 10 g bacto-yeast extract, and 5 g NaCl with 1 L water. Adjust to pH 7.0 with 5 N NaOH. Sterilize by autoclaving. Dilute 30 mL medium with 5 mL sterile 20% glucose and 15 mL sterile 50% glycerol solutions.
8. 50% glycerol solution: Mix 50 mL glycerol and 50 mL deionized water. Autoclave.

2.4. Trial Induction of SICLOPPS Library Members

1. 100 m*M* IPTG solution: Dissolve 0.238 g of isopropylthio-β-D-galactoside (IPTG, Sigma, St. Louis, MO) in 10 mL deionized water. Filter sterilize.
2. SDS gel-loading buffer: Mix 100 m*M* Tris·HCl, pH 6.8, 200 m*M* dithiothreitol (DTT), 4% sodium dodecyl sulfate (SDS), 0.2% bromophenol blue, and 20% glycerol.
3. SDS-PAGE minigel: A 16% resolving SDS-PAGE gel and 5% stacking gel ***(12)*** were cast in a Mini Protean 3 electrophoresis module as instructed by the manufacturer (Bio-Rad Laboratories).
4. Tris-glycine electrophoresis buffer: Dissolve 15.1 g Tris base, 94 g glycine, and 5 g electrophoresis grade SDS to 1 L with deionized water. Dilute 100 mL to 500 mL with water to obtain working buffer.
5. Stain solution: Dissolve 0.25 g Coomassie brilliant blue R250 in 45 mL methanol, 45 mL water, and 10 mL glacial acetic acid.
6. Destain solution: Mix 45 mL methanol, 45 mL water, and 10 mL glacial acetic acid.

2.5. Purification of SICLOPPS Library Members

1. 1 *M* IPTG solutions: Dissolve 2.38 g IPTG in 10 mL water. Filter sterilize.
2. Chitin buffer: Dilute 25 mL of 1 *M*, Tris·HCl, pH 7.0, and 29.2 g NaCl with water to 1 L final volume.
3. 100 m*M* PMSF solution: Dissolve 17 mg of phenylmethyl sulfonyl fluoride (PMSF, Sigma) in 1 mL EtOH.
4. Chitin beads (New England BioLabs)

5. 97% α-cyano-4-hydroxycinnamic acid (Aldrich, Milwaukee, WI): Recrystallize before using.
6. 10 m*M* Trifluoroacetic solution (Pierce Chemical, Rockford, IL, sequencing grade). Dilute 76.5 μL with water to 100 mL final volume.

3. Methods

3.1. Preparation of Competent Cells

This protocol is derived from the method of Seidman et al. ***(11)***.

1. Inoculate 3 mL LB broth with *E. coli* cells. Grow overnight at 37°C.
2. Add 2.5 mL overnight culture to 500 mL LB broth. Incubate at 37°C, with vigorous shaking, until culture reaches an OD_{600} of 0.5–0.6.
3. Chill bacterial culture at 4°C for 15 min (*see* **Note 8**).
4. Pellet cells by centrifugation (6000*g*, 10 min). Discard supernatant.
5. Resuspend cells in 20 mL cold water with gentle swirling. Dilute with water to 500 mL final volume. Repeat **steps 4–5**.
6. Resuspend cells in 10 mL water and transfer to 50-mL Falcon tube.
7. Pellet cells in clinical centrifuge (1500*g*, 10 min). Resuspend cells in 0.5 mL water (*see* **Note 9**).
8. Resuspend cells in 40 mL of a 10% glycerol solution. Pellet as before.
9. Resuspend cells in 0.5 mL of a 10% glycerol solution.
10. Aliquot 50 μL cells in 0.7-mL Eppendorf tubes. Freeze and store at –70°C.

3.2. Construction of SICLOPPS Vector (see Note 10)

3.2.1. Cloning of I_C Gene

3.2.1.1. Preparation of I_C Insert

1. Mix 5 μL PCR Buffer (10X), 3 μL $MgCl_2$ solution, 1 μL dNTP mix, 1 μL *Synechocystis* sp. PCC6803 genomic DNA, 5 μL I_Cf primer, 5 μL I_Cr primer, 1 μL Taq DNA polymerase, and water to 50 μL final volume.
2. PCR using the following protocol: 94°C (5 min), 55°C (2 min), 72°C (1.5 min) [1 cycle] and, 94°C (1 min), 55°C (1 min), 72°C (1.5 min) [25 cycles], 72°C (8 min) [1 cycle].
3. Treat PCR product with QiaQuick PCR purification kit as instructed by manufacturer.
4. Mix 7 μL digestion buffer (10X), 7 μL BSA (10X), 50 μL I_C PCR fragment, 1 μL restriction enzyme {A}, 1 μL restriction enzyme {B}, and water to 70 μL final volume. Incubate overnight at 37°C.
5. Recover PCR product with QiaQuick PCR purification kit as above.

3.2.1.2. Preparation of Cloning Vector

1. Mix 1 μg cloning vector, 2 μL digestion buffer (10X), 2 μL BSA (10X), 1 μL restriction enzyme {A}, 1 μL restriction enzyme {B}, and water to 20 μL final volume. Incubate overnight at 37°C.

2. Add 1 µL shrimp alkaline phosphatase and continue incubation at 37°C for 1 h.
3. Inactivate enzymes by incubating at 65°C for 30 min.
4. Run digested vector on 2% Seakem agarose gel in 1X TAE buffer. Excise the vector backbone.
5. Purify excised band using QiaQuick gel extraction kit as instructed by manufacturer.

3.2.1.3. Ligation and Electroporation

1. Determine DNA concentrations by running insert and vector fragments on a 2% agarose gel in 1X TAE buffer. Use 10 µL each PhiX174 DNA/HaeIII and Lambda DNA/HinDIII markers as standards.
2. Combine 100 ng digested and dephosphorylated vector, 100 ng digested I_C fragment, 1 µL ligation buffer (10X), 1 µL T4 DNA ligase, and water to 10 µL final volume. As control, prepare a ligation mixture containing all components except insert. Incubate overnight at room temperature.
3. Inactivate T4 ligase by incubating at 65°C for 30 min.
4. Thaw competent cells on ice. Add 1.5 µL ligation mixture.
5. Transfer mixture to electroporation cuvet. Incubate on ice for 1 min.
6. Apply electric pulse with electroporation apparatus as instructed by manufacturer.
7. Add 1 mL SOC medium and transfer to 1.7-mL Eppendorf tubes. Recover cells for 1 h at 37°C.
8. Pellet cells in microfuge. Resuspend in 100 µL SOC medium. Plate in LB agar plates containing appropriate antibiotic. Incubate overnight at 37°C.

3.2.1.4. Identification of Vectors Containing I_C Gene

1. Inoculate randomly picked colonies into separate 3 mL LB broth containing appropriate antibiotic. Grow culture to stationary phase. Reserve 200 µL.
2. Purify plasmids with Wizard Minipreps kit as instructed by manufacturer.
3. Mix 10 µL plasmid, 2 µL digestion buffer (10X), 2 µL BSA (10X), 1 µL restriction enzyme {A}, 1 µL restriction enzyme {B}, and water to 20 µL final volume. Incubate at 37°C for 6 h.
4. Analyze plasmids for I_C insertion by running digestion mixtures on a 2% agarose gel with 1X TAE buffer.
5. Inoculate 50 mL LB broth with reserved cells. Incubate overnight at 37°C.
6. Purify plasmid with QIAGEN Plasmid Midi kit as instructed by the manufacturer.
7. Confirm I_C vector identity by DNA sequencing.

3.2.2. Cloning of I_N

1. Follow protocols in **Subheading 3.2.1.1.** with the following modifications. Use I_Nf and I_Nr primers for gene amplification in **step 1**. Digest I_N gene fragment with restriction enzymes {C} and {D} in **step 4**.
2. Digest I_C vector with enzymes {C} and {D} using protocol from **Subheading 3.2.1.2.**
3. Clone I_N fragment into the I_C vector following protocols in **Subheadings 3.2.1.3–3.2.1.4.**

4. Using the Quick Change site-directed mutagenesis kit with the AFLf and AFLr primers introduce an *Afl*II restriction site in the vector obtained in **step 3** (*see* **Note 11**).
5. Using the Quick Change site-directed mutagenesis kit with the KPNf and KPNr primers introduce a *Kpn*I restriction site in the vector obtained in **step 4** (*see* **Note 12**).
6. Confirm SICLOPPS vector identity by DNA sequencing.

3.2.3. Cloning of Chitin Binding Domain

1. Follow protocols described in **Subheading 3.2.1.1.** with the following modifications. Use CBDf primer, CBDr primer, and pCYB1 plasmid vector in **step 1**. Digest DNA fragments using restriction enzymes *Kpn*I and {D} in **step 4**.
2. Digest SICLOPPS vector with *Kpn*I and {D} using protocol from **Subheading 3.2.1.2.**
3. Insert CBD gene into SICLOPPS vector using protocols in **Subheadings 3.2.1.3–3.2.1.4.**
4. Confirm SICLOPPS-CBD vector identity by DNA sequencing.

***3.3. SICLOPPS Library Construction* (see *Note 13*)**

3.3.1. Preparation of SICLOPPS Library Insert

1. Mix 5 μL PCR Buffer (10X), 3 μL $MgCl_2$ solution, 1 μL dNTP mix, 100 ng SICLOPPS-CBD plasmid, 5 μL S+5 or Δ4+5 library primer, 5 μL CBDr primer, 1 μL Taq DNA polymerase, and water to 50 μL final volume.
2. PCR using the following protocol: 94°C (5 min), 65°C (2 min), 72°C (1.5 min) [1 cycle] and, 94°C (1 min), 65°C (1 min), 72°C (1.5 min) [30 cycles], 72°C (8 min) [1 cycle], (*see* **Note 14**).
3. Treat PCR fragment with QiaQuick PCR purification kit as above.
4. Determine DNA fragment concentrations using protocol from **Subheading 3.2.1.3., step 1**.
5. Mix 5 μL PCR Buffer (10X), 3 μL $MgCl_2$ solution, 1 μL dNTP mix, 100 ng amplified library fragment, 5 μL zipper primer, 5 μL CBDr primer, 1 μL Taq DNA polymerase, and water to 50 μL final volume (*see* **Note 15**).
6. PCR using the following protocol: 94°C (5 min), 65°C (2 min) , 72°C (1.5 min) [1 cycle] and, 94°C (1 min), 65°C (1 min), 72°C (1.5 min) [15 cycles].
7. Digest gene fragment with *Bgl*I (or *Mfe*I) and {D} following protocol from **Subheading 3.2.1.1., steps 3–5**.

3.3.3. Preparation of Cloning Vector

1. Digest SICLOPPS-CBD vector with restriction enzymes *Bgl*I (or *Mfe*I) and {D}, following protocol described in **Subheading 3.2.1.2.**

3.3.4. Optimization of Library Ligation Conditions (see ***Note 16***)

1. Combine 100, 150, 200, 250, and 300 ng digested and dephosphorylated vector, with 30, 60, and 120 n*M* digested library insert (total of 15 ligation reactions).

Add 1 μL ligation buffer (10X), and water to a final volume of 9 μL for every sample. As controls, prepare ligation mixtures for every vector concentration without insert.
2. Incubate samples at 42°C for 5 min. Allow to cool to room temperature for 5 min.
3. Add 1 μL T4 DNA ligase and incubate overnight at room temperature.
4. Inactivate enzyme by incubating at 65°C for 30 min.
5. Ethanol precipitate DNA using Pellet Paint kit as instructed by manufacturer.
6. Transform samples following protocol described in **Subheading 3.2.1.3., steps 4–8.**
7. Plate 10^{-4} and 10^{-6} dilutions for every ligation condition. Incubate overnight at 37°C.
8. Subtract CFU in control plates from CFU in library plates to determine best ligation conditions.

3.3.5. Transformation and Recovery of SICLOPPS Library

1. To obtain a 10^8 member library, prepare enough ligation mixture for 12 transformations using the best conditions determined in **Subheading 3.3.4.**
2. Transform samples following protocol described in **Subheading 3.2.1.3., steps 4–7.**
3. Pool all transformations and pellet cells in microfuge (18000*g*, 1 min). Resuspend in 1 mL LB broth.
4. Plate cells in library agar plates. Plate 10^{-6} dilution in small agar plate to determine CFU. Incubate overnight at 37°C.
5. Scrap library colonies into 10 mL recovery media with a bent glass Pasteur pipet.
6. Pellet cells in clinical centrifuge (1500*g*, 10 min).
7. Resuspend cells in 2 mL recovery media. Place 0.5-mL aliquots in 1.7-mL Eppendorf tubes. Freeze in liquid nitrogen and store at –70°C.
8. Inoculate 50 mL LB broth with 10 μL frozen library cells. Incubate overnight at 37°C.
9. Extract and purify DNA using QIAGEN plasmid Midi kit as above.
10. Confirm SICLOPPS library identity by DNA sequencing.

3.4. Trial Induction of SICLOPPS Library Members

3.4.1. Selection of Random Colonies for SICLOPPS Peptide Expression

1. Transform library plasmid into an *E. coli* protein expression strain (*see* **Note 17**).
2. Plate approximately 100 colonies in LB agar plate. Incubate overnight at 37°C.
3. Using sterile toothpicks, inoculate 12 colonies into separate 2 mL LB broth containing appropriate antibiotic. Grow overnight at 37°C.
4. Dilute 500 μL of each culture with 500 μL of 50% glycerol solution. Freeze in liquid nitrogen. Store at –70°C. Reserve remaining cell suspensions.

3.4.2. Trial Induction

1. Inoculate 6 mL LB broth with reserved cultures from **Subheading 3.4.1., step 4.** Incubate at 37°C to an OD_{600} in the range 0.5–0.6.
2. Separate each culture into 2 × 3 mL aliquots. Induce one aliquot from each sample with 3 μL IPTG (100 m*M*). The remaining aliquots will be used as uninduced controls.

3. Grow all samples for 20 h at room temperature with vigorous shaking (*see* **Note 18**).
4. Aliquot 1 mL cultures into separate 1.7-mL Eppendorf tubes.
5. Pellet cells in a microfuge (18,000*g*, 1 min). Discard supernatant.
6. Resuspend cells in 100 µL SDS-PAGE loading buffer.
7. Lock tubes securely. Incubate for 15 min in boiling water bath.
8. Centrifuge samples in microfuge (18,000*g*, 15 min).
9. Load 2 µL from each sample in a 16% SDS-PAGE minigel.
10. Run under Tris-glycine buffer at 45 mA until tracking dye is 5 mm from gel bottom.
11. Cover SDS-PAGE gel with stain solution. Incubate with agitation until gel is stained.
12. Decant stain solution and cover gel with destain solution. Incubate with agitation until protein bands are visible.
13. Select constructs showing incomplete precursor protein processing (*see* **Note 19**).

3.5. Purification of SICLOPPS Library Members

3.5.1. Induction of Selected Constructs

1. Inoculate individual 500 mL LB broth with 2 mL uninduced samples reserved in **Subheading 3.4.1., step 4**.
2. Incubate at 37°C, with agitation, until cell suspension reaches an OD_{600} between 0.5–0.6. Reserve 1 mL for analysis and incubate along as uninduced control.
3. Cool cells to room temperature for 30 min. Induce with 50 µL IPTG solution (1 *M*).
4. Continue incubation at room temperature for 16–20 h. Reserve 1 mL for analysis.
5. Pellet cells by centrifugation (6000*g*, 10 min). Store cells at –70°C.

3.5.2. In vitro SICLOPPS Peptide Synthesis

1. Resuspend cell pellet obtained in **Subheading 3.5.1., step 5** in 30 mL chitin buffer. Add 1 mL PMSF solution.
2. Lyse cells with 0.5 in tip sonicator (5 × 20 s) at 50% output and power setting 10.
3. Pellet cell debris with by centrifugation (30000*g*, 1 h).
4. Pipet 2 mL chitin beads into a 10-mL column. Wash with 100 mL chitin buffer.
5. Load cell lysate onto chitin column at 0.5 mL/min. Reserve 100 µL lysate aliquot.
6. Wash with 100 mL chitin buffer at fastest flow rate. Reserve 100 µL wash aliquot.
7. Leave 2 mL buffer above chitin beads. Incubate at room temperature for 16 h.
8. Elute columns. Reserve 100 µL eluate and 100 µL chitin beads aliquots.

3.5.3. Analysis of SICLOPPS Peptide Formation

3.5.3.1. SDS-PAGE Gel Electrophoresis

1. Pellet samples from **Subheading 3.5.1., steps 2** and **4** in microfuge (18000*g*, 1 min). Decant supernatant. Resuspend pellet in 100 µL of SDS-PAGE loading buffer.
2. Resuspend reserved aliquots from **Subheading 3.5.2., steps 5**, **6**, and **8** in 100 µL of SDS-PAGE loading buffer.

3. Follow procedure described in **Subheading 3.4.2., steps 7–12**.
4. Check for protein intermediate processing in the chitin column.

3.5.3.2. DNA Sequencing

1. Dilute 10 µL frozen cell stocks obtained in **Subheading 3.4.1., step 4** into 50 mL LB broth. Incubate overnight at 37°C.
2. Extract and purify plasmids with QIAGEN Midi kit.
3. Determine SICLOPPS peptide identity by DNA sequencing.

3.5.3.3. Matrix-Assisted Laser Desorption Ionization Mass Spectrometry (MALDI)

1. Cocrystallize 0.5 µL aliquot of chitin column eluate with α-cyano-4-hydroxycinnamic acid on a MALDI sample plate. Allow solvent to evaporate at room temperature.
2. Wash five times with 5 µL cold trifluoroacetic acid solution.
3. Acquire positive ion MALDI mass spectra in linear mode as the summed signal from 256 shots of 337 nm radiation from a nitrogen laser.

4. Notes

1. Media and buffers were prepared according to standard methods described in ***(12)***.
2. Any intein expressed in an active, soluble form may be compatible with SICLOPPS. Because the residues immediately adjacent to the intein domains can modulate activity, the choice of inteins will be dependent on target identity. We utilized the intein associated with the *DnaE* polymerase of *Synechocystis* sp. PCC6803 because is the only naturally occurring intein where I_C and I_N are expressed as two separate polypeptides ***(13)*** and cyclization shows little dependence on target sequence ***(10)***.
3. The SICLOPPS construct can be cloned into any expression vector containing at least four contiguous restriction sites arranged in the form 5'-{A}{B}{C}{D}-3' (*see* **Fig. 2**).
4. All primers were synthesized by the phosphoramidite method. The oligonucleotides were purified by G-25 gel chromatography and 10 µ*M* working solution prepared.
5. The chitin binding domain from *Bacillus circulans WL-12* was added *C*-terminal to the SICLOPPS construct to aid in precursor protein purification.
6. Two different library primers have been used. Primer S+5 encodes hexapeptides with an invariable serine and five variable positions. Primer Δ4+5 encodes nonapeptides of the form c[SGXXXXXPL]. Both constructs yield cyclic peptides: S+5 peptides have lower molecular weights, and 7 out of 10 of library members tested yielded cyclic products. The four amino acid scaffold of the Δ4+5 construct resulted in cyclic products in all tested constructs.
7. The zipper primer is identical to the constant 5'- region of the library primers.
8. All reagents and equipment for making electrocompetent cells must be sterile and pre-chilled to 4°C.
9. Use ElectroMAX DH5α-E at this step to maximize library diversity. Frozen cells show lower transformation efficiency and should be used for routine cloning only.

10. The procedure to prepare SICLOPPS vectors is represented in **Fig. 2**.
11. The I_C gene has unique *Mfe*I and *Bgl*I restriction sites at its 3'-end. Introducing an *Afl*II restriction site at the 5'-end of the I_N gene allows targets to be cloned *Mfe*I/*Afl*II or *Bgl*I/*Afl*II. This allows the expression of cyclic peptides without extra amino acids derived from restriction site usage.
12. Adding a unique *Kpn*I restriction site at the 3'-end of I_N allows affinity tags to be fused to the *C*-terminus of the SICLOPPS construct.
13. The procedure to prepare SICLOPPS libraries is represented in **Fig. 3**.
14. Annealing at lower temperatures results in constructs containing extensive frameshifts downstream of the library sequence.
15. The "zipper" PCR reaction ensures that all DNA fragments are annealed to their complementary sequence (*see* **Fig. 3**).
16. Small changes in vector to insert ratio have large effects on transformation efficiency. Best conditions must be determined for every plasmid/insert pair. Transformation efficiency is improved by pre-warming SOC medium to 37°C and adding it rapidly after the electric pulse. Samples should be placed immediately in 15-mL falcon tubes and recovered at 37°C for 1 h.
17. Best protein expression and peptide recovery was obtained with vectors containing a T7 promoter (e.g., pET) and the *E. coli* Tuner (DE3) strain. Expression of the SICLOPPS library from an arabinose-inducible promoter resulted in a marked decrease in CFU.
18. Expressing SICLOPPS at 37°C resulted in large amounts of insoluble proteins.
19. The analytical detection of cyclic peptides relies on the cyclization reaction proceeding in vitro. Incomplete in vivo processing result in accumulation of SICLOPPS intermediates that can splice in the affinity column.

Acknowledgments

We want to thank Prof. A. Daniel Jones for mass spectrometry and Dr. Deborah S. Grove for DNA sequencing.

References

1. Rosamond, J. and Allsop, A. (2000) Harnessing the power of the genome in the search of new antibiotics. *Science* **287,** 1973–1976.
2. Gururaja, T. L., Narasimhamurthy, S., Payan, D. G., and Anderson, D. C. (2000) A novel artificial loop scaffold for the noncovalent constraint of peptides. *Chem. Biol.* **7,** 515–527.
3. Jermutus, L., Honegger, A., Schwesinger, F., Hanes, J., and Pluckthun, A. (2001) Tailoring in vitro evolution for protein affinity or stability. *Proc. Natl. Acad. Sci. USA* **98,** 75–80.
4. Scott, C. P., Abel-Santos, E., Wall, M., Wahnon, D. C., and Benkovic, S. J. (1999) Production of cyclic peptides and proteins in vivo. *Proc. Natl. Acad. Sci. USA* **96,** 13,638–13,643.
5. Perler, F. B. and Adam, E. (2000) Protein splicing and its applications. *Curr. Opin. Biotechnol.* **11,** 377–383.

6. Evans T. C., Jr., Martin, D., Kolly, R., et al. (2000) Protein trans-splicing and cyclization by a naturally split intein from the DnaE gene of *Synechocystis species PCC6803. J. Biol. Chem.* **275,** 9091–9094.
7. Iwai, H., Lingel, A., and Pluckthun, A. (2001) Cyclic green fluorescent protein produced in vivo using an artificially split *PI-PfuI* intein from *Pyrococcus furiosus. J. Biol. Chem.* **276,** 16,548–16,554.
8. Stern, B. and Gershoni, J. M. (1998) Construction and use of a 20-mer phage display epitope library in *Methods in Molecular Biology: Combinatorial peptide library protocols*, (Cabilly, S., ed.), Humana Press, Totowa, NJ, Vol. 87, pp. 137–154.
9. Luzzago A., and Felici, F. (1998) Construction of disulfide constrained random peptide libraries displayed on phage coat protein VIII in *Methods in Molecular Biology: Combinatorial peptide library protocols*, (Cabilly, S. ed.), Humana Press, Totowa, NJ, Vol. 87, pp. 155–164.
10. Scott, C. P., Abel-Santos, E., Jones, A. D., and Benkovic, S. J. (2001) Structural requirements for the biosynthesis of backbone cyclic peptide libraries, *Chem. Biol.* **8,** 801–815.
11. Seidman, C. E., Struhl, K., and Sheen, J. (1997) *Short Protocols in Molecular Biology.* 3rd edition, (Ausubel, F. M., Brent, R., Kingston, R. E., et al., eds.), John Wiley & Sons, New York, NY.
12. Sambrook, J., Fritsch, E. F., and Maniatis, I. (1989) *Molecular Cloning: A Laboratory Manual,* 2nd edition, Cold Spring Harbor Laboratory, Cold Spring Harbor, NY.
13. Wu, H., Hu, Z., and Liu, X. Q. (1998) Protein trans-splicing by a split intein encoded in a split DnaE gene of *Synechocystis* sp. PCC6803. *Proc. Natl. Acad. Sci. USA* **95,** 9226–9231.

20

Hyperphage

Improving Antibody Presentation in Phage Display

Olaf Broders, Frank Breitling, and Stefan Dübel

1. Introduction

Since its invention in the early 1990s, phage display has revolutionized the generation and engineering of monoclonal antibodies (for review, *see* **ref.** ***1***). Without the need for laboratory animals or hybridomas, it was now possible to create antibodies binding to almost any antigen of choice. All this is accomplished in a system that completely by-passes our bodies immune system.

Antibody phage display is done by fusing antigen-binding antibody fragments to the phage minor coat protein pIII. Incorporation of this fusion protein into the mature phage coat results in the presentation of the antibody on the phage surface, while the genetic material of the fusion resides within the phage particle. This physical linkage between the antibody gene and its product allows the enrichment of antigen specific phage antibodies by employing immobilized or labeled antigen. While non-adherent phages will be removed by washing, phage that display the relevant antibody will be retained on an antigen-coated surface. Bound phages can then be recovered from the surface, re-infected into bacteria, and thus amplified for further enrichment. Each re-grown colony represents a single molecular interaction event, thus allowing an enormous sensitivity. By using large combinatorial antibody fragment repertoires (10^8–10^{11} independent clones), antigen-specific antibodies to almost any chosen antigen can be selected. These highly specific antibodies can then be recloned into various expression vectors and/or be further modified to optimize their diagnostic or therapeutic capabilities. A recent breakthrough to enhance the performance of antibody phage display was the development of hyperphage technology *(2)*. By using this method, antigen binding activity was increased

From: *Methods in Molecular Biology, vol. 205, E. coli Gene Expression Protocols*
Edited by: P. E. Vaillancourt © Humana Press Inc., Totowa, NJ

approximately 400-fold by enforcing oligovalent antibody display on every phage particle. The use of hyperphage for packaging a universal human scFv library improved the specific enrichment factor: after two rounds of panning, more than 50% of the isolated antibody clones bound to the antigen, compared to 3% when conventional M13KO7 helper phage was used. Thus, hyperphage are particularly useful in stoichiometrical situations, where the chances of a single phage capable of locating the wanted antigen are known to be low. In particular, new tumor markers may be detected by allowing panning on cell surfaces with higher sensitivity. In the search for novel targets to deliver genes or drugs in a tissue- or cell-specific manner, in vivo panning can be expected to benefit from hyperphage packaging.

1.1. Concept of the Hyperphage

An overview of various phage display systems can be found in ***(3)***. The approach described in this protocol is applicable to all these phagemid display systems, which employ full length pIII. It uses a novel helper phage design to improve the presentation of antibody fragments on the phage surface.

In commonly used phagemid-based systems ***(4–6)***, only a small percentage of the total phage population carries an antibody fragment on its surface. This problem arises from the presence of two copies of the gene III which encodes the pIII gene product (g3p). One copy resides on the antibody expression phagemid and is fused to the antibody gene. However, the phagemid lacks the other structural genes required for phage assembly. To provide these, infection with a helperphage is employed. This helperphage, however, brings in a second pIII gene. This is a wildtype pIII gene and cannot easily be deleted from the helperphage genome, since functional pIII is an essential surface protein for infection, by providing F-pilus binding. This wildtype pIII is favored during assembly of the phage particle, resulting in a minor fraction of phage carrying any antibody:pIII fusion protein at all. The problem was finally overcome by avoiding the delivery of wild-type pIII during the phage antibody packaging (*see* **Fig. 1**). Hyperphage are helper phages with a deletion in the pIII gene, but with wildtype pIII phenotype, thus capable of infecting F^+ *E. coli* cells with high efficiency. During phagemid packaging to create an antibody expression phage library, they render the phagemid encoded antibody-pIII fusion as the only source of pIII in phage assembly. This results in both an increase of the fraction of phage carrying antibodies and the number of antibodies displayed per phage, the latter providing a significant increase of the apparent affinity by the avidity effect.

Until now all reported approaches to generate a respective helperphage combined a pIII deleted helperphage genome and a pIII supplementing plasmid. These approaches, however, were impeded by packaging of the pIII supplementing plasmid into helper phage particles even though the plasmid lacks signals for

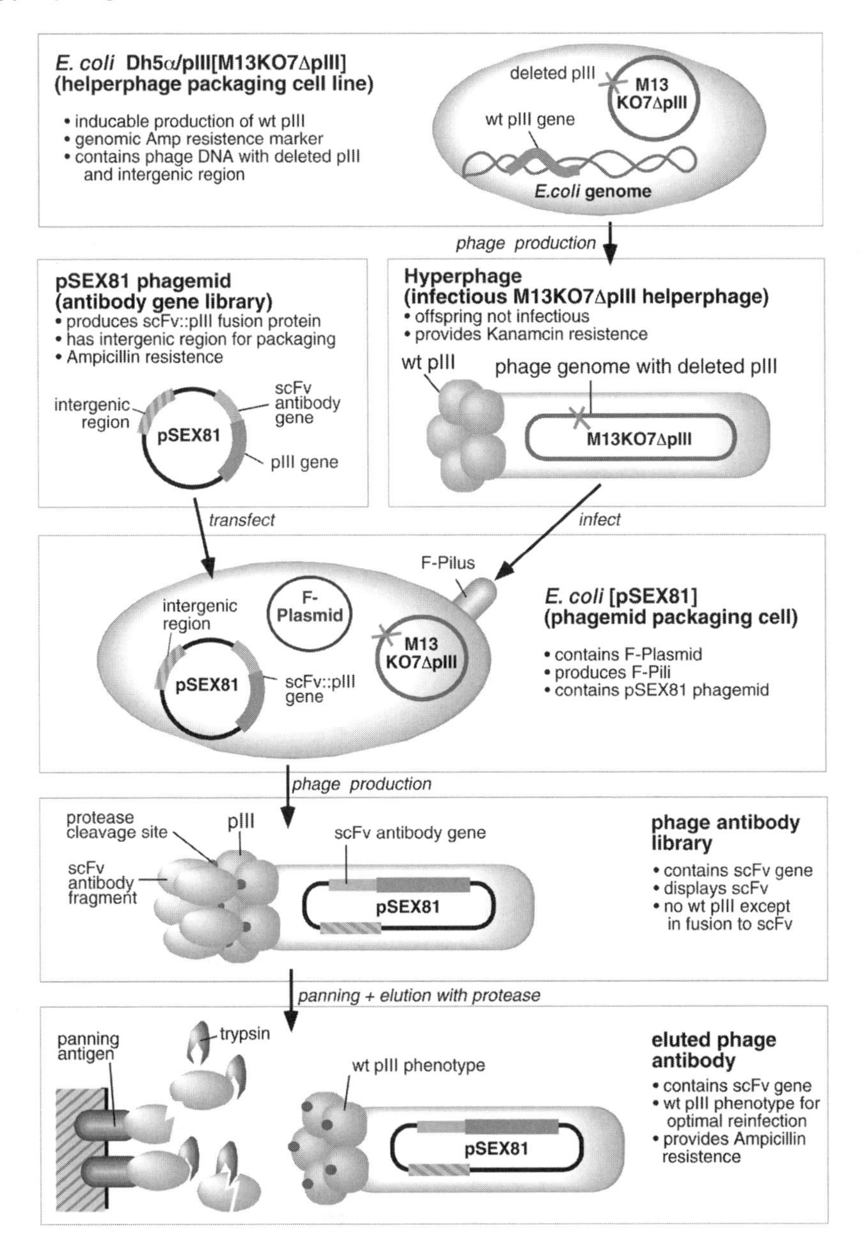

Fig. 1. The concept of hyperphage: a gene pIII-deleted helper phage with wild-type infection phenotype. Note that elution during panning can be done with proteases (e.g., trypsin) when using the pSEX81 phagemid encoding a trypsin cleavage site between the pIII and the antibody fragment (lowest panel). This allows the use of physiological pH throughout the panning procedure to get optimal reinfectivity but ensures elution of very high affinity binders.

the phage assembly machinery. These "wrongly packaged" plasmids then contaminate the phage with significant amounts of wildtype pIII genes and protein, thereby decreasing the quality of the phage-antibodies. To avoid the presence of a plasmid during helper phage production, hyperphage are produced without a supporting plasmid, by employing an *E. coli* packaging cell line with a copy of the gene III integrated into the bacterial genome (*see* **Fig. 1**). A genomically integrated pIII gene is expressed under the control of a strong, but tightly repressable synthetic promotor allowing an inducable expression of the pIII protein during helper phage generation.

The following protocol describes a typical phagemid "packaging" and two convenient titration methods for it's control, Nitrocellulose Plating and Phage ELISA.

2. Materials

2.1. Packaging of a Single-Chain Fv Antibody Fragment Library Employing M13K07ΔpIII Hyperphage

1. 2X YT *(7)* medium (1 L): 16 g peptone, 10 g yeast, 5 g NaCl. Bring to 1 L with didistilled water. Adjust pH to 7.0 if necessary. Autoclave and store at room temperature for up to several weeks.
2. 2 *M* glucose. Filter to sterilize.
3. 100 mg/mL Ampicillin and 70 mg/mL kanamycin. Filter to sterilize.
4. Phage dilution buffer: 10 m*M* Tris-HCl, pH 7.5, 20 m*M* NaCl, 2 m*M* ethyleediamine tetraacetic acid (EDTA).
5. PEG/NaCl: 16.7% w/v PEG-6000, 3.3 m*M* NaCl. Autoclave and store at 4°C.
6. 450-mL Sterile plastic rotor flasks and conventional Erlenmeyer flasks.
7. Hyperphage stock (Progen Biotechnik GmbH, Heidelberg, Germany).
8. F^+ bacteria transfected with an antibody expression phagemid library.

2.2. Titration on Nitrocellulose Filters to Determine the Number of Infective Particles (CFU Calculation)

1. Round nictrocellulose filters: BA 85, 0.45 mm, 82 mm diameter (Schleicher & Schuell, Dassel, Germany).
2. Luria Bertani (LB) agar plates: 20 g peptone, 10 g yeast, 20 g NaCl and 15 g agar for 1 L, adjust pH to 7.0 and autoclave) supplemented with 100 m*M* glucose and 70 µg/mL kanamycin.
3. *E. coli* TG1 bacteria (Stratagene, Amsterdam, Netherlands).
4. Phage dilution buffer: *see* **Subheading 2.1.**

2.3. Phage Enzyme-Linked Immunosorbent Assay (ELISA) for the Estimation of Total Particle Number

1. MaxiSorp ELISA plates (Nunc, Naperville, USA).
2. 100 m*M* $NaHCO_3$, pH 8.6.
3. 1 *M* H_2SO_4.

4. 2% skim milk/PBS/0.05% Tween (SERVA, Heidelberg, Germany). Always prepare freshly prior to use, or store frozen until use. Do not store at 4°C for more than 2 h.
5. Antibody B62-FE2~HRP m*M*ab to filamentous phage major coat protein (pVIII) (Progen Biotechnik GmbH, Heidelberg, Germany).
6. Developer solution for ELISA: Mix 4.5 mL H_2O, 0.5 mL sodium acetate (1 *M*, pH 6.0), 12.5 µL TMB substrate (Promega, Madison, USA), and 6 µL 30% H_2O_2. Prepare freshly before use. Alternatively use TMB substrate from Progen Biotechnik GmbH (Heidelberg).

3. Methods

3.1. Packaging of a Single-Chain Fv Antibody Fragment Library Employing M13K07ΔpIII Hyperphage

1. Grow an overnight culture of bacteria transfected with an antibody expression phagemid library (for details on phagemid vectors, *see* **ref. *3***) in 150 mL 2X YT medium supplemented with 100 µg/µL ampicillin and 100 m*M* glucose (*see* **Note 1**).
2. Supplement 500 mL of fresh 2X YT medium with ampicillin and glucose as before and incoculate with 1/100 volume of the overnight culture. Let the bacteria grow to an OD_{600} of 0.1.
3. Infect the bacteria with hyperphage at a multiplicity of infection (MOI) of 20 and incubate the culture at 37°C for 15–20 min without shaking.
4. Shake for 45 min with 230 rpm/37°C (*see* **Note 2**)
5. Pellet the bacteria in 250-mL centrifuge tubes at 1500–2000*g* for 10 min at 4°C.
6. Resuspend the pelleted bacteria in 500 mL of 2X YT medium supplemented with 100 µg/µL ampicillin and 70 µg/µL kanamycin, but without glucose (*see* **Note 3**).
7. Shake overnight with 230 rpm at 37°C for antibody-phage production.
8. Pellet the bacteria with 6000*g* for 20 min at 4°C and recover the supernatant.
9. Precipitate the produced phage particles with 1/5 vol of PEG/NaCl for >5 h on ice.
10. Pellet the antibody phages by centrifugation with 13,000*g* at 4°C for 1 h. Discard the supernatant. Remove all traces of medium carefully.
11. Resuspend the white phage pellet in 1/100 of the initial culture volume (5 mL) of phage dilution buffer and aliquot into 1.5 mL Eppendorf tubes.
12. Remove bacterial debris by two times centrifugation with 16,000*g* for 5 min at 4°C in a table-top centrifuge.
13. Titrations can be done by two methods (*see* **Note 4**): cfu (colony forming units) calculation (on conventional plates or on nitrocellulose, *see* **Subheading 3.2.** or **ref. *8***), or for estimation of particle numbers by phage ELISA.

3.2. Titration on Nitrocellulos Filters to Determine the Number of Infective Particles (CFU Calculation)

Usually, phage titration is done by infecting *E. coli* plating bacteria with dilution series of phage *(7)*. Approximately 16 h after embedding the infected bacteria in top-agar, plaques can be counted. Since these plaques are transient, a more convenient method is to plate infected bacteria on agar plates and selecting for the antibiotics resistance gene provided by the phage genome. The resulting colonies can be identified and counted easily. To save material, the simplified method presented below uses plating of multiple samples on a nitrocellulose filter *(8)*.

1. Mark 16 fields on round nictrocellulose filters (Schleicher & Schuell) with a ballpen and place onto Luria Broth agar plates containing 70 µg/mL kanamycin.
2. Infect 2 mL *E. coli* TG1 bacteria at an OD_{600} of 0.6 with 100 µL of serial dilutions of phage (depending on expected titer, start with 10^{-2}–10^{-9}, but use at least three different dilutions) in phage dilution buffer for 20 min at 37°C.
3. Pipet 10 µL aliquots of each infection onto the nitrocellulose filters into the middle of each field and incubate overnight at 27°C (*see* **Note 5**)

Count colonies and calculate the titer of the initial phage suspension (*see* **Note 6**).

3.3. Phage ELISA for the Estimation of Total Particle Number

The number of phage particles can be determined by ELISA using an antibody specifically recognizing the pVIII phage outer surface protein *(9)*. This ELISA determines the particle number in comparison to a dilution series of a phage suspension of a known titre (standardization curve). The number of phage particle is usually not identical to the number of colony forming units (cfu) (*see* **Note 4**).

1. Coat MaxiSorp ELISA plates with serial dilutions of a reference phage of known titer and your new phage in parallel (e.g., 10^{-1} – 5×10^{-4} dilutions in 100 m*M* $NaHCO_3$, pH 8.6). Apply 100 µL of each dilution per well and coat 2–3 h at room temperature or overnight at 4°C (*see* **Note 7**).
2. Block with 2% skim milk/phosphate buffered saline (PBS)/0.05% Tween. Apply 200 µL/well and incubate for 1–2 h at room temperature.
3. Wash 5 times with PBS/0.05% Tween. Apply 200 µL/well (*see* **Note 8**).
4. Apply 100 µL of the mouse monoclonal antibody B62-FE2~HRP in 2% milk/PBS/0.05% Tween according to the manufacturer. Incubate for 1 h at room temperature.
5. Wash 5 times with 200 µL PBS/0.05% Tween.
6. Prepare the developer solution consisting of 4.5 mL H_2O, 0.5 mL sodium acetate (1 *M*, pH 6), 12.5 µL 3,3',5,5'-tetramethylbenzidine (TMB) substrate (Promega, Madison, USA) and 6 µL 30% H_2O_2. Apply 100 µL per well (*see* **Note 9**).
7. To stop the color development, add 50 µL of 1 *M* H_2SO_4 to each well and measure the absorption at OD_{450} with an ELISA reader. Calculate the total number of phage particles by comparison with the titration curve of your reference phage of known titre.

4. Notes

1. Glucose at this step of the protocol is required in order to prevent expression of the antibody:pIII fusion protein during the initial amplification of the library. In pSEX-based phagemid libraries, successfully transfected clones can be selected through the phagemid encoded ampicillin resistance.
2. During this step, successfully infected cells will acquire a kanamycin resistance provided by the helperphage genome.
3. Glucose removal leads to scFv expression by activation of the scFv:pIII promotor.
4. The obtained cfu are not necessarily identical to the number of phage particles since typically a fraction of the produced particles is not infective. We have, however, observed that the ratio of cfu/particle is close to constant for a given combination of phage and phagemid, so that the assays can be interchanged in routine applications once this ratio is established.
5. Higher temperatures might be used if the developing colonies are too small at the next morning. Standard (37°C) incubation usually yields colonies which cannot be counted anymore.
6. In case the retrieved colonies are too small to count after overnight incubation, prolong the growth time for 1–2 h at 37°C to increase colony size.
7. For each step of serial dilution use a new pipet tip. Mix well by pipetting up and down several times. Do not use polystyrene vessels for dilution series (e.g., polystyrole ELISA plates).
8. After addition of the washing solution wait for 5 s. Shake well and take care to remove washing solution completely after each washing step.
9. A color reaction should be visible after 5 min. In case no signal appears, check the quality of the detecting antibody and the detection reagents.

References

1. Breitling, F. and Dübel, S. (1999) Recombinant Antibodies. John Wiley and Sons, New York, NY.
2. Rondot, S., Koch, J., Breitling, F., and Dübel, S. (2001) A helper phage to improve single-chain antibody presentation in phage display. *Nature Biotechnol.* **19,** 75–78.
3. Kontermann, R. and Dübel, S. (eds.) (2001) Antibody Engineering. Springer Verlag; Heidelberg, New York, NY.
4. Barbas III, C. F., Kang, A. K., Lerner, R. A., and Benkovic, S. J. (1991) Assembly of combinatorial antibody libraries on phage surfaces: the gene III site. *Proc. Natl. Acad. Sci. USA* **88,** 7978–7982.
5. Breitling, F., Dübel, S., Seehaus, T., Klewinghaus, I., and Little, M. (1991) A surface expression vector antibody screening. *Gene* **104,** 147–153.
6. Hoogenboom, H. R., Griffith, A. D., Johnson, K. S., Chiswell, D. J., Hudson, P., and Winter, G. (1991) Multi-subunit proteins on the surface of filamentous phage: methodologies for displaying antibody (Fab) heavy and light chains. *Nucleic Acids Res.* **19,** 4133–4137.

7. Sambrook, J., Fritsch, E. F., and Maniatis, T. (1989) *Molecular Cloning: A Laboratory Manual.* Cold Spring Harbor Laboratory Press, Cold Spring Harbor, New York, Second Edition.
8. Koch, J., Breitling, F. and Dübel, S. (2000) Rapid Titration of Multiple Samples of Filamentous Bacteriophage (M13) on Nitrocellulose Filters. Benchmarks, *BioTechniques* **29,** 1196–1202.
9. Micheel, B., Heymann, S., Scharte, G., et al. (1994) Production of monoclonal antibodies against epitopes of the main coat protein of filamentous fd phages. *J. Immunol. Methods* **171,** 103–109.

21

Combinatorial Biosynthesis of Novel Carotenoids in *E. coli*

Gerhard Sandmann

1. Introduction

Among secondary metabolites, many interesting compounds including flavors, fragrances, and those with pharmaceutical potential can be found. Once the biosynthetic pathway of an interesting compound or group of compounds has been elucidated and the genes encoding the enzymes of the reaction sequence cloned, they can be used for heterologous production in suitable hosts. Combinations of selected genes from organisms which synthesize different end products of a branched pathway makes it possible to design and produce novel products. The potential of this combinatorial biosynthesis has been demonstrated for the synthesis of novel polyketide antibiotics *(1)* and novel carotenoids *(2)*.

Carotenoids are important as nutriceutical compounds and natural lipophilic antioxidants. In the cell, carotenoids protect against oxidative damage by quenching photosensitizers, interacting with singlet oxygen, and scavenging of peroxy radicals *(3)*. The antioxidative potential of carotenoids depends on their chemical properties, such as the number of conjugated double bonds, structural end groups, and oxygen-containing substituents *(4)*. Evidence is accumulating that carotenoids play an important role in human health by prevention of degenerative diseases. Carotenoids with unsubstituted β-ionone end groups are precursors of vitamin A. Hundreds of carotenoids with diverse chemical structures have been identified in bacteria, fungi, algae, and plants. However, most of them are biosynthetic intermediates which accumulate only

From: *Methods in Molecular Biology, vol. 205, E. coli Gene Expression Protocols*
Edited by: P. E. Vaillancourt © Humana Press Inc., Totowa, NJ

in trace amounts, making it very difficult to extract and purify sufficient material. The commercial demand of carotenoids as food and feed supplements, for pharmaceutical purposes, and as food colorants is mainly met by chemical synthesis and to a minor extent by extraction from natural sources. Therefore, the supply of carotenoids is restricted to a very few derivatives. One possibility to overcome this limitation, is the heterologous expression of carotenoid genes in suitable microorganisms. The non-carotenogenic yeasts *Candida utilis* and *Saccharomyces cerevisiae* **(5)** and especially the bacterium *Escherichia coli* *(2)* have been used for the synthesis of rare derivatives. In this combinatorial biosynthesis approach, a carotenogenic pathway is assembled in a non-carotenogenic host in a modular way by transformation with the appropriate genes which encode the enzymes responsible for the individual catalytic steps. By combining carotenoid genes from different host species which synthesize different carotenoids even novel carotenoids, which have not previously been discovered can be generated.

Carotenoid production is limited by the supply of precursors. Their formation can be increased by metabolic engineering of the early terpenoid pathway *(6)*. Additive effects on stimulation of carotenoid formation were observed by overexpression of the *dxs*, *dxr,* or *idi* genes.

2. Materials

Once the production of a desired carotenoid is anticipated, the several steps have to be followed: selection of the necessary genes which cover the whole pathway, construction of expression plasmids, transformation of a suitable *E. coli* strain with a combination of plasmids, and cultivation under optimized carotenoid production conditions. Finally, carotenoid extraction and analysis by HPLC must be adapted to the nature of the synthesized products.

2.1. Plasmids

E. coli can be transformed with several plasmids as long as they all possess a different origin of replication. Furthermore, it is essential that each plasmid carries a different antibiotic resistance marker, and that selection pressure is maintained at a high level to prevent spontaneous plasmid loss. In **Table 1**, several useful vectors belonging to different incompatibility groups are compiled. They have all been used successfully for expression of carotenogenic genes. They can all be introduced simultaneously in *E. coli* for carotenoid synthesis. However, it is convenient to combine several genes on one plasmid which mediate the formation of certain carotene intermediates, e.g., of a C_{40} carbon skeleton with a certain degree of desaturation, and co-transformation of *E. coli* with additional plasmids carrying genes which are needed for the modification of this structure.

Table 1
Compatible Plasmids for Simultanous Transformation of *E. coli*

Plasmids	Origin of replication	Antibiotic resistance	Reference
pUC (or other pBR322-related plasmids)	pMB1	ampicillin	*(17)*
pACYC184	p15A	chloramphenicol	*(18)*
pRK404	RK2	tetracyclin	*(19)*
pBBR1MCS2	SC101	kanamycin	*(20)*

2.2. Carotenogenic Genes for Establishment of Biosynthetic Pathways

Carotenogenic genes from higher plants, fungi, algae, and bacteria have been used for carotenoid synthesis in *E. coli*. The bacterial genes originated largely from gram-negative bacteria with a GC content of ca. 50%. **Table 2** lists carotenogenic genes which have been successfully utilized for production of various carotenoid structures (part A) or genes which stimulate overall carotenoid synthesis (part B). A negative example for not functional-genes in *E. coli* are the *crtU* genes from *Streptomyces* griseus *(7)* and *Brevibacterium linens (8)*, which both encode a β-carotene desaturase with GC contents of >60%. After transcription, carotenogenic enzymes from algae and higher plants possess an N-terminal extention for transfer into plastids, where carotenoid biosynthesis is located. Upon plastid import, this portion is cleaved. Using genes from algae and plants, deletion of this transit sequence may improve the catalytic activity of the expressed enzyme *(9)*.

2.3. E. coli *Strains*

Several *E. coli* strains have been analyzed for carotenoid synthesis (unpublished). The most productive ones were JM101 and HB101. To some extent also DH5α and NM554 were suitable.

2.4. Carotenoid Extraction Procedures and HPLC Systems for Product Analysis

E. coli cells should be freeze-dried prior to extraction of the highly lipophilic carotenoids. The best solvent to penetrate the unbroken cell powder is methanol at 60°C. It also offers the advantage of simultaneous saponification of acyl lipids, by adding KOH to a final concentration of 6%. However, carotenes with an extended double-bond system like lycopene or 3,4-didehydrolycopene are not

Table 2
Examples of Usefull Carotenogenic Genes for Carotenoid Production in *E. coli* (A) and for Metabolic Engineering of the Pathway (B)

Enzyme/Gene	Substrate	Reaction product	Ref.
A. For Carotenoid Production in *E. coli*			
C_{30} chain			
Diapophytoene synthase/*crtM*	FPP	Diapophytoene	***(21)***
Diapophytoene desaturase/*crtN*	Diapophytoene	Diaponeurosporene	***(21)***
C_{40} chain			
GGPP synthase/*crtE*	FPP	GGPP	***(22)***
Phytoene synthase/*crtB*	GGPP	Phytoene	***(22)***
Desaturases/*pds*	Phytoene	ζ-Carotene	***(23)***
crtIRc	Phytoene	Neurosporene	***(23)***
crtIEu	Phytoene	Lycopene	***(23)***
al-1	Phytoene	3,4-Didehydrolycopene	***(24)***
crtQa	ζ-Carotene	Lycopene	***(25)***
crtQb	ζ-Carotene	Lycopene	***(26)***
crtD	Hydroxyneurosporene	Demethylspheroidene	***(14)***
Lycopene β-cyclase/*crtY*	Lycopene	β-Carotene	***(13)***
Lycopene ε-cyclase/*lcy-ε*	Lycopene	δ-Carotene	***(27)***
Hydroxylase/*crtZ*	β-Carotene	Zeaxanthin	***(15)***
Epoxydase/*zep*	Zeaxanthin	Violaxanthin	unpublished

Ketolases/*crtA*	Spheroidene	Spheroidenone	unpublished
crtW, *bkt*	β-Carotene	Canthaxanthin	***(28)***
crtO	β-Carotene	Echinenone	***(29)***
Hydratase/*crtC*	Neurosporene	1-Hydroxyneurosporene	***(14)***
Glycosilase/*crtX*	Zeaxanthin	Zeaxanthin diglucoside	***(30)***
$C_{45/50}$ chain			
Lycopene elongase/*crtEb*	Lycopene	Nonaflavuxanthin	***(11)***
B. For Metablic Engineering of Pathway			
1-Deoxyxylulose-5-P Synthase/*dxs*	Glyceraldehyde/ pyruvate	1-Deoxyxylulose-5-P	***(6)***
1-Deoxyxylulose-5-P reductoisomerase/*dxr*	1-Deoxyxylulose-5-P	2-C-methyl-D-erythritol-4-P	***(6)***
Isopentenyl pyrophosphate Isomerase/*idi*	Isopentenyl pyrophosphate	Dimethylallyl pyrophosphate	***(6)***

Abbreviations: FPP, farnesyl pyrophosphate; GGPP, geranylgeranyl pyrophosphate.

quantitatively recovered with methanol. In these cases, acetone at 50°C is a much better extraction solvent. Before HPLC analysis, a simple partition into 10% ether in petrol removes many other unwanted metabolites from the carotenoids. The latter are concentrated in the upper phase. When carotenoid glycosides are present, the percentage of ether should be increased to at least 50%.

HPLC separation of carotenoids on a Nucleosil C_{18}, 3-μm column with acetonitrile/methanol/ 2-propanol (85:10:5, v/v) is very convenient. This simple and fast routine system works very well isocratically, but it has some limitations which should be considered (*see* **Notes 1–3**).

3. Methods

The following example is taken from a case study involving the production of 1-hydroxy acyclic carotenoids. Unique 1-hydroxy acyclic carotenoids which are powerful antioxidants have been produced in *E. coli* by combining carotenogenic genes from various bacteria *(12)*. The individual experimental steps resulting in the formation of 1-HO-3,4-didehydrolycopene, 1,1'-$(HO)_2$-3,4,3',4'-tetradehydrolycopene and demethylspheroidene will be explained. These products and their biosynthetic pathway are shown in **Fig. 1**.

3.1. Gene and Plasmid Combinations

The genes necessary for the formation of the carotenoids mentioned above are indicated by their gene products in **Figure 1**. Typically one plasmid with several genes is used for the synthesis of the carotene precursor. In our case pACCRT-EBI_{Eu} *(13)* with the *crtE, crtB,* and *crtI* genes from the carotenogenic gene cluster of *Erwinia uredovora* mediates the formation of lycopene. The remaining *crtC* and *crtD* genes were on individual compatible plasmids, pRKCRT-C and pQECRT-D. Alternatively, the *idi* gene was added on a fourth plasmid *(12)*. The overexpression of this isopentenyl pyrophosphate gene increased the carotenoid yield due to a better precursor supply.

E. coli JM101 was transformed according to standard procedures. Plasmids were brought in by transformation of competent JM101 cells with two plasmids simultaneously as the first step. Additional plasmids were introduced by making the resulting transformant competent and transformation with a single plasmid. The latter steps were repeated in the case when a fourth plasmid was introduced.

3.2. Growth Conditions

The growth medium mainly determines the cell density rather than the carotenoid contents per cell. Media of choice are Luria-Bertani (LB) or those with up to 2.5% casein hydrolysate *(2)*. Carotenoid production is best at the end of the log phase *(14)*. Furthermore, carotenoid yields are generally higher at sub-

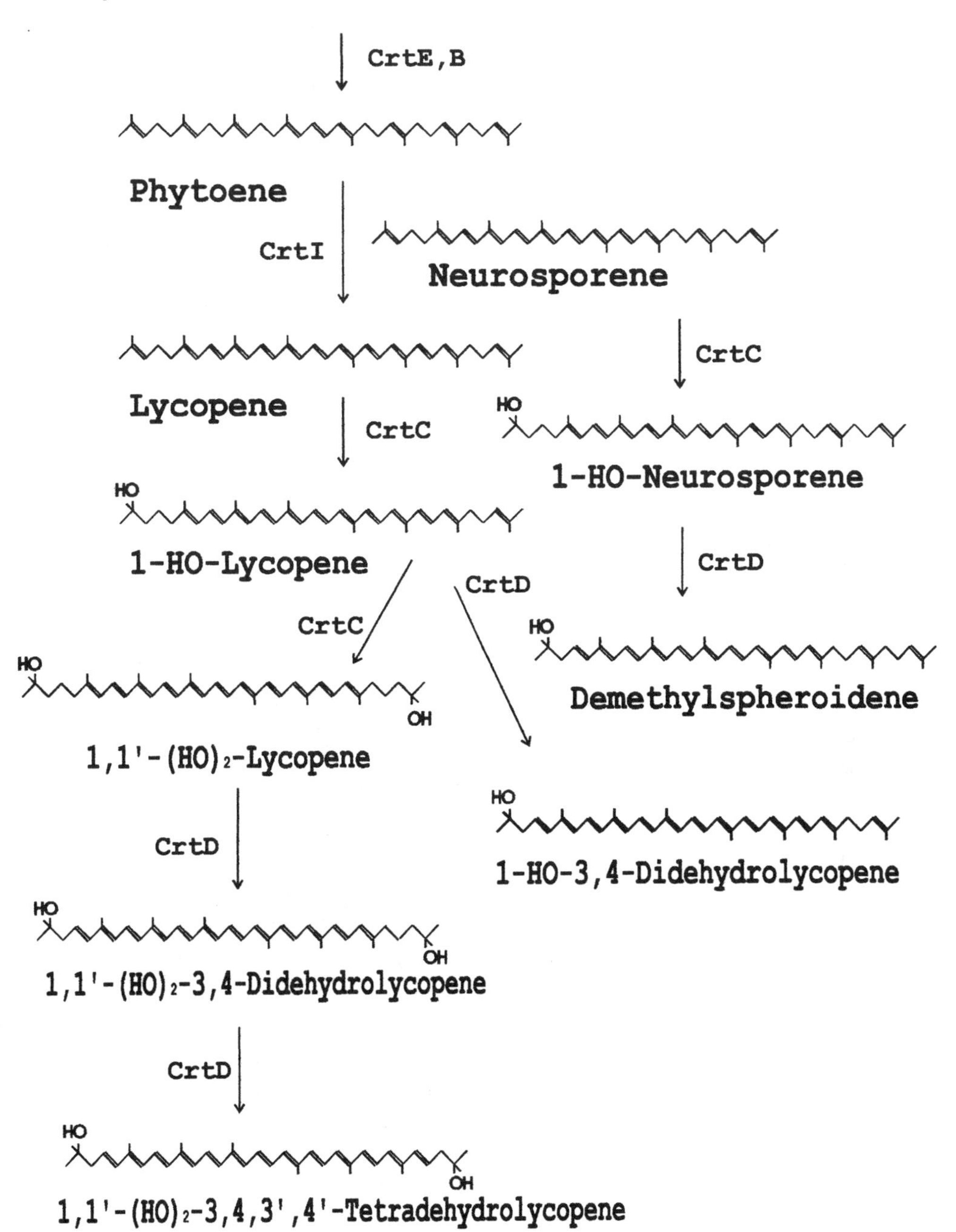

Fig. 1. Carotenogenic pathway to 1-HO-3,4-didehydrolycopene, 1,1'-$(HO)_2$-3,4,3',4'-tetradehydrolycopene and demethylspheroidene. The gene products which catalyze the individual reactions are indicated at the arrows.

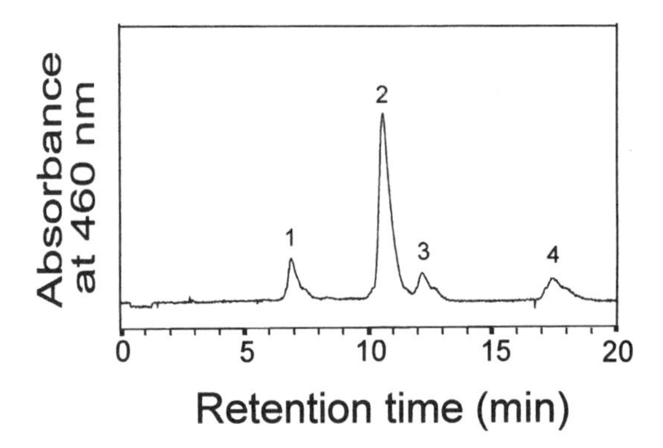

Fig. 2. HPLC separation of carotenoids from the *E. coli* transformants JM101/pACCRT-EBI$_{EU}$/pRKCRT-C/pQECRT-D. The separated carotenoids were identified as 1,1'-$(HO)_2$-3,4,3',4'-tetradehydrolycopene (peak 1), 1-HO-3,4-didehydrolycopene (peak 2), demethylspheroidene (peak 3) and lycopene (peak 4).

optimal growth temperatures around 28°C and after a 48 h growth period ***(15)***. Selection pressure by combination of the appropriate antibiotics has to be maintained throughout growth. Otherwise the plasmids will be lost rapidly.

3.3. Extraction, Analysis, and Yields

Carotenoids from lyophilized cells of *E. coli* were extracted with methanol containing 6% KOH by heating for 20 min to 60°C and partitioned into diethylether/petrol (bp 35–60°C) (1:9, v/v). The upper phase was collected and the solvent evaporated under a stream of nitrogen. After resuspension into acetone, carotenoids were separated by HPLC on a 25-cm Nucleosil C_{18}, 3-µm column with acetonitrile/methanol/ water (48:50:2, v/v) at a flow rate of 1 mL/min. The HPLC separation is documented in **Fig. 2**. Two major peaks, 1 and 2, and two minor ones were obtained. Spectra were recorded on-line from the elution peaks by a photodiode array detector. The following major absorbance maxima were obtained and used for carotenoids identification: 466, 492, and 525 nm for $(HO)_2$-3,4,3',4'-tetradehydrolycopene with 13 conjugated double bonds (peak 1); 456, 481, and 513 nm for HO-3,4-didehydrolycopene with 11 (peak 2); 427, 452, and 482 nm for demethylspheroidene with 10 (peak 3); and 442, 468, and 499 nm for lycopene with 11 conjugated double bonds (peak 4). This assignment of the carotenoid products was confirmed by mass spectroscopy ***(12)***. The molecular masses were determined after collection of the individual carotenoid fractions.

Carotenoids can be quantitated by integration of the HPLC peaks and calibration with defined amounts of an authentic standard. In the case of new caro-

Table 3
Production of Carotenoid (μg/g dw) in Transgenic *E. coli*

Plasmid combinations	Carotenoids			
	1,1'-$(HO)_2$-TDHL	1-HO-3,4-DDL	DMS	Lyc
pACCRT-EBI_{EU} + pRKCRT-C	141	99	23	23
+ pQECRT-D + pBBRK-idi	153	150	46	59

Lyc, lycopene; DDL, 3,4-didehydrolycopene; Car, carotene; TDHL, 3,4,3',4'-tetradehydrolycopene; DMS, demethylspheroidene

tenoids another carotenoid with similar absorbance maxima (= same conjugated double bond system) can be used instead. Alternatively, an estimate of the carotenoids can be obtained by spectrometric determination of total carotenoids in the extract at the wavelength of the major carotenoid using the extinction coefficient of a carotenoid with similar absorbance maxima ***(16)***, like 1-(HO)-3,4-didehydrolycopene and 3,4-didehydrolycopene or 1,1'-$(HO)_2$-3,4,3',4'-tetradehydrolycopene and anhydrorhodovibrin. The amounts of the individual carotenoids can then be calculated using the percentage distribution from the integrated HPLC peaks.

The concentrations of the carotenoids formed in *E. coli*/pACCRT-EBI_{Eu}/pRKCRT-C/pQECRT-D are given in **Table 3**. The simultaneous expression of *idi* resulted in a higher yield of all carotenoids.

4. Notes

1. This solvent is unable to resolve hydroxy and keto derivatives. If this is anticipated, a solvent of acetonitrile/methanol/water (48:50:2, v/v) should be used instead ***(10)***.
2. Hydroxy carotenoids carrying one ε-end group instead of a β-end group (i.e., lutein vs zeaxanthin) are not separated from each other.
3. Polar carotenoids (i.e., acyclic hydroxy derivatives) differing only by one double-bond may be poorly resolved. In such cases, HPLC on a C_{30} stationary phase with methanol/methyl-*tert*-butyl ether mixtures containing a fixed water content (typically 4%) as mobile phase is the matter of choice (*see* **ref. *11*** for examples). However, the separation works only with a solvent gradient and takes much longer than the other C_{18} systems. Furthermore, the resolution of many cis/trans isomers may be confusing. Fast information for a preliminary assignment of the produced carotenoids can be obtained from the retention behavior on HPLC and the absorbance spectra recorded on-line from the elution peaks by a photodiode array detector. However, their final identification should be confirmed by mass spectroscopy or if possible by NMR spectroscopy.

References

1. McDaniel, R., Ebert-Khosla, S., Hopwood, D. A., and Khosla, C. (1993) Engineered biosynthesis of novel polyketides. *Science* **262,** 1546–1550.
2. Sandmann, G., Albrecht, M., Schnurr, G., Knörzer, O., and Böger, P. (1999) The biotechnological potential and design of novel carotenoids by gene combination in *Escherichia coli. Trends Biotechnol.* **17,** 233–237.
3. Edge, R., McGarvey, D. J., and Truscott, T. G. (1997) The carotenoids as antioxidants—a review. *J. Photochem. Photobiol. B* **42,** 189–200.
4. Britton, G. (1995) Structure and properties of carotenoids in relation to function. *FASEB J.* **9,** 1551–1558.
5. Misawa, N. and Shimada, H. (1998) Metabolic engineering for the production of carotenoids in non-carotenogenic bacteria and yeasts. *J. Biotechnol.* **59,** 169–181.
6. Albrecht, M., Misawa, N., and Sandmann, G. (1999) Metabolic engineering of the terpenoid biosynthetic pathway of *Escherichia coli* for production of the carotenoids β-carotene and zeaxanthin. *Biotechnol. Lett.* **21,** 791–795.
7. Krügel, H., Krubasik, P., Weber, K., Saluz, H. P., and Sandmann, G. (1999) Functional analysis of genes from *Streptomyces griseus* involved in the synthesis of isorenieratene, a carotenoid with aromatic end groups, revealed a novel type of carotenoid desaturase. *Biochim. Biophys. Acta* **1439,** 57–64.
8. Krubasik, P. and Sandmann, G. (2000) A carotenogenic gene cluster from *Brevibacterium linens* with novel lycopene cyclase genes involved in the synthesis of aromatic carotenoids. *Molec. Gen. Genet.* **263,** 423–432.
9. Misawa, N., Truesdale, M. R., Sandmann, G., et al. (1994) Expression of a tomato cDNA coding for phytoene synthase in *Escherichia coli*, phytoene formation *in vivo* and in *vitro*, and functional analysis of the various truncated gene products. *J. Biochem.* **116,** 980–985.
10. Albrecht, M., Steiger, S., and Sandmann, G. (2001) Expression of a ketolase gene mediates the synthesis of canthaxanthin in *Synechococcus* leading to resistance against pigment photodegradation and UV-B sensitivity of photosynthesis. *Photochem. Photobiol.* **73,** 551–555.
11. Breitenbach, J., Braun, G., Steiger, S., and Sandmann, G. (2001) Chromatographic performance on a C_{30}-bonded stationary phase of monohydroxycarotenoids with variable chain length or degree of desaturation and of lycopene isomers synthesized by different carotene desaturases. *J. Chromatogr. A* **936,** 59–69.
12. Albrecht, M., Takaichi, S., Steiger, S., Wang, Z.-Y., and Sandmann, G. (2000) Novel hydroxycarotenoids with improved antioxidative properties produced by gene combination in *Escherichia coli. Nature Biotechnol.* **18,** 843–846.
13. Cunningham, F. X., Chamovitz, D., Misawa, N., Gantt, E., and Hirschberg, J. (1993) Cloning and functional expression in *Escherichia coli* of a cyanobacterial gene for lycopene cyclase, the gene that catalyzes the biosynthesis of β-carotene. *FEBS Lett.* **328,** 130–138. **6,** 130–138.
14. Albrecht, M., Takaichi, S., Misawa, N., Schnurr, G., Böger, P., and Sandmann, G. (1997) Synthesis of atypical cyclic and acyclic hydroxy carotenoids in *Escherichia coli* transformants. *J. Biotechnol.* **58,** 177–185.

15. Ruther, A., Misawa, N., Böger, P., and Sandmann, G. (1997) Production of zeaxanthin in *Escherichia coli* transformed with different carotenogenic plasmids. *Appl. Microbiol. Biotechnol.* **48,** 162–167.
16. Britton, G. (1995) UV/visible spectroscopy, in *Carotenoids, Volume 1B: Spectroscopy* (Britton, G., Liaaen-Jensen, S., Pfander, H., eds.) Birkhäuser Verlag, Basel, Switzerland, pp. 13–62.
17. Vieira, J. and Messing, J. (1982) The pUC plasmids, an M13mp7-derived system for insertion mutagenesis and sequencing with synthetic universal primers. *Gene* **19,** 259–268.
18. Rose, R. E. (1988) The nucleotide sequence of pACYC184. *Nucleic. Acids Res.* **16,** 355.
19. Ditta, G., Schmidhauser, T., Yakobson, E., et al. (1985) Plasmids related to the broad range vector, pRK290, useful for gene cloning and for monitoring gene expression. *Plasmid* **13,** 149–153.
20. Kovach, M. E., Elzer, P. H., Hill, D. S., et al. (1994) Four new derivatives of the broad-host range cloning vector pBBR1MCS, carrying different antibiotic resistance cassettes. *Gene* **166,** 175–176.
21. Wieland, B., Feil, C., Gloria-Maercker, E., Thumm, G., Lechner, M., Bravo, J., Porolla, K., and Götz, F. (1994) Genetic and biochemical analysis of the biosynthesis of the yellow carotenoid 4,4' diaponeurosporene of *Staphylococcus aureus. J. Bacteriol.* **176,** 7719–7726.
22. Sandmann, G. and Misawa, N. (1992) New functional assignment of the carotenogenic genes *crtB* and *crtE* with constructs of these genes from *Erwinia* species. *FEMS Microbiol. Lett.* **90,** 253–258.
23. Linden, H., Misawa, N., Chamovitz, D., Pecker, I., Hirschberg, J., and Sandmann, G. (1991) Functional complementation in *Escherichia coli* of different phytoene desaturase genes and analysis of accumulated carotenes. *Z. Naturforsch.* **46c,** 160–166.
24. Hausmann, A. and Sandmann, G. (2000) A single 5-step desaturase is involved in the carotenoid biosynthesis pathway to β-carotene and torulene in *Neurospora crassa. Fung. Gen. Biol.* **30,** 147–153.
25. Linden, H., Vioque, A., and Sandmann, G. (1993) Isolation of a carotenoid biosynthesis gene coding for a ζ-carotene desaturase from *Anabaena* PCC7120 by heterologous complementation. *FEMS Microbiol. Lett.* **106,** 99–104.
26. Albrecht, M., Klein, A., Hugueney, P., Sandmann, G., and Kuntz, M. (1995) Molecular cloning and functional expression in *E. coli* of a novel plant enzyme mediating ζ-carotene desaturation. *FEBS Lett.* **372,** 199–202.
27. Cunningham, F. X., Jr, Pogson, B., Sun, Z., McDonald, K. A., DellaPenna, D., and Gantt, E. (1996) Functional analysis of the β and ε lycopene cyclase enzymes of *Arabidopsis* reveals a mechanism for control of cyclic carotenoid formation. *Plant Cell* **8,** 1613–1626.
28. Misawa, N., Satomi, Y., Kondo, K., et al. (1995) Structure and functional analysis of a marine bacterial carotenoid biosynthesis gene cluster and astaxanthin biosynthetic pathway proposed at the gene level. *J. Bacteriol.* **177,** 6575–6584.

29. Fernandez-Gonzalez, B., Sandmann, G., and Vioque, A. (1997) A new type of asymmetrically acting β-carotene ketolase is required for the synthesis of echinenone in the cyanobacterium *Synechocystis* sp. PCC 6803. *J. Biol. Chem.* **272,** 9728–9733.
30. Sandmann, G., Woods, W. S., and Tuveson, R. W. (1990) Identification of carotenoids in *Erwinia herbicola* and in a transformed *Escherichia coli* strain. *FEMS Microbiol. Lett.* **71,** 77–82.

22

Using Transcriptional-Based Systems for In Vivo Enzyme Screening

Steven M. Firestine, Frank Salinas, and Stephen J. Benkovic

1. Introduction

The advent of combinatorial approaches to problems at the interface between chemistry and biology has had a profound impact on areas ranging from drug discovery to protein chemistry. One area of intense work has been the application of combinatorial libraries of proteins to the discovery of proteins with novel functions or properties *(1)*. Numerous methods exist for creating these libraries, examples being DNA shuffling and incremental truncation *(2,3)*. However, library creation is only one phase of a combinatorial solution to protein discovery. The second equally important phase is the screening of the library for proteins displaying the desired property or function. This chapter focuses on a method for examining libraries for enzymatic function.

The literature on screening for enzymatic function is as old as the field of enzymology. However, many of the methods employed to analyze for the presence of an enzyme are not applicable for the screening of thousands to millions of samples. The interrogation of large libraries requires high-throughput methods and efforts in this area have increased in recent years *(4)*. The interested reader is encouraged to read two excellent reviews by Fastrez and Georgiou on enzyme screening methods *(4,5)*.

This chapter will focus on an in vivo screening method called QUEST (QUerying for EnzymeS using the Three-hybrid system) *(6)*. QUEST functions by coupling substrate turnover to a transcriptional event (*see* **Fig. 1**), a common theme in metabolic regulation where the expression of genes is controlled by a small molecule, which in turn is regulated by enzymatic turnover. The single most important criteria to establishing a QUEST system is a synthetic transcriptional protein that is responsive to two molecules (*see* **Note 1**).

From: *Methods in Molecular Biology, vol. 205, E. coli Gene Expression Protocols*
Edited by: P. E. Vaillancourt © Humana Press Inc., Totowa, NJ

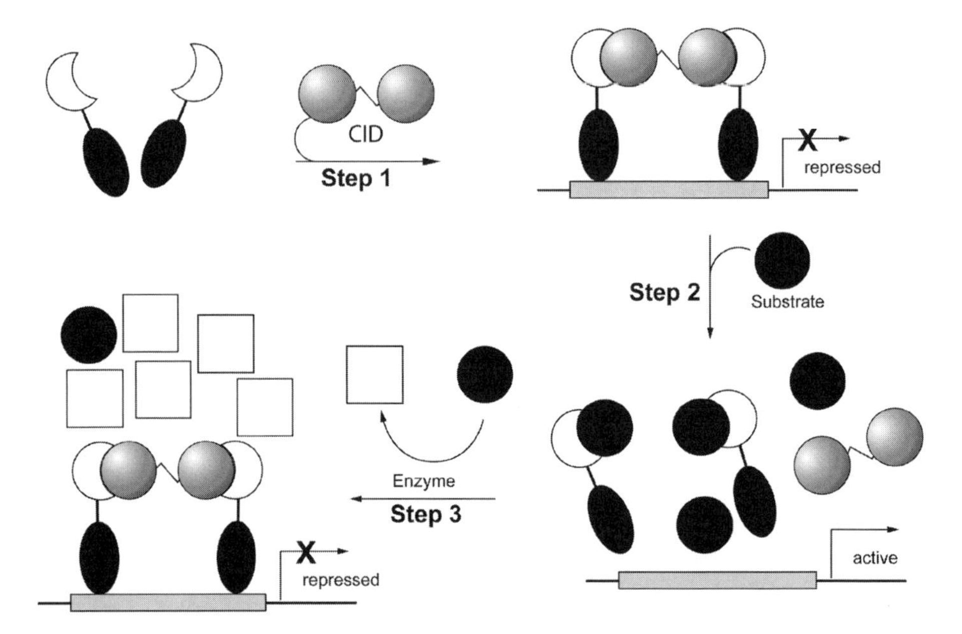

Fig. 1. Overview of the QUEST system. A synthetic transcriptional protein, constructed from a DNA-binding domain (black oval) and a ligand-binding domain (crescent-shaped), is produced inside of a bacterial cell. When a chemical inducer of dimerization is added (the two-domain, dumbbell shaped molecule), the two transcriptional proteins dimerize resulting, in this case, in repression (activation is also possible). Addition of substrate (black circles) results in competition and dissociation of the transcriptional protein from the DNA. If an enzyme is present, conversion of the substrate into product (open squares) results in a decrease in the in vivo concentration of the substrate and the formation of a functional repressor. Each of the steps results in a distinct phenotype depending upon the system and assay utilized (repressor or activator, *see* **Table 1**).

The first is the substrate of the reaction and the second is a chemical inducer of dimerization (CID) (*see* **Note 2**). The CID serves as a molecular switch to facilitate a protein-protein interaction between two transcriptional proteins, and thereby controls the expression of a gene or operon. This system, called a three-hybrid system is related to the two-hybrid system but formation of a functional dimer is mediated by a third molecule rather than an intrinsic protein-protein interaction *(7)*. When cells containing the transcriptional protein are treated with the CID, dimerization of the transcriptional protein occurs resulting in a functional transcriptional protein (*see* **Fig. 1, step 1**). The function of the transcriptional protein could be activator (AraC-based system) or repressor (lambda-based system) depending upon the DNA-binding domains used in the

Table 1
Phenotype Displayed by the Cell at Each Step of QUEST as Measured by the Assays Described in the Chapter

Step	MacConkey Assay	LysoSensor Green Assay	Fluorescein Assay	Phage Assay
Step 1	Red	High fluorescence	Low fluorescence	Low titer
Step 2	Colorless	Low fluorescence	High fluorescence	High titer
Step 3	Red	High fluorescence	Low fluorescence	Low titer

construction of the transcriptional protein (*see* discussion next paragraph). Regardless of the type of transcriptional protein used, dimerization results in a certain phenotype (*see* **Table 1**). Addition of substrate to the cells results in competition for binding sites on the transcriptional protein, inducing dissociation of the dimeric protein and a reversal of the phenotype in the cell. However, if the cell contains an enzyme which converts substrate to product, the substrate will be removed, shifting the equilibrium towards CID binding. Dimerization of the transcriptional protein once again changes the phenotype displayed by the cell. Thus, QUEST functions by connecting the phenotype displayed by the cell to substrate turnover inside of the cell (*see* **Note 3**).

At the heart of the QUEST system is the utilization of the three-hybrid system to establish a transcriptional switch responsive to substrate levels. The three-hybrid system has been previously described for yeast and mammalian cells *(7,8)*. However, since QUEST was designed to analyze large protein libraries, a three-hybrid system functional in *E. coli* was needed, in order to take advantage of the high transformation efficiency displayed by bacteria. Since the three-hybrid system is related to two-hybrid systems, QUEST relies upon previously described bacterial two-hybrid systems, several of which are described in this volume (*see* Chapters 16 and 17). This chapter will detail the utilization of two systems for construction of a QUEST screening system, one being the AraC-based system and the other being the lambda-based system *(9,10)*. **Table 2** outlines some of the properties of these systems. The two systems are complementary in that they represent the two modes of transcription in bacteria (activation and repression) and cloning of the ligand-binding domain onto the DNA-binding domain can be done at either the N- or C-terminus. Both methods offer the possibility of screening for enzyme function. For the AraC-based system, activation is detected by inducing expression of the arabinose operon, which in turn, converts arabinose into intermediates in glycolysis. This results in a decrease in the pH of the media, detected by pH indicators such as MacConkey media or pH-sensitive fluorescent dyes. The lambda-based system has the advantage of having the ability to either select or screen for the function of interest. This is accomplished by the fact that

Table 2
Properties of the Bacterial Hybrid Systems Used to Construct a QUEST System

Properties	AraC-Based	Lambda-Based
Transcriptional system	Activation system	Repression system
Cloning onto DNA-binding protein	N-terminus	C-terminus
Selection	Possible, but requires a reporter vector	Yes. Phage selection
Screening	Yes, pH dependent assays	Yes, phage titer assays

the lambda system relies upon protection from phage to detect enzymatic activity. Thus, phage infection can be used as a selection method, or phage titers can be used to screen for enzymatic function.

2. Materials

2.1. Bacterial Strains

1. For common manipulation and amplification of vectors, any common strain of bacteria can be used; however, we routinely used DH5α.
2. For the AraC-based system, we utilized strain MC1163 (F-*hsdR*-X1488 *Δ(araCO)1109 Δ(lacIPOZY)74 galE15 galK16 mcrA mcrB1 relA1 spoT1 rpsL150(strAR)*, (M. Casadaban, University of Chicago) for our studies. The strain can be stored in the same manner as common laboratory strains. The strain can be made competent either by a $CaCl_2$ method or by treatment for use in electroporation.
3. For the lambda-based system, we utilized the same strain as outlined in Chapter 16. The strain is AG1688 (MC1061, F128, *lacI^q^ lacZ*::Tn5, J. Hu, Texas A&M University). This strain was stored in the same manner as common laboratory strains. The strain is made competent by the $CaCl_2$ method.

2.2. Phage

For the lambda-based system, the phage used is the same as outlined in Mariño-Ramirez et al. (Chapter 16 of this volume). The phage, KH54 (cI-, J. Hu, Texas A&M University) is prepared according to the procedures outlined in Chapter 17, and the stock titer is determined using standard methods ***(11)***.

2.3. Vectors

1. pGB017 (R. Schlief, Johns Hopkins University) is a kanamycin-resistant plasmid that contains the C-terminal DNA binding domain of the AraC protein. Restriction sites, *Nco*I and *Bam*HI, allow for cloning of the desired ligand-binding proteins. The resulting chimeric gene creates a new transcriptional regulatory protein that can control any AraC-based system.

2. Plasmid JH391 (J. Hu, Texas A&M University) is an ampillicin-resistant vector that contains the DNA-binding domain of the lambda repressor. For details on cloning into this vector, *see* Chapter 16 of this book.
3. Compatible vectors for the expression of the enzyme of interest (*see* **Note 4**).

2.4. Chemicals used in QUEST

1. Chemical inducer of dimerization (CID) (*see* **Note 5**).
2. Substrate of enzymatic reaction.
3. 20% L-arabinose, filter sterilized and prepared freshly before each use (Sigma, St. Louis, MO, cat. no. A3256).
4. 20% Maltose, filter sterilized and prepared freshly before each use (Sigma, cat. no. M9171).
5. Antibiotics, according to vectors required, prepared according to standard conditions.
6. LysoSensor Green DND-189, diluted to 10 μ*M* in DMSO from the standard stock solution purchased from the company (Molecular Probes, Eugene OR, cat. no. L-7535). Stock solution should be stored at –20°C.
7. Fluorescein prepared as a 100 μ*M* stock solution in DMSO (Sigma, cat. no. F7505). The stock solution should be stored at 4°C.

2.5. Media

1. Luria-Bertani (LB) medium is used to grow all bacteria: 10 g Tryptone, 5 g Yeast Extract, 5 g NaCl, sterilize by autoclaving. For pH-dependent measurements of bacteria in LB, the pH of the medium was adjusted to 7.0 (typically with 1.0 *M* NaOH) and the media was then filter sterilized before use.
2. MacConkey base agar (Difco, Detroit, MI, cat. no. 0818-17-3). Other MacConkey agar formulations that contain sugars will not work, as they will give false positives.
3. Medium for phage infection: LB media supplemented with 1.0 m*M* $MgSO_4$ and 1.0% maltose.

3. Methods

3.1. Cloning of the Substrate and Ligand-Binding Domain to the DNA-Binding Domain

This is the single most critical aspect of the project. Before the QUEST system can be established, a protein capable of binding a CID and the substrate or product of the reaction must be available (*see* **Note 2**).

3.2. Testing the Effect of the CID on Transcription

This section outlines the primary detection methods for the two systems outlined in this chapter. The section below assumes that the cloning of the appropriate ligand and substrate-binding domain have been accomplished (*see* **Subheading 3.1.**) and that the CID has been prepared.

3.2.1. MacConkey Agar

1. An individual colony of MC1163 containing the desired ligand-binding domain fused to the AraC DNA-binding domain is grown overnight in LB with kanamycin (50 μg/mL, this concentration is used throughout this chapter) at 37°C. A second colony expressing only the DNA-binding domain should also be grown for use as a negative control.
2. CID-containing plates (*see* **Note 6**) are prepared by diluting various concentrations of the CID into MacConkey media supplemented with kanamycin (kan) and 0.5% arabinose (*see* **Note 7**). A control containing no CID should also be used (*see* **Note 8**).
3. The overnight culture is diluted such that no more than 500 colonies/plate are present (*see* **Note 9**). The plates are then incubated until the colonies develop a red/purple color (typically 12 h, *see* **Notes 10–12**). The colonies should turn this color before the negative control (which normally turns the plate brown colored) and the plate lacking any CID. A functional CID should show a concentration dependence on the number of colored colonies present on the plate (*see* **Note 13**).

3.2.2. Fluorescent Assays

The fluorescent assays allow quantitation of the activation of the pathway. The method also removes some of the subjective assessment of the MacConkey agar. The assay is based upon the same principle of the MacConkey agar assay, namely the measurement of the decrease in the pH of the media. The assay can be done on Petri dishes; however, it is far more efficient to utilize microtiter plates and scan the plates using a fluorescent plate reader.

1. Using a sterilized toothpick, inoculate into each well of a microtiter plate containing LB/kan individual colonies of MC1163 containing the ligand-binding domain fused to the DNA-binding domain (*see* **Note 14**). Seal the plate and grow overnight at 37°C with shaking. The resulting plate will be the master plate.
2. To test the effects of the CID, prepare a clear microtiter plate using 100 μL total volume of pH-adjusted LB/kan containing 0.1–0.5% arabinose, 1.0 m*M* $MnCl_2$, and various concentrations of the CID (*see* **Notes 15–17**). To each well, add a 1% inoculum from the master plate. Each well from the master plate should be duplicated. The plates are sealed and incubated without shaking at the appropriate temperatures for the time necessary to get a good density of cells (*see* **Notes 18, 19**).
3. Scan the plate using a Vis-plate reader at OD_{600} to calculate the cellular density in each well. This value is needed to correct for differences in cellular growth in each well (*see* **Note 20**).
4. To each well, a pH-sensitive fluorescent dye is added to its final concentration and the well is mixed using a pipet (*see* **Note 21**). Each well is then transferred to a black microtiter plate for reading by a plate reader (*see* **Note 22**). The data generated is then corrected for differences in the amount of bacteria (as determined in **step 3**) in each well (*see* **Note 23**).

3.2.3. Phage Plaque Assays (see **Note 24**)

1. Transform the vector containing the ligand-binding domain and the lambda DNA-binding domain into strain AG1688. Plate the resulting transformation onto LB/amp (100 μg/mL) plates.
2. Inoculate an individual colony into LB/amp and grow overnight at 37°C.
3. To the phage infection media supplemented with ampicillin and various concentrations of the CID (including no CID) add a 1% inoculum (*see* **Note 25**). Incubate with shaking at 37°C for 2 h.
4. Add phage KH54 to the culture such that the multiplicity of infection is less than 1 (*see* **Note 26–27**). Incubate at 37°C for 1 h.
5. Centrifuge (15,000*g*) the culture to pellet the bacteria. Remove the supernatant and centrifuge again. Transfer the supernatant from the second spin to a fresh culture tube.
6. Titer the phage produced from **step 4**. Dilute the supernatant from **step 5** into 0.1 mL of a fresh culture of AG1688 and incubate at 37°C for 20 min. The resulting culture is added to 10 mL of top agar and then immediately poured onto a LB agar plate. After the top agar has cooled, the plates are incubated until clear plaques are seen (typically 10–12 h). The amount of plaques are counted and the titer of the phage stock from **step 4** is determined. The percent of infection is calculated by taking the phage titer for the strain with the various concentrations of CID and dividing by the phage titer for the strain alone (no CID).

3.3. Testing the Effect of the Substrate and CID on Transcription

Any of the above methods can be utilized to detect the effect of the substrate and CID on transcription. Assays are run identical to those described in **Subheading 3.2.**, with the addition of various concentrations of substrate to the media (*see* **Note 28**). Increasing concentrations of substrate should result in a decrease in CID-mediated transcription.

3.4. Testing the Effect of the Substrate, CID, and Enzyme on Transcription (with Library Vector)

Any of the assays described in **Subheading 3.2.** can be used to test the effect of substrate, CID and enzyme on transcription. The key difference between this section and **Subheading 3.3.** is the presence of a second vector encoding for the enzyme or library. The cloning of the enzyme or library into a vector of choice has already been discussed in **Note 4**. Once the enzyme vector and the transcriptional vector have been transformed into the appropriate strain, all media must contain the selection antibiotic necessary to maintain the enzyme vector.

3.5. Screening the Combinatorial Library

Analysis of a combinatorial library of potential enzymes can be conducted in a manner similar to **Subheading 3.4.** with any of the assays described in

Subheading 3.2. All members of the library should be verified by duplication. For the AraC-based system, false positives occur at a reasonable frequency. False positives can be eliminated by repeating the assay on all of the potential positives identified in the first screen.

3.5.1. Selection of a Combinatorial Library Using Phage

For the lambda-based system, potential functional enzymes from the library can be selected using the phage as the selection pressure.

1. Transform the library into AG1688 along with the synthetic transcriptional protein. Determine the transformation efficiency and library size by plating dilutions of the transformed culture onto LB/ampicillin plates supplemented with the required antibiotic for the library vector. Plate the remaining library culture onto a large, square petri dish (245 × 20 mm, Nunc) and grow overnight.
2. Remove the colonies from the large petri dish by adding 5 mL of LB media to the plate, scrapping the plate with a bent Pasture pipet and removing the resulting suspension using a sterile pipet. The process is repeated and the combined bacterial suspension is centrifuged. The cells are then suspended in 2.0 mL of LB to which 60% glycerol is added and the culture is stored at –70°C.
3. To subject the library to selection, a small amount of the library from **step 2** is added to LB media containing the required antibiotics, the CID and substrate. The culture is grown at 37°C until log phase is reached, at which time phage KH54 is added to a MOI < 1.0.
4. The culture is allowed to incubate at 37°C for 1 h, then centrifuged. The resulting cells are suspended in LB and then centrifuged. This process is repeated three times in order to remove any remaining phage.
5. The cells remaining are grown overnight in LB supplemented with the appropriate antibiotics. The selection process can be repeated until the desired level of selection is obtained.

4. Notes

1. The transcriptional protein must be a monomer since an intrinsic protein-protein interaction will prevent the determination of substrate turnover in the cell.
2. The identification of an appropriate protein-ligand pair for the construction of the QUEST system is the most difficult and challenging aspect of the method. If a binding protein for the substrate exists (for example, by a related enzyme), then this protein could be engineered to function as a transcriptional protein. The most obvious ligand to use would be the substrate of the reaction, although known competitive inhibitors should also be considered. If there are no known proteins available for construction of the transcriptional protein, a protein will have to be engineered. To accomplish this, an appropriate scaffold protein (a small structural protein with the desired three-dimensional shape) is cloned into a phage display system and the protein is randomly mutated. Selection is then conducted using a substrate affinity column to identify proteins capable of binding the sub-

strate. The CID could then be constructed using the substrate. If a second molecule were desired for the construction of the CID (to avoid problems with enzyme action on the CID), competition studies using small molecules structurally related to the substrate could be conducted to identify molecules capable of removing the protein from the substrate affinity column. Proteins eluted from the column using this procedure would have the ability to recognize the substrate and the other molecule. Similar studies have been conducted to identify ligand-binding proteins ***(12–14)***. The resulting protein would also provide an excellent starting point for the engineering of enzymes, as the substrate-binding pocket would already have been created.

The above approach, although based upon well-documented procedures, is not guaranteed to produce a protein with the desired properties. To circumvent this problem, several variations of QUEST can be envisioned, all of which could take advantage of well-known protein-ligand pairs (*see* **Fig. 2**). The version of QUEST outlined in the chapter is the most general since it relies solely upon competitive binding of the substrate and CID (*see* **Fig. 1**). However, any action that affects the CID would result in a detectable signal in the cell. For example, enzymatic cleavage or synthesis of the CID would be detected by changes in the phenotype of the cell in a manner similar to competition with substrate. Enzymatic release of a competitor molecule would also result in a detectable signal.

3. While the discussion has focused on substrate binding to the transcriptional protein, it is important to realize the system would work equally well with competition by the product of the enzyme.
4. Besides the desired restriction sites, vectors must be chosen based upon the following key features: (a) Antibiotic resistance compatible with the vectors encoding for the synthetic transcriptional protein, (b) "Orthogonal" inducers of expression of the protein. Since many vectors are induced by the same molecule (e.g., isopropyl-thio-β-D-galactopyranoside [IPTG]), identifying vectors that allow controlled expression of the enzyme of interest and not the transcriptional protein can be difficult. This problem is highlighted in the lambda-based system, where induction by IPTG results in false positives. Therefore, we find that AraC-based expression systems are good for the lambda-based transcription systems and IPTG-based expression systems are good for AraC-based transcription. "Leaky" expression vectors are also good, since high levels of protein are not necessarily the best for this assay.
5. The cartoon representation of the CID presented in **Fig. 1** offers a good representation for the design of this molecule, namely a bifunctional linker molecule covalently attaches two ligands that bind to the synthetic transcriptional protein. Linker molecules can be chosen from an array of commercially available bifunctional molecules that display a wide range of functional groups that allow for attachment of the ligand of interest. Consideration should be given to potentially liable functional groups, such as esters that could affect the lifetime of the CID. The distance between the two ligands can be varied, but, in general, linkers containing 8–10 atoms have been utilized with success in creating CID molecules.

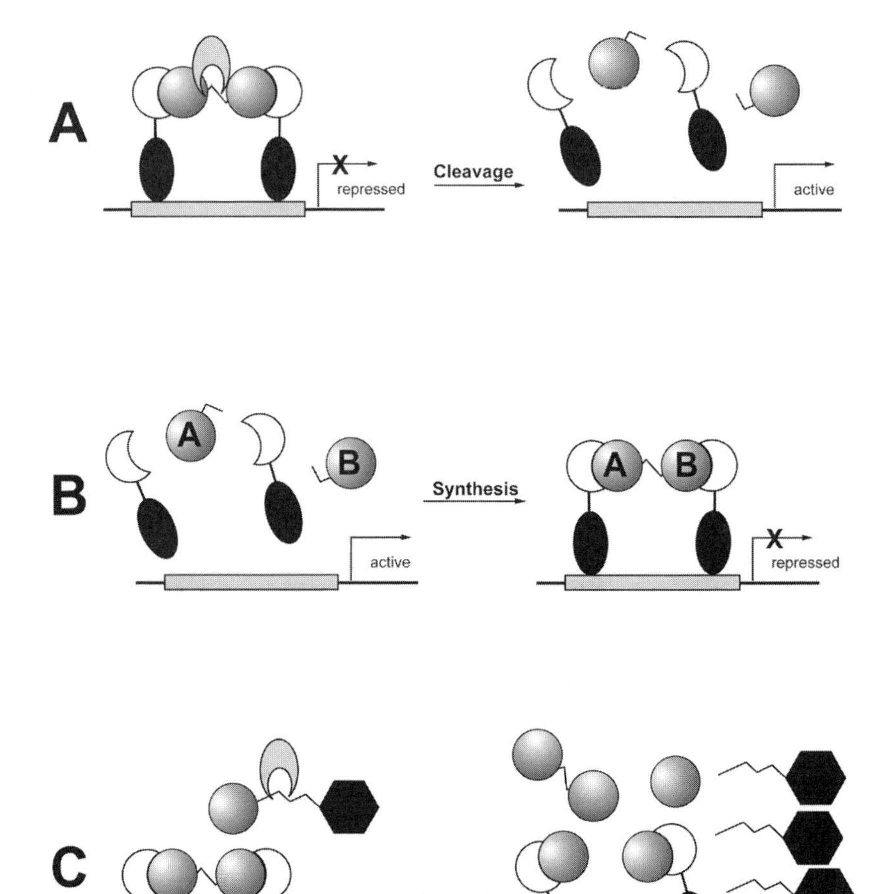

Fig. 2. Variations of the QUEST system. The presence of an enzyme can be detected by other ways besides strict competition between substrate and the CID. For example, enzymatic cleavage or synthesis of the CID (**A** and **B**) results in a detectable transcriptional signal. Enzymatic release of a competitor molecule (**C**) could also be utilized. While repression is shown in the figure, each of these systems could also be established for an activator protein.

6. If the researcher has only a small amount of CID to test, small petri dishes (35 × 10 mm, Fisher) can be used. Each plate utilizes between 2–3 mL of media, making the total amount of CID needed for the experiment quite low. The small plates dry out very easily in an incubator. To prevent this, plates should be stored inside of a plastic bag with a couple of wet paper towels.

7. The concentration of the CID needed can be estimated from either the K_i or K_m of the ligand used in its construction. However, transport and other phenomena can complicate this estimate. Thus, investigators should explore a wide range of CID concentrations and then narrow the concentration range to a region acceptable for the experiment.
8. Investigators should always run a control plate lacking any CID. Oftentimes the ligand-binding domain may have some background activity, resulting in false signals.
9. MacConkey media is sensitive to overcrowding of colonies on the plate. Thus, keeping the number of colonies low is critical for success. For small plates (*see* **Note 6**), we recommend only 20 colonies/plate.
10. The temperature of incubation is critical. A wide range of temperatures should be explored to test for the optimal function.
11. The time of incubation can vary depending upon the temperature and the level of induction of the arabinose operon.
12. The colonies themselves must be red/purple and should not be confused with translucent colonies displaying the background color of the media.
13. Not all colonies on the positive plate will be red. We have found that typically about 5–10% of the colonies display the background color of the plate regardless of the amount of CID added or the origin of the starting colony.
14. We have only used a 96-well plate, but the method could be extended to plates with higher well density (e.g., 384-well).
15. The volume of the well is dependent upon the density of wells for the microtiter plate used (e.g., 384).
16. Manganese chloride is supplemented into the media, since the ribulose isomerase enzyme in the arabinose pathway requires this metal for catalysis.
17. The arabinose levels may be modified to adjust for any background problems associated with the ligand-binding protein itself.
18. The time and temperature are dependent upon one another and should be investigated to determine the optimal conditions. The density of the cells can be estimated by eye and in general, each well should appear to be at least OD_{600} of 0.3 to the well-trained investigator before proceeding.
19. It is recommended that the plates NOT be shaken at this stage. This will allow for the build-up of CO_2 and increase the acidity of the media.
20. This step is critical to the success of the fluorescent assays. Growth can vary as much as 10% between the wells. Failure to correct for this difference could prevent the determination of positive results in the assay.
21. We have used two pH-sensitive dyes. One is LysoSensor Green which has the property of increasing fluorescence as pH decreases. LysoSensor Green is used at a concentration of 500 n*M*, is analyzed at an excitation wavelength of 430 nm, and has an emission wavelength of 500 nm. We have found that this dye is not very stable and loses signal over time. The second dye that we have used is fluorescein, at a final concentration of 1.0 µ*M*. This dye displays a decreased fluorescence as the pH decreases. It has an excitation wavelength of 490 nm and an emission wavelength of 520 nm.

22. Black microtiter plates are recommended to decrease scatter and lower the background signal. A clear-bottom black microtiter plate may be necessary depending upon the configuration of the plate reader. For studies conducted in our laboratory, we utilized a MicroMax plate reader attachment on a Fluoromax-2 fluorimeter, in which the excitation and emission signals are generated and detected from above the plate. For this purpose, we used black microtiter plates from PGC Scientific (Gaithersburg, MD, cat. no. 05-6114-25).
23. Since changes in pH are dependent upon the amount of bacteria present, the greater the amount of bacteria present, the lower the pH. In the case of LysoSensor Green, the greater the amount of bacteria, the greater the signal. Since fluorescein decreases fluorescence as the pH decreased, the greater the amount of bacteria, the more the signal should decrease.
24. This assay is used to detect dimerization of the lambda-based transcriptional protein. In this assay, function dimers bind to DNA injected by the phage and prevent the lytic cycle of the phage. Thus, the more dimers present (by action of the CID) in the cell, the lower the phage titer.
25. The amount of bacteria added can be adjusted; however, large amounts of bacteria are not necessary and cells should be in log phase during the phage infection.
26. The amount of the bacteria can be estimated by using a sample of the culture to determine the absorbance at OD_{600}. The titer of the phage stock should be determined before use.
27. The MOI should be below 1 to prevent multiple infections of the cell. This is especially necessary in the case where dimerization mediated by the CID is weak and therefore the amount of functional dimmers in the cell is low.
28. Care should be taken regarding the stability of the substrate. Some materials may be too unstable to be utilized in the QUEST system. For example, substrates with short half-lives (~minutes) would likely decompose before the bacteria grow to a size great enough to detect the desired changes. The fluorescence and phage assays have been used to test an unstable substrate. Another mechanism to circumvent this problem would be to use more stable, alternative substrate. For substrates with marginal stability, an alternative nitrocellulose-based MacConkey assay can be used as detailed *(6)*. This procedure reduces the total time that the substrate is present to only a few hours.

References

1. Nixon, A. E. and Firestine, S. M. (2000) Rational and "irrational" design of proteins and their use in biotechnology. *IUBMB Life* **49,** 181–187.
2. Stemmer, W. P. (1994) DNA shuffling by random fragmentation and reassembly: in vitro recombination for molecular evolution. *Proc. Natl. Acad. Sci. USA* **91,** 10,747–10,751.
3. Ostermeier, M., Nixon, A. E., and Benkovic, S. J. (1999) Incremental truncation as a strategy in the engineering of novel biocatalysts. *Bioorg. Med. Chem.* **7,** 2139–2144.
4. Olsen, M., Iverson, B., and Georgiou, G. (2000) High-throughput screening of enzyme libraries. *Curr. Opin. Biotechnol.* **11,** 331–337.

5. Fastrez, J. (1997) In vivo versus in vitro screening or selection for catalytic activity in enzymes and abzymes. *Mol. Biotechnol.* **7,** 37–55.
6. Firestine, S. M., Salinas, F., Nixon, A. E., Baker, S. J., and Benkovic, S. J. (2000) Using an AraC-based three-hybrid system to detect biocatalysts in vivo. *Nature Biotechnol.* **18,** 544–547.
7. Licitra, E. J. and Liu, J. O. (1996) A three-hybrid system for detecting small ligand-protein receptor interactions. *Proc. Natl. Acad. Sci. USA* **93,** 12,817–12,821.
8. Rivera, V. M., Clackson, T., Natesan, S., et al. (1996) A humanized system for pharmacologic control of gene expression. *Nature Med.* **2,** 1028–1032.
9. Bustos, S. A. and Schleif, R. F. (1993) Functional domains of the AraC protein. *Proc. Natl. Acad. Sci. USA* **90,** 5638–5642.
10. Zeng, X., Herndon, A. M., and Hu, J. C. (1997) Buried asparagines determine the dimerization specificities of leucine zipper mutants. *Proc. Natl. Acad. Sci. USA* **94,** 3673–3678.
11. Sambrook, J. and Russell, D. (2001) *Molecular cloning: A laboratory manual.* Cold Springs Harbor Laboratory Press, Cold Springs Harbor, NY, pp. 2.25–2.31.
12. Smith, G. P., Patel, S. U., Windass, J. D., Thornton, J. M., Winter, G., and Griffiths, A. D. (1998) Small binding proteins selected from a combinatorial repertoire of knottins displayed on phage. *J. Mol. Biol.* **277,** 317–332.
13. McConnell, S. J. and Hoess, R. H. (1995) Tendamistat as a scaffold for conformationally constrained phage peptide libraries. *J. Mol. Biol.* **250,** 460–470.
14. Beste, G., Schmidt, F. S., Stibora, T., and Skerra, A. (1999) Small antibody-like proteins with prescribed ligand specificities derived from the lipocalin fold. *Proc. Natl. Acad. Sci. USA* **96,** 1898–1903.

23

Identification of Genes Encoding Secreted Proteins Using Mini-O*phoA* Mutagenesis

Mary N. Burtnick, Paul J. Brett, and Donald E. Woods

1. Introduction

Protein fusions are invaluable tools for the genetic studies involving the mechanisms of protein export in bacteria. In 1985, Hoffman and Wright developed an in vitro fusion approach that allowed for fusions of the gene encoding *Escherichia coli* alkaline phosphatase to a variety of cloned genes *(1)*. The modified *phoA* gene employed in these studies, designated *'phoA*, resulted in the production of a highly active alkaline phosphatase protein missing its signal sequence *(1)*. This approach is based on the fact that bacterial alkaline phosphatase is normally periplasmic and must be located extracytoplasmically to be active, i.e., export is essential for high levels of alkaline phosphatase activity *(1)*. Through the fusion of *'phoA* to portions of heterologous genes containing signal sequences, export from the cytoplasm and subsequent PhoA activity can be observed *(1)*.

The utility of the *phoA* fusion approach was extended with the construction of Tn*phoA* (7733 bp), a Tn*5* based transposon with a truncated *phoA* gene at one end *(2)*. Tn*phoA* randomly generates *'phoA* fusions upon integration into a recipient bacterial chromosome *(2,3)*. Isolation of mutants harboring active PhoA fusions can be identified easily as they appear blue on agar plates containing the chromogenic substrate 5-bromo-4-chloro-3-indolyl phosphate (XP). Such gene fusions result in the expression of hybrid proteins with PhoA activity if the gene forming the fusion encodes an extracytoplasmic product, i.e., a membrane, periplasmic, outer membrane, or secreted protein.

In 1990, De Lorenzo et al. constructed mini-Tn*5phoA*, a mini-Tn*5* derivative with the *'phoA* gene from Tn*phoA* *(4)*. Mini-Tn*5phoA* possesses a transposase (*tnp**) external to the mobile element and is about half the size of

From: *Methods in Molecular Biology, vol. 205, E. coli Gene Expression Protocols*
Edited by: P. E. Vaillancourt © Humana Press Inc., Totowa, NJ

Tn*phoA* *(4)*. These features simplify genetic analysis by ensuring stability of the mini-Tn*5phoA* integration in the recipient chromosome and increasing the ease of cloning. More recently, a broad host range, self cloning plasposon containing the *'phoA* gene has been constructed *(5)*. The *'phoA* gene from Tn*phoA* *(3)* was PCR amplified and ligated it into the plasposon pTnmodOGm *(6)*. resulting in a construct designated mini-O*phoA* (*see* **Fig. 2**) *(5)*. Similar to the mini-Tn5 derivatives, plasposons include the presence of a cognate transposase outside of the inverted repeats allowing for integration into the recipient chromosome without the transposase thereby avoiding additional genetic rearrangements *(6)*. In addition, plasposons possess a pMB1 conditional origin of replication and multiple cloning sites within its inverted repeats that allow for the rapid cloning of DNA flanking the integration site *(6)*. The mini-O*phoA* fusion system works in the same manner as Tn*phoA* and mini-Tn*5phoA*, however, the presence of an origin of replication that allows for self cloning of the DNA flanking mini-O*phoA* confers a significant advantage *(5)*. This system simplifies and expedites the identification, cloning and sequence analysis of genes encoding extracytoplasmic products. **Figure 1** shows the steps in formation of an active gene fusion using a self-cloning mini-transposon carrying the *'phoA* gene. **Figure 2** schematically represents the pmini-O*phoA* plasposon.

We have optimized the mini-O*phoA* system for use in phosphatase-negative strains of three *Burkholderia* spp. *(5,7)*, and this system should prove useful in most gram negative bacteria. Described here are the materials and methods used for the identification and characterization of genes encoding extracytoplasmic products using mini-*OphoA*. *See* **Fig. 3** for an overview of this approach.

2. Materials

Unless otherwise stated chemicals were purchased from Sigma-Aldrich Canada (Oakville, ON, Canada). Tryptone and yeast extract were purchased from Difco (Detroit, MI, USA).

2.1. Conjugation and Screening

1. XP stock solution: 40 mg/mL in deionized water, filter sterilize through a 0.22 μm filter (Millipore Corp., Mississauga, ON, Canada). Light sensitive: wrap in tin foil and store at –20°C.
2. Gentamicin (Gm) stock solution: 20 mg/mL in deionized water, filter sterilize through 0.22 μm filter (*see* **Note 1**). Store at –20°C.
3. Streptomycin (Sm) stock solution: 100 mg/mL in deionized water, filter sterilize through 0.22 μm filter (*see* **Note 2**). Store at –20°C.
4. Luria-Bertani (LB) Broth: 10 g tryptone, 5 g yeast extract and 5 g sodium chloride (NaCl), dilute to 1 L with deionized water (*see* **Note 3**). Store at room temperature.

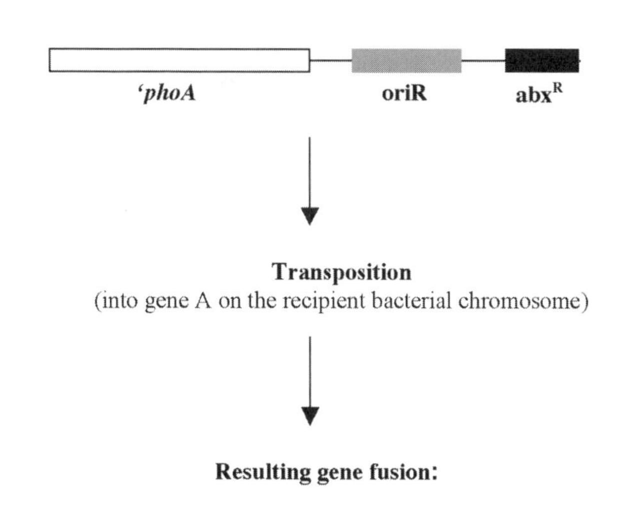

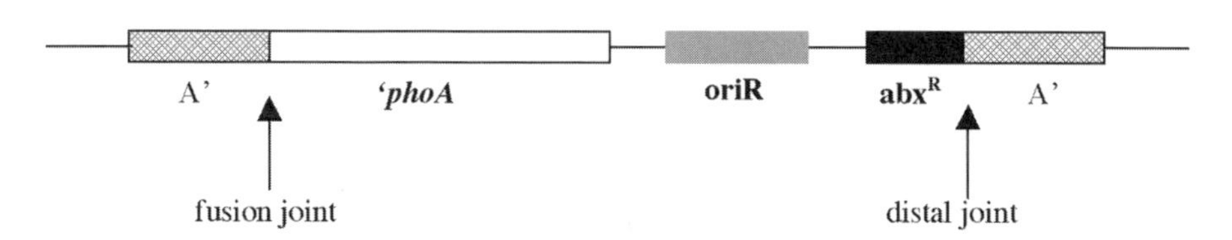

Fig. 1. Formation of an active *phoA* gene fusion using a self-cloning mini-transposon carrying a truncated *phoA* gene. *'phoA*: modified *E. coli* alkaline phosphatase gene minus the signal sequence; oriR: origin of replication that allows for self-cloning; abx^R: antibiotic resistance cassette appropriate for selection of transposition events in the recipient bacterial species in question. An active insertion into gene A results in interruption of the gene and the production of a hybrid protein from A-*phoA* fusion.

5. LB agar: LB broth plus 1.5% agar (15 g/L). Autoclave. Cool agar before pouring plates. Store at 4°C.
6. LBGmXP agar: Prepare low salt LB agar, cool to 55°C. Add Gm to a final concentration of 20 µg/mL, and XP to a final concentration of 40 µg/mL. Wrap in tin foil and store at 4°C.
7. Fresh LBGm agar plate of the donor bacterial strain, SM10 (pmini-O*phoA*).
8. Fresh LBSm agar plate of the recipient bacterial strain.
9. LBSmGmXP agar: Prepare low salt LB agar, cool to 55°C. Add Sm to a final concentration of 100 µg/mL, gentamicin to a final concentration of 20 µg/mL, and XP to a final concentration of 40 µg/mL. Wrap in tin foil and store at 4°C.
10. 40% Glycerol. Autoclave to sterilize. Store at room temperature.

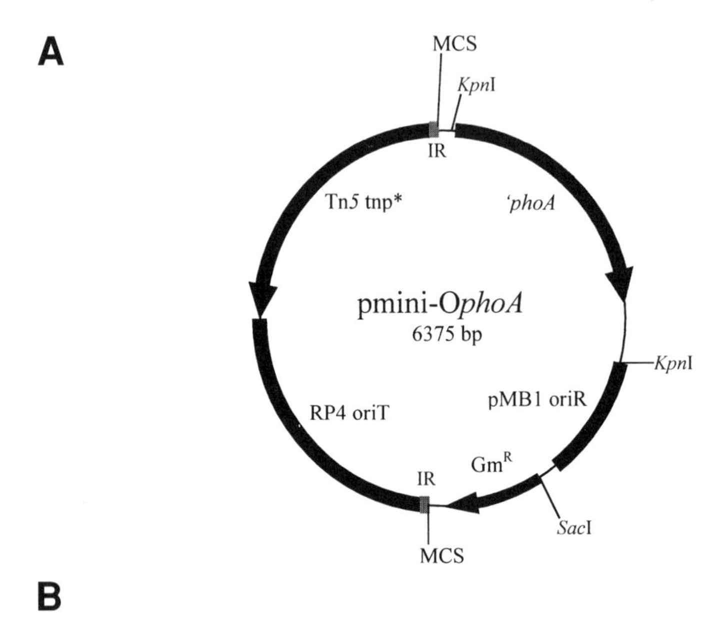

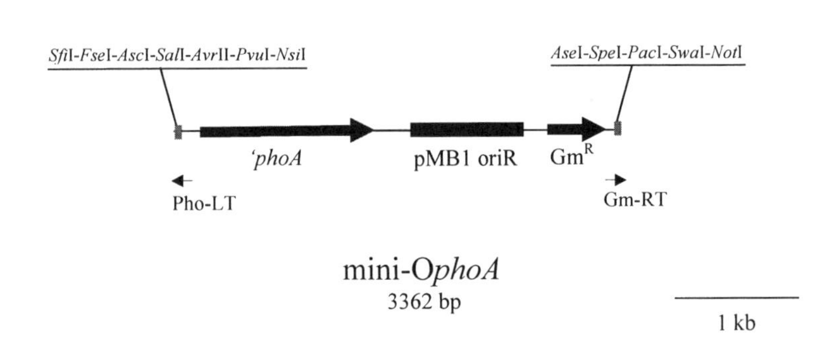

Fig. 2. (**A**) Schematic of the pmini-O*phoA* plasposon. *'phoA, E. coli* alkaline phosphatase gene lacking the signal sequence; pMB1 oriR, origin of replication; GmR, gentamicin resistance gene; RP4 oriT, origin of transfer; Tn5 tnp*, Tn5 transposase; IR, Tn5 inverted repeats; MCS, multiple cloning sites. (**B**) mini-O*phoA*, the portion of the plasposon that integrates into the chromosomal DNA of the bacterial recipient. Restriction endonuclease cleavage sites of the MCSs are shown. Pho-LT and Gm-RT sequencing primers are indicated as arrows. Adapted from **Ref. *5***.

2.2. Analysis of Transconjugates with PhoA Activity: Cloning of the DNA Flanking Mini-OphoA Integrations and DNA Sequencing

Restriction endonucleases and T4 DNA ligase were obtained from Gibco-BRL (Rockville, MD, USA) or New England Biolabs (Mississauga, ON, Canada) and were stored at –20°C and used as per manufacturer's instructions.

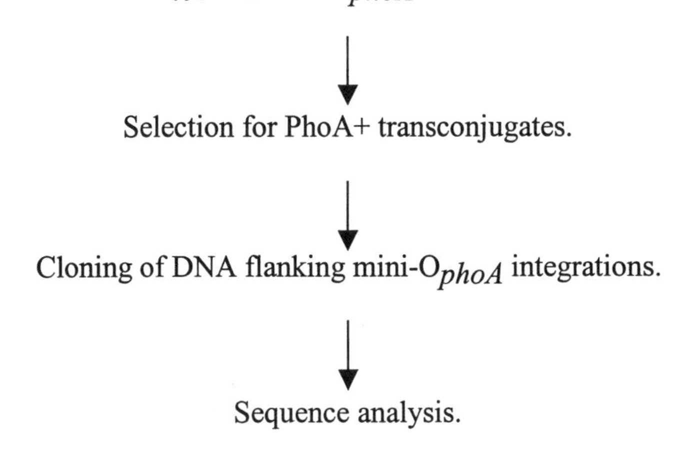

Fig. 3. A summary of the steps in mini-O*phoA* mutagenesis and analysis of transconjugates with PhoA activity (PhoA+).

1. LBSmGmXP agar (*see* **Subheading 2.1., item 9**).
2. LBGm and LBSm: LB broth with gentamicin 20 µg/mL or streptomycin 100 µg/mL, respectively.
3. 15-mL Polypropylene round bottom "snap-cap" tubes (Starstaedt).
4. Wooden toothpicks. Autoclave to sterilize.
5. Genomic DNA isolation protocol of your choice. We use Wizard™ Genomic DNA Isolation kit (Promega, Madison, WI, USA).
6. Restriction endonucleases. *See* **Fig. 2**, mini-O*phoA* map for positions restriction endonuclease cleavage sites.
7. 3 *M* Sodium acetate (NaOAc) pH 4.6. Autoclave. Store at room temperature.
8. 100% Absolute ethanol, store at –20°C and 70% ethanol, store at room temperature.
9. Sterile deionized water.
10. T4 DNA Ligase and 5X Ligase Buffer.
11. Chemically competent or electrocompetent *E. coli* cells. We use Top10 *E. coli* (Invitrogen) or electrocompetent High Efficiency *E. coli* DH5α (GibcoBRL).
12. LBGmXP agar (*see* **Subheading 2.1., item 6**).
13. Plasmid DNA isolation protocol of your choice. We use the QIAprep plasmid miniprep kit (QIAGEN, Mississauga, ON, Canada).
14. 0.8% Agarose gel, appropriate buffers and gel running apparatus. See Sambrook et al. for standard procedures ***(8)***.
15. Sequencing primers ***(5)***:
    a. Pho-LT 5'-CAGTAATATCGCCCTGAGCAGC-3'
    b. Gm-RT 5'-GCCGCGGCCAATTCGAGCTC-3'
16. DNA sequence analysis software and BLASTX program ***(9)***: www.ncbi.nlm.nih.gov/blast/index.html.

3. Methods

The following procedures have been optimized for use with *Burkholderia* spp. It may be necessary, however, to modify some of the steps described below in order to achieve optimal results in other organisms.

3.1. Conjugation and Screening

1. Day 1: Inoculate an LBGmXP plate with SM10 (pmini-O*phoA*) from a frozen glycerol stock. Additionally, inoculate an LBSm plate with the recipient bacterial strain. Be sure to streak for isolated colonies. Invert and incubate plates at 37°C overnight.
2. Day 2: Using a sterile toothpick, inoculate 2 mL of LBGm broth in a snap cap tube with a single white SM10 (pmini-O*phoA*) colony. Again, using a sterile toothpick, inoculate 2 mL of LBSm broth in a snap cap tube with a single colony of the recipient bacterial strain and. Incubate at 37°C with aeration (250 rpm) for 18 h.
3. Day 3: Divide an LB agar plate into eight sections with a marker, label one section as the donor control and one section as the recipient control, label the other six sections with an "X" for the donor plus recipient conjugations (*see* **Fig. 4**). Pipet 5 µL from the overnight culture of SM10 (pmini-O*phoA*), i.e., the donor, onto the donor control section and 5 µL onto each of the "X" sections of the LB agar plate. Next, pipet 5 µL from the recipient strain overnight culture onto the recipient control section and onto each of the "X" sections of the agar plate. Make sure that the cultures spotted onto to the "X" sections are mixed. Incubate the plate at 37°C overnight. For alternate conjugation methods (*see* **Note 4**).
4. Day 4: Using a sterile scraper or glass spreader, scrape the cells from each conjugation ("X" section) and spread them onto selective media, LBSmGmXP. Additionally scrape the cells from the control sections onto selective media. Incubate at 37°C for 24–48 h.
5. Days 5 and 6: Examine LBSmGmXP plates for the presence of transconjugates. Blue (PhoA+) colonies represent transconjugates that have acquired mini-O*phoA* and have 'phoA fusions. Retain the PhoA+ colonies for further analysis. There should not be any growth present on the control plates.
6. Purify the PhoA+ colonies by streaking them onto LBSmGmXP plates to ensure a homogeneous culture. At this point, it is suggested that frozen glycerol stocks be prepared by adding saturated bacterial culture to 40% glycerol in a 1:1 ratio, store at –70°C. The PhoA+ colonies for further analysis should be maintained on selective media.

3.2. Analysis of Transconjugates with PhoA Activity: Cloning of the DNA Flanking Mini-OphoA Integrations and DNA Sequencing

1. Inoculate a single PhoA+ transconjugate colony into 3 mL of LBSmGm in a snap cap tube. Incubate overnight at 37°C with aeration (250 rpm).
2. Isolate the chromosomal from the overnight cultures of each PhoA+ transconjugate using the method of your choice, we use the Wizard™ Genomic DNA

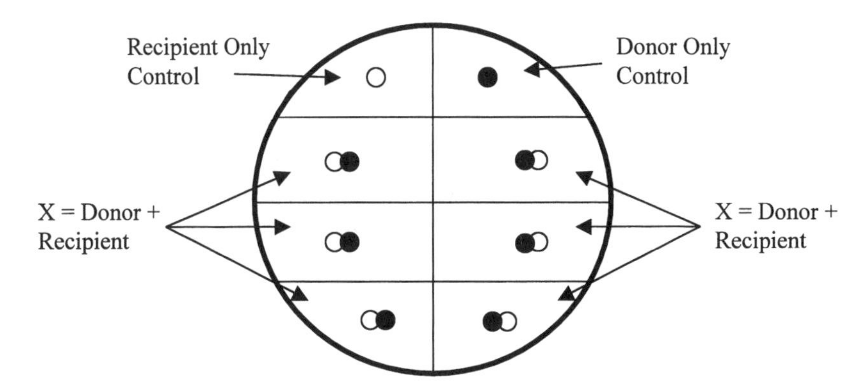

Fig. 4. Agar plate conjugation procedure used for transfer of mini-O*phoA* from SM10 λ pir (donor strain) to a gram negative bacterial recipient strain. Two control sections containing 5 µL of donor *or* recipient strain alone, *and* six conjugation ("X") sections containing 5 µL of both donor and recipient strains mixed together are shown.

Isolation kit (Promega). Quantitate the concentration of the chromosomal DNA obtained using OD260/280 method *(8)*.

3. Self-cloning of the DNA flanking mini-O*phoA* integrations: digest 1–2 µg of chromosomal DNA from each PhoA+ strain with an appropriate restriction endonuclease. For example, to clone the 'phoA fusion joint, *Not*I, *Swa*I, *Pac*I, *Spe*I or *Ase*I can be used. Alternatively, to clone the DNA flanking the opposite (Gm^R gene) side of mini-O*phoA*, *Sfi*I, *Fse*I, *Asc*I, *Sal*I, *Avr*II, *Pvu*I, or *Nsi*I can be used (*see* **Fig. 2**). Set up a 20 µL restriction endonuclease reaction in a 1.5 mL microfuge tube as per manufacturer's instructions. Generally, incubation at 37°C for 1 h is appropriate.
4. Heat inactivate the restriction digest by boiling for 5 min (*see* **Note 5**). Briefly centrifuge to recover any condensation.
5. Ethanol precipitate the DNA. Add 1/10 volume 3 *M* NaOAc, pH 4.6, and 2.5 vol of ice cold 100% Absolute ethanol. Place this reaction at –20°C for at least 30 min (*see* **Note 6**). Centrifuge at top speed in a microfuge for 15 min. Carefully remove the supernatant and wash pellet with 70% ethanol. Centrifuge for 5 min. Carefully remove the supernatant and air dry the pellet.
6. Set up a ligation reaction as follows. Thoroughly resuspend the dried DNA pellet from **step 5** in 50 µL of sterile deionized water (*see* **Note 7**). Use 19 µL of the resuspended DNA, 5 µL of 5X ligase buffer and 1 µL of T4 DNA ligase (Gibco-BRL). Incubate at 16°C overnight.
7. Use 2–5 µL of the ligation mixture from **item 6** to transform high efficiency competent *E. coli* cells (*see* **Note 8**). For chemical transformations, we use *E. coli* Top 10 cells (Invitrogen) and for electroporations, we use Max Efficiency *E. coli* DH5α cells (Gibco-BRL) as per manufacturer's instructions. Select for transformants on LBGmXP plates, incubate overnight at 37°C.

8. Using a sterile toothpick, inoculate individual blue transformants into LBGm broth. Incubate at 37°C with aeration (250 rpm).
9. Isolate plasmids from each transformant using your method of choice, we use the QIAprep plasmid miniprep kit (QIAGEN). Check each plasmid by digesting with the same enzyme that was used to clone it. Load the digested plasmids onto a 0.8% agarose gel. The appropriate clones will have only one band when visualized with ethidium bromide under a UV light source. The plasmid size can then be estimated; additional double digestions can be performed to more accurately determine the size of the cloned flanking DNA fragment.
10. Sequence the appropriate plasmids using the Pho-LT and Gm-RT primers (*see* **Note 9**).
11. Analyze the sequence obtained for homology to known gene sequences using the BLASTX program.

4. Notes

1. Gentamicin is used to select for the presence of mini-O*phoA* in the donor strain prior to conjugation, and to select for the integration of mini-O*phoA* into the chromosome of the recipient strain following conjugation. If Gm is not a desirable selectable marker for a specific recipient bacterial strain, the Gm^R cassette on the mini-O*phoA* can easily be replaced. This cassette can be excised using *Sac*I or *Sst*I and a resistance cassette of the user's choice can be ligated into mini-O*phoA*.
2. Streptomycin is used for selection against the donor strain, SM10 λ pir (pmini-O*phoA*), following conjugations procedures. When using certain gram negative bacterial strains, it may be necessary to use an antibiotic other than Sm if the recipient strain does not display a streptomycin resistant phenotype. For example, when *B. mallei* was used as a recipient for mini-O*phoA*, naladixic acid was used in place of Sm due to the fact that a stable Sm^R derivative of *B. mallei* could not be obtained *(5,7)*.
3. Low salt Luria-Bertani broth (10 g tryptone, 5 g yeast extract and 5 g NaCl, not 10 g NaCl) was used throughout this protocol as high salt concentrations may interfere with the activity of the gentamicin.
4. The conjugation method described in **Subheading 3.1., item 3** has been used specifically for *B. pseudomallei* and *B. thailandensis*. For other gram negative bacteria, the incubation time at 37°C may need to be adjusted depending on the conjugation efficiency of the recipient strain. Additionally, the bacterial growth from the conjugations can be scraped off of the LB plate and diluted as necessary in 0.85% NaCl and then plated. This step may be taken if the density of single transconjugates on the selective media is too high. Other methods of conjugation may be used instead of using the plate method described here. A broth method may be employed as follows: inoculate a snap cap tube containing 2 mL LB broth with 100 μL of overnight cultures of each of the donor and the recipient strains, incubate at 37°C 250 rpm for a few hours to overnight. Following the incubation, 100 μL aliquots (or less if necessary) of the conjugation mixture should be plated

on selective media (LBSmGmXP) and incubated at 37°C. The duration of the conjugation step may differ for different bacterial species and should be optimized for the specific gram negative recipient strain in question.

5. The heat inactivation step may be altered depending on the restriction endonuclease used, for example heating to 65°C for 10 min is sufficient for inactivation of certain restriction endonucleases. See the manufacturer's heat inactivation specifications for the particular enzyme being used.
6. During the ethanol precipitation step, we have found that placing the reaction in a –20°C or –70°C freezer overnight works efficiently, and may in fact increase the amount of DNA recovered following this step.
7. It is important at this stage to ensure that the dried DNA pellet is thoroughly resuspended. It is suggested that the DNA be allowed to resuspend at room temperature for at least 10 min prior to preparation of the ligation reaction.
8. The use of high efficiency competent *E. coli* cells was necessary to obtain mini-O*phoA* flanking clones. User prepared cells did not have a high enough transformation efficiency to obtain clones on the first cloning attempt. Additionally, it is often helpful to microconcentrate the ligation reactions prior to transformation in order to increase the chances of obtaining the desired transformants.
9. Nucleotide sequencing was performed with the ABI PRISM DyeDeoxy Termination Cycle Sequencing System and analyzed using an ABI 1373A DNA Sequencer by University Core DNA Services (University of Calgary). Both the Pho-LT and Gm-RT primers can be used for sequencing of plasmid DNA isolated from a single mini-O*phoA* flanking clone regardless of which side of the integration was cloned. For example, if the *phoA* fusion joint was cloned, PhoA-LT would provide the sequence immediately adjacent to *'phoA*, while the Gm-RT primer would provide the sequence of the DNA upstream. Alternately, if the Gm^R gene joint was cloned, Gm-RT would provide the sequence immediately adjacent to the integration and Pho-LT would provide the sequence of the DNA downstream of the mini-O*phoA* integration. This feature of self-cloning expedites the sequencing process and allows the user to quickly and efficiently assess the interrupted gene as well as neighboring genes.

Acknowledgments

This work was supported by a Canadian Institutes for Health Research grant MOP36343. M.N.B. is the recipient of an Alberta Heritage Foundation for Medical Research studentship.

References

1. Hoffman, C. and Wright, A. (1985) Fusions of secreted proteins to alkaline phosphatase: an approach for studying protein secretion. *Proc. Natl. Acad. Sci. USA* **82,** 5107–5111.
2. Manoil, C. and Beckwith, J. (1985) Tn*phoA*: a transposon probe for protein export signals. *Proc. Natl. Acad. Sci. USA* **82,** 8129–8133.

3. Taylor, R. K., C. Manoil, and Mekalanos, J. J. (1989) Broad-host-range vectors for delivery of Tn*phoA*: use in genetic analysis of secreted virulence determinants of *Vibrio cholerae*. *J. Bacteriol.* **171,** 1870–1878.
4. de Lorenzo, V, Herrero, M., Jakubzik, U., and Timmins, K. N. (1990) Mini-Tn*5* transposon derivatives for insertion mutagenesis, promoter probing, and chromosomal insertion of cloned DNA in gram-negative eubacteria. *J. Bacteriol.* **172,** 6568–6572.
5. Bolton, A. and Woods, D. E. (2000) Self-Cloning Minitransposon *phoA* Gene-Fusion System Promotes the Rapid Genetic Analysis of Secreted Proteins in Gram-Negative Bacteria. *Biotechniques* **29,** 470–474.
6. Dennis, J. J. and Zylstra, G. J., (1998) Plasposons: modular self-cloning minitransposon derivatives for rapid genetic analysis of gram-negative bacterial genomes. *Appl. Environ. Microbiol.* **64,** 2710–2715.
7. Burtnick, M. N., Bolton, A. J., Brett, P. J., Watanabe, D., and Woods, D. E. (2001) Identification of the acid phosphatase (*acpA*) gene homologues in pathogenic and non-pathogenic *Burkholderia* spp. facilitates Tn*phoA* mutagenesis. *Microbiology* **147,** 111–120.
8. Sambrook, J., Fritsch, E., and Maniatis, T. (1989) *Molecular cloning: a laboratory manual 2nd ed.* Cold Spring Harbor Laboratory Press, Cold Spring Harbor, NY.
9. Altshul, S., Madden, T. L., Schaffer, et al. (1997) Gapped BLAST and PSI-BLAST: a new generation of protein database search programs. *Nucleic Acids Res.* **25,** 3389–3402.

Index

From: *Methods in Molecular Biology, vol. 205, E. coli Gene Expression Protocols*
Edited by: P. E. Vaillancourt © Humana Press Inc., Totowa, NJ